Manual of Cross-Connection Control Tenth Edition

Published by the
Foundation for Cross-Connection Control and Hydraulic Research
a Division of the University of Southern California

University of Southern California
Research Annex 219
Los Angeles, California 90089-7700

October, 2009

Foundation for Cross-Connection Control and Hydraulic Research

a Division of the University of Southern California

University of Southern California
Research Annex 219
Los Angeles, California 90089-7700

First Edition, 1960
Second Edition, 1965
Third Edition, 1966
Fourth Edition, 1969
Fifth Edition, 1974
Sixth Edition, 1979
Revised Sixth Edition, 1982
Seventh Edition, 1985
Eighth Edition, 1988
Ninth Edition, 1993

2009 © University of Southern California
Fourth Printing, 2017
ISBN: 978-0-9638912-5-9

ALL RIGHTS RESERVED

All portions of this publication, including text, graphics and electronic media may not be reproduced in any manner whatsoever without written permission of the publisher

Preface to the Tenth Edition

The Tenth Edition is the most extensive revision of the *Manual of Cross-Connection Control*, to date. The Manual Review Committee has been working on this edition of the Manual off and on for over a decade. Various technical issues held up the publication of the Manual on several occasions. The Tenth Edition includes a chapter on the history of cross-connection control. Although of interest to many people, this information has not been published before. A chapter on hydraulics will be invaluable to the students of backflow in helping them understand the hydraulics behind backflow as well as how the backflow preventers operate.

Chapters 6 and 7 include information on facilities and equipment to help the Specialist prepare for site surveys. Illustrations of various water using equipment is included to help the reader understand how certain pieces of equipment use water and may contaminate the water supply. There are some modifications to the field test procedures. The illustrations have also been improved. The illustrations use two colors, making the presentation of the field test procedures very clear.

Chapter 10 includes the standards for backflow prevention assemblies, along with a standard for a field test kit. This leads the way for the Foundation's Approval of field test kits. There are new standards for a double check detector assembly-type II and a reduced pressure principle detector assembly-type II. This edition of the manual includes a second color throughout to help the reader understand hydraulic conditions as well as the field-testing procedures. A CD-ROM is also included on the inside back cover which provides the model ordinances, forms and letters as well as a table of backflow incidents.

The Manual Review Committee was comprised of representatives from water agencies, health agencies, certified testers, Foundation staff and a non-voting representative of the Backflow Prevention Manufacturer's Association. The committee members volunteered a considerable amount of time and effort over the past decade considering many suggestions, changes and additions to the Manual. Backflow prevention assembly manufacturers were asked to comment on Chapter 10 of the Manual and there was a general concurrence. The Committee is confident these changes will result in better backflow prevention assemblies. We recognize the Manual Review Committee for their efforts in making this manual even more useful than previous editions to those involved in cross-connection control. We greatly appreciate the comments and feedback received from the many people who contributed information and made suggestions for improvements in this Tenth Edition of the *Manual of Cross-Connection Control*.

We trust that you will find this Manual invaluable in your efforts in cross-connection control. We believe this new edition will be used much more effectively and by a wider audience than previous editions. As always, should you have any questions or comments, please feel free to contact the Foundation Office.

Paul H. Schwartz, P.E.
Chief Engineer

J. J. Lee, Ph.D., P.E.
Director

Dedication

Alfred W. "Al" Jorgensen

Al Jorgensen was born December 1, 1927 in downtown Los Angeles at the Lutheran Hospital, which is still there. He was raised and worked in southern California most of his working years. He attended UCLA from 1945-47 and graduated from U. C. Berkeley in 1949 with a Bachelor of Science in Process Engineering. At this same graduation he received his commission as a Second Lieutenant in the U. S. Army Corps of Engineers. In 1967, Al also earned a Master of Science degree in Public Administration from the University of Southern California.

He was called to active duty in 1951 and spent nearly two years in Korea as the commander of the 91st Engineer Water Supply Company furnishing water from Taegu in the South through Seoul and the Panmunjom Peace Camp in the North. In 1961 his Reserve Unit, the 916th Engineer Construction Group, was called to duty during the Berlin Wall Crisis and he spent a year at Fort Meade, MD. Al retired from the Army Reserve in 1973 with the rank of Lieutenant Colonel.

Al Jorgensen devoted his entire career to the water industry in some fashion—from operations to planning, design and construction as well as management in the water industry. He has always been a supporter and consumer of public water supplies (no bottled water). In 1950 he was the first full time employee of the newly formed Crescenta Valley Water District and after the Korean War he was the General Manager of the Mountain Water Company, both of La Crescenta. From there he became the Director of Utilities for the City of Monrovia from 1956 to 1967. During this time he was President of the Upper San Gabriel Valley Water Association for four years (1963-1967). He became a Rotarian in 1956 and was President of the Monrovia Club in 1966-1967.

In 1967 Al became Director of Operations and Planning for the international consulting firm of Engineering-Science, which involved many consulting trips to Rio de Janeiro. In 1971 he became a partner in Lowry and Associates, Consulting Engineers in Orange County, California, with offices in the San Francisco Bay Area and Arizona. In 1986 two firms merged to form NBS/Lowry that grew to some 300 people. Al became the Executive Vice President and Chief Operating Officer. He retired in 1990 and moved to Prescott, Arizona.

During his working years he was very active in the American Water Works Association (AWWA) and served as Chairman of the California-Nevada Section in 1980. He was the National Director representing the CA/NV Section from 1984 through 1987.

Al Jorgensen became a member of the Southern California Water Utilities Association (the Walter Weight Group) in the early 1950's. He was instrumental in forming the financial backing and support of the water industry for the Foundation for Cross-Connection Control and Hydraulic Research at the University of Southern California. He was present when the first check for $10,000 was presented to Dr. Norman Topping, President of the University of Southern California, in 1967. Al has continued from that date to the present (2012) as a member of the Advisory Board of Directors of the Foundation for Cross-Connection Control and Hydraulic Research.

The Manual Review Committee

The Tenth Edition of the Manual was prepared by the members of the Manual Review Committee, who spent many, many hours considering the changes incorporated herein. It is the sincere hope of this committee that the water agencies, health agencies, plumbing officials and all who enjoy the use of clean water will understand better the problems created by cross-connections and will work toward the control or elimination of all cross-connections.

Mr. Michael C. Ahlee
Ahlee Backflow Service, Inc.
Santee, CA

Mr. Kenneth Anderson
(retired)
City of Riverside
Riverside, CA

Mr. Richard Bird
Sweetwater Authority
Chula Vista, CA

Mr. Carlos Borja
County of Los Angeles
Department of Public Health
Los Angeles, CA

Mr. Richard Carlson
Water Specialist Consultants
La Mesa, CA

Mr. Henry Chang
Mechanical Engineer/Program Manager
Foundation for Cross-Connection Control
and Hydraulic Research
University of Southern California
Los Angeles, CA

Mr. Marty Friebert
Environmental Health Specialist
Orange County Health Care Agency
Santa Ana, CA

Mr. Bill Gedney
VP Asset Management
Golden State Water Company
Ontario, CA

Mr. Ernest Havlina
(retired)
City of Los Angeles, Department of Water and Power
Los Angeles, CA

Mr. Daniel Jimenez
Graphic Artist
Foundation for Cross-Connection Control
and Hydraulic Research
University of Southern California
Los Angeles, CA

Mr. Sam Johnson
(retired)
City of Riverside
Riverside, CA

Dr. J. J. Lee, Ph.D., P.E.
Director
Foundation for Cross-Connection Control
and Hydraulic Research
University of Southern California
Los Angeles, CA

Mr. Charles Nena
(retired)
California Water Service Company
San Pedro, CA

Ms. Grazyna Newton
California Department of Public Health
Los Angeles, CA

Mr. Brad Noll
Backflow Prevention Manufacturer's Association
Paso Robles, CA

Mr. Robert Purzycki
Backflow Apparatus and Valve Company
Long Beach, CA

Mr. Paul Schwartz, P.E.
Chief Engineer
Foundation for Cross-Connection Control
and Hydraulic Research
University of Southern California
Los Angeles, CA

Mr. Patrick Sylvester
Business Manager
Foundation for Cross-Connection Control
and Hydraulic Research
University of Southern California
Los Angeles, CA

Table of Contents

Chapter 1 Definitions
Definitions 1

Chapter 2 History
History of Cross-Connection Control..20
    Ancient Civilizations 20
    The Modern Age 20
    Backflow Preventers...21
    Development of Regulations 24
    University of Southern California 25
    The State of the Industry..32

Chapter 3 Hydraulics
Hydraulics 38
    Water...38

Backflow 43
    Backsiphonage 43
    Backpressure 46

Cross-Connections 46

Degree of Hazard 47

The Backflow Incident 48

How to Prevent Backflow 48
    Check Valve...49
    Double Check Valve Assembly (DC)..50
    Reduced Pressure Principle Assembly (RP) 52
    Atmospheric Vacuum Breaker (AVB) 54
    Pressure Vacuum Breaker (PVB) 55
    Spill-Resistant Pressure Vacuum Breaker (SVB) 56

Three Questions 57

Hydraulics of the Backflow Prevention Assemblies 59
    Double Check Valve Assembly 59
    Reduced Pressure Principle Assembly 65
    Pressure Vacuum Breaker Backsiphonage Prevention Assembly........74
    Spill-Resistant Vacuum Breaker Backsiphonage Prevention Assembly...............78

Chapter 4 Elements of a Program

Jurisdiction 86

Responsibilities 86
Responsibility of the Health Agency 86
Responsibility of the Water Supplier 87
Responsibility of the Plumbing Official 87
Responsibility of the Consumer 87
Responsibility of the Certified Backflow Prevention Assembly Tester 88
Responsibility of the Repair and Maintenance Technician 88

The ABC's of a Cross-Connection Control Program 88
Authority 89
Backflow Preventers 89
Certified Testers and Specialists 92
Defensible and Detailed Records 93
Education and Training 94

Policies & Procedures 95
Administration of the Program 96
Authority 96
Auxiliary Water Systems 96
Certified Backflow Prevention Assembly Testers 97
Change of Occupancy or Use 97
Combined Services 98
Critical Services 98
Dual Services 98
Equipment 98
Existing Assemblies 98
Fire Sprinkler Systems 99
Incident Response 99
Irrigation Systems 100
Low Water Pressure 102
Multiple Services 103
Non-Compliance 103
Plumbing Code 103
Recycled Water Systems 103
Restricted or Classified Services 104
Sump and Lift Stations 104
Single and Multiple Family Dwellings 104
Service Across Political or Water Agency Boundaries 105
Typical High Hazard Services 105

Summary 105

Chapter 5 Cross-Connection Control Surveys

Cross-Connection Control Programs 110
    Preparing for a Cross-Connection Control Survey 111
        General 111
        Notification 111
        Existing Systems vs. New Construction 111
        Typical Methods of Backflow Prevention 112
        Before the Survey 112

The Site Survey 114
        Documentation 115
        Compliance 115

Chapter 6 Facilities

Typical Facilities: Cross-Connections or Water Uses Which May Endanger the Public Water System 120
        Services 120
        Manufacturing 121
        Food Processing/Service 123
        Medical Facilities 124
        Restricted 126
        Other Facilities 127

Chapter 7 Equipment and Systems

Equipment and Systems 132
        Air Scrubber (Wet Scrubbing) 132
        Aspirators 133
        Autoclaves 134
        Autopsy/Mortuary Tables 135
        Boiler 136
        Can and Bottle Washing Machines 137
        Car Washing Machine (automatic type) 138
        Carbonators 139
        Chemical Dispenser 140
        Cookers 141
        Cooling Tower 142
        Dental Vacuum Pump 143
        Fire Sprinkler Systems 144
        Heat Exchangers 145
        Irrigation Systems 146
        Kidney Dialysis Machine 147
        Laboratory Equipment 148
        Laundry Machines 149
        Photographic Film Processing Machines 150
        Portable Cleaning Equipment 151
        Pumps 152
        Recycled Water Systems 153
        Sewage Ejector 154

　　　　　Tanks/Vats 155
　　　　　Toilet 156
　　　　　Urinal 157
　　　　　Water Softener 158

Chapter 8 Samples and Forms

Approved Backflow Prevention Assemblies (service protection) 162

Installation and Maintenance Requirements 164

Guidelines for Parallel Installation of
　　Backflow Prevention Assemblies 173

Periodic Field Test and Maintenance Report with Report Form 175

Non-compliance Notice to
　　Install Backflow Prevention Assembly 177

Non-compliance of Periodic Field Test and Maintenance 178

Notice of Discontinuance of Water Supply 179

Air Gap Periodic Test and Maintenance Report 180

Notice of Appointment of Water Supervisor 182

Resolution Relative to Backflow Prevention Assembly Testers 183

Application for a Backflow Prevention Assembly
　　Tester Certificate 185

List of Certified Backflow Prevention Assembly Testers 187

Notice of Shutdown 188

Model Ordinance 189

Minimum Requirements for Backflow Prevention
　　Assembly Tester Certification Program 193

Minimum Requirements for Cross-Connection Control
　　Specialist Certification Programs 195

Field Survey Form 196

Backflow Incident Report 197

Approved Backflow Prevention Assemblies (internal protection) 199

Chapter 9 Field Test Procedures

Preliminary Steps 204

Field Test Reporting 205

Maintenance and Repairs 206

Manual of Cross-Connection Control Tenth Edition Modifications 206

Reduced Pressure Principle Backflow Prevention Assembly (RP)
    Using Five Needle Valve Test Kit 210
    Field Test Procedure 210
    Diagnostics 224

Reduced Pressure Principle Backflow Prevention Assembly (RP)
    Using Two Needle Valve Field Test Kit 230
    Field Test Procedure 231
    Diagnostics 246

Reduced Pressure Principle Backflow Prevention Assembly (RP)
    Using Three Needle Valve Field Test Kit 253
    Field Test Procedure 254
    Diagnostics 270

Double Check Valve Backflow Prevention Assembly (DC) 277
    Field Test Procedure 278
    Diagnostics 284

Pressure Vacuum Breaker Assembly (PVB) 290
    Field Test Procedure 291
    Diagnostics 301

Spill-Resistant Pressure Vacuum Breaker Assembly (SVB) 307
    Field Test Procedure 308
    Diagnostics 315

Reduced Pressure Principle-Detector
    Backflow Prevention Assembly (RPDA) 318

Double Check-Detector Backflow Prevention Assembly (DCDA) 320

Reduced Pressure Principle-Detector Backflow
    Prevention Assembly-Type II (RPDA-II) 322

Double Check-Detector Backflow
    Prevention Assembly-Type II (DCDA-II) 324

Chapter 10 Standards
Backflow Prevention Assemblies 330
    General Design and Material Requirements and Laboratory Testing 330
    Reduced Pressure Principle Backflow Prevention Assemblies(RP) 349
    Double Check Valve Backflow Prevention Assemblies(DC) 361
    Pressure Vacuum Breaker Backsiphonage
        Prevention Assemblies (PVB) 368
    Atmospheric Vacuum Breaker Backsiphonage
        Prevention Assemblies(AVB) 376
    Double Check Detector Backflow Prevention Assemblies(DCDA) 381
    Reduced Pressure Principle Detector Backflow
        Prevention Assemblies(RPDA) 385
    Spill-Resistant Pressure Vacuum Breaker
        Backsiphonage Preventions Assemblies (SVB) 390
    Double Check-Detector Backflow Prevention
        Assemblies-Type II (DCDA-II) 397
    Reduced Pressure Principle-Detector Backflow Prevention
        Assemblies-Type II (RPDA-II) 402

Differential Pressure Gage Field Test Kits Used for
    Field Testing of Backflow Prevention Assemblies 407
    Scope 407
    Definitions 407
    General Design Requirements 408
    Design Requirements 409
    Material Requirements 410
    Evaluation of Design and Performance 410

Chapter 11 Backflow Incidents
Backflow Incidents 417

Appendix A Field Test Guidance Documents
General Information 484
    Line Pressure 484
    Bleed-off Valve Arrangement 484
    Field Test Results 486
Reduced Pressure Principle Backflow Prevention Assembly 486
    Attaching hoses to self-actuating test cocks 486
    Check Valve No. 2: Direction-of- flow 487
Pressure Vacuum Breakers (PVB & SVB) 488
    Backpressure Evaluation 488
    Limited Space 488
Detector Assemblies: Operation of bypass 489
    RPDA: Verify Detection of Flow through Bypass 489
    RPDA Diagnostics: Operation of Bypass 490
    DCDA: Verify Detection of Flow through Bypass 491
    DCDA Diagnostics: Operation of Bypass 491

RPDA-II: Verify Detection of Flow through Bypass 493
DCDA-II: Verify Detection of Flow through Bypass 493
Abbreviated Field Test Procedures: Training Aid 493
RP 493
DC 498
PVB 499
SVB 500
Periodic Test of Differential Pressure Field Test Kit 501

Appendix B Evaluation Documents and Procedures
Evaluation Forms 505
Laboratory Submittal Form 505
Field Site Application Form 508
Letter of Acknowledgement for Field Evaluation Sites 512
Field Evaluation Phase of Approval Program 513
3 PSI Buffer 513
Field Test Procedure 513

Index 515

Chapter 1: Definitions

1.1 Absolute Pressure
Pressure measured on a scale where a perfect vacuum is zero. Absolute pressure is the sum of gage pressure and atmospheric pressure.

1.2 Accessible
Capable of being reached for testing and maintenance, when referring to a backflow prevention assembly. However, it first may require the removal of an access panel, door or similar obstruction. (See 1.54 Readily Accessible.)

1.3 Administrative Authority
The individual official, board, department, or agency established and authorized by a state, county, city or other political entity created by law to administer and enforce the provisions of the cross-connection control program.

1.4 Air Gap
A physical separation between the free flowing discharge end of a potable water supply pipeline and an open or non-pressure receiving vessel. An "approved air gap" shall be at least twice the diameter of the supply pipe measured vertically above the overflow rim of the receiving vessel; in no case less than 1 inch (2.54 cm). (Additional reference: ASME A112.1.2 - 2004 Air Gaps in Plumbing Systems) See figure 1.1.

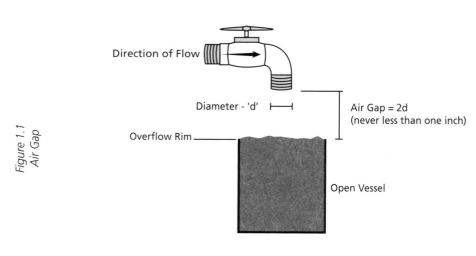

Figure 1.1
Air Gap

1.5 Approved Backflow Prevention Assembly
An assembly that has been investigated and approved by the administrative authority having jurisdiction. The approval of backflow prevention assemblies by the administrative authority shall be on the basis of a favorable laboratory and field evaluation report by an approved testing laboratory recommending such approval (see 1.7).

1.6 Approved Check Valve
A check valve that is drip-tight in the normal direction of flow when the inlet pressure is at least 1.0 psi (pound per square inch) and the outlet pressure is zero. The check valve shall permit no leakage in a direction reverse to the normal flow. The closure element (e.g., clapper or poppet) shall be internally loaded to promote rapid and positive closure. An approved check valve is only one component of an approved backflow prevention assembly (i.e., pressure vacuum breaker {PVB and SVB}, double check valve assembly {DC} or reduced pressure principle assembly {RP}). (See Chapter 3.)

1.7 Approved Testing Laboratory
The Foundation for Cross-Connection Control and Hydraulic Research of the University of Southern California or other laboratory having equivalent capabilities for both the laboratory and field evaluation of backflow prevention assemblies.

1.8 Approved Water Supply
Any public potable water supply, which has been investigated and approved by the health agency having jurisdiction. The system must be operating under a valid health permit. In determining what constitutes an approved water supply, the health agency has final judgment as to its safety and potability.

1.9 Aspirator
A device used for creating suction, specifically by flowing water through a venturi or restricted area of flow. At this restricted area of flow the pressure drops to sub-atmospheric, thus suction is created. Usually a tube is attached at this location for aspiration or suction purposes.

1.10 Aspirator Effect
The effect created by an aspirator, restricted area of flow or undersized piping.

1.11 Atmospheric Pressure
The pressure (or weight per unit area) exerted by the atmosphere on a surface. At sea level the atmospheric pressure is 14.7 psia (pounds per square inch, absolute).

1.12 Atmospheric Vacuum Breaker Backsiphonage Prevention Assembly (AVB)

An assembly containing an air inlet valve, a check seat and an air inlet port(s). (Also known as the non-pressure type vacuum breaker.) The flow of water into the body causes the air inlet valve to close the air inlet port(s). When the flow of water stops the air inlet valve falls and forms a check valve against backsiphonage. At the same time it opens the air inlet port(s) allowing air to enter and satisfy the vacuum. A shutoff valve immediately upstream may be an integral part of the assembly, but there shall be no shutoff valves or obstructions downstream. The assembly shall not be subjected to operating pressure for more than twelve (12) hours in any twenty-four (24) hour period. An atmospheric vacuum breaker is designed to protect against a non-health hazard (i.e., pollutant) or a health hazard (i.e., contaminant) under a backsiphonage condition only. See figure 1.2.

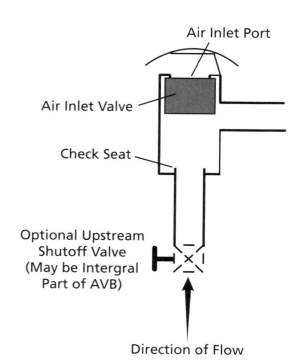

Figure 1.2
Atmospheric Vacuum Breaker Backsiphonage Prevention Assembly (AVB)

1.13 Auxiliary Water Supply

Any water supply on or available to the premises other than the water purveyor's approved public potable water supply. These auxiliary waters may include water from another purveyor's public potable water supply or any natural source such as a well, spring, river, stream, harbor, etc. They may be polluted or contaminated or they may be objectionable and constitute an unacceptable water source over which the water purveyor does not have sanitary control.

1.14 Backflow

The undesirable reversal of flow of water or mixtures of water and other liquids, gases or other substances into the distribution pipes of the potable supply of water from any source or sources. See terms Backpressure (1.16) and Backsiphonage (1.17).

1.15 Backflow Prevention Assembly

Any effective assembly used to prevent backflow into a potable water system. The type of assembly used shall be based on the existing or potential degree of hazard and backflow condition. The types of backflow prevention assemblies are:

- Atmospheric Vacuum Breaker Backsiphonage Prevention Assembly (see 1.12)
- Double Check Valve Backflow Prevention Assembly (see 1.29)
- Double Check Detector Backflow Prevention Assembly (see 1.30)
- Double Check Detector Backflow Prevention Assembly-Type II (see 1.31)
- Pressure Vacuum Breaker Backsiphonage Prevention Assembly (see 1.52)
- Reduced Pressure Principle Backflow Prevention Assembly (see 1.57)
- Reduced Pressure Principle Detector Backflow Prevention Assembly (see 1.58)
- Reduced Pressure Principle Detector Backflow Prevention Assembly-Type II (see 1.59)
- Spill-Resistant Pressure Vacuum Breaker Backsiphonage Prevention Assembly (see 1.64)

1.16 Backpressure

Any elevation of pressure in the downstream piping system (by pump, elevation of piping, steam pressure, air pressure, etc.) above the supply pressure at the point of consideration, which would cause or tend to cause a reversal of the normal direction of flow.

1.17 Backsiphonage

A form of backflow due to a reduction in system pressure, which causes a sub-atmospheric pressure to exist in the water system.

1.18 Certified Backflow Prevention Assembly Tester

A person who has proven ability in field testing backflow prevention assemblies to the satisfaction of the administrative authority having jurisdiction, either directly or through a third party certification program. Each person who is certified to perform field tests and prepare reports on backflow prevention assemblies shall be conversant in applicable laws, rules and regulations and have had experience in plumbing or pipe fitting or have other equivalent qualifications in the opinion of the administrative authority having jurisdiction.

1.19 Column of Water

A vertical tube of water usually used to create a specific pressure or used to measure pressure by the elevation of the water in the tube. (A column of water 27 3/4 inches (2.31 feet) high generates a pressure of one pound per square inch.) See Chapter 3.

1.20 Consumer

The owner or operator of an on-site water system(s) having a service from a public potable water system.

1.21 Consumer's Potable Water System

The portion of the privately owned potable water system lying between the point of delivery and the point of use. This system includes all pipes, conduits, tanks, receptacles, fixtures, equipment and appurtenances used to produce, convey, store or utilize the potable water.

1.22 Consumer's Water System(s)

Any water system located on the consumer's premises whether supplied by a public potable water system or an auxiliary water supply. The system or systems may be either a potable water system or an industrial piping system.

1.23 Containment
See Service Protection, 1.63

1.24 Contaminant
Any substance that shall impair the quality of water, in such a way as to create an actual hazard to the public health through poisoning, the spread of disease, etc.

1.25 Critical Level
The minimum elevation above the flood level rim of the fixture or receptacle served, downstream piping and water uses on atmospheric vacuum breakers, pressure vacuum breakers and spill-resistant vacuum breakers, at which the unit may be installed. This is indicated by the marking "C-L" or "C/L." When an AVB, PVB, or SVB does not bear a critical level marking, the bottom of the assembly shall constitute the critical level.

1.26 Critical Service
A water service that can never be interrupted due to the critical nature of facility involved.

1.27 Cross-Connection
Any actual or potential connection or structural arrangement between a public or a consumer's potable water system and any other source or system through which it is possible to introduce into any part of the potable system any used water, industrial fluid, gas, or substance other than the intended potable water with which the system is supplied. Bypass arrangements, jumper connections, removable sections, swivel or change-over devices and other temporary or permanent devices through which or because of which backflow can occur are considered to be cross-connections.

- A *direct cross-connection* is a cross-connection which is subject to both backsiphonage and backpressure.
- An *indirect cross-connection* is a cross-connection which is subject to backsiphonage only.

1.28 Degree of Hazard
Either a pollutant (non-health hazard) or contaminant (health hazard); derived from the assessment of the materials, which may come in contact with the distribution system through a cross-connection.

1.29 Double Check Valve Backflow Prevention Assembly (DC)
An assembly composed of two independently acting, approved check valves, including tightly closing resilient seated shutoff valves attached at each end of the assembly and fitted with properly located resilient seated test cocks. (See Chapter 10 for additional details.) This assembly shall only be used to protect against a non-health hazard (i.e., pollutant). See figure 1.3.

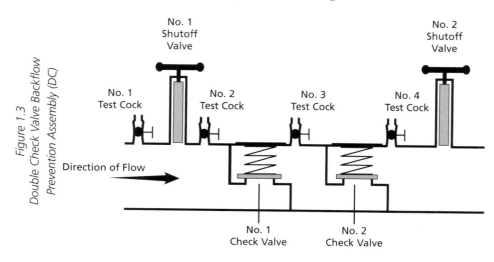

Figure 1.3 Double Check Valve Backflow Prevention Assembly (DC)

1.30 Double Check Detector Backflow Prevention Assembly (DCDA)

A specially designed assembly composed of a line-size approved double check valve assembly with a bypass containing a specific water meter and an approved double check valve assembly. The meter shall register accurately for rates of flow up to 2 gpm (gallons per minute) and shall show a registration for all rates of flow. (See Chapter 10, for additional details.) This assembly shall only be used to protect against a non-health hazard (i.e., pollutant). The DCDA is primarily used on fire sprinkler systems. See figure 1.4.

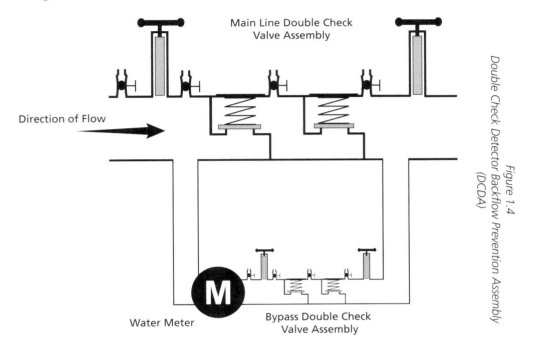

Figure 1.4
Double Check Detector Backflow Prevention Assembly (DCDA)

1.31 Double Check Detector Backflow Prevention Assembly-Type II (DCDA-II)

A specially designed assembly composed of a line-sized approved double check valve assembly with a bypass around the second check containing a specific water meter and a check valve. The meter shall register accurately for rates of flow up to 2 gpm and shall show a registration for all rates of flow. (See Chapter 10, for additional details.) This assembly shall only be used to protect against a non-health hazard (i.e., pollutant). The DCDA-II is primarily used on fire sprinkler systems. See Fig 1.5.

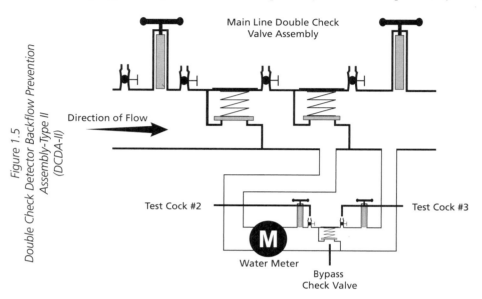

Figure 1.5 Double Check Detector Backflow Prevention Assembly-Type II (DCDA-II)

1.32 Gage Pressure
The pressure above atmospheric pressure.

1.33 Grey Water
Wastewater other than toilet contaminated waste. Wastewater generated by kitchen sinks and dishwashers are not considered grey water.

1.34 Health Hazard
See Contaminant, 1.24

1.35 Health Agency
The health authority having jurisdiction.

1.36 Hospital
Any institution, place, building, or agency which maintains and operates facilities for one or more persons for the diagnosis, care and treatment of human illness, including convalescence and care during and after pregnancy or which maintains and operates organized facilities for any such purpose, and to which persons may be admitted for overnight stay or longer. The term hospital includes sanitarium, nursing home and maternity home.

1.37 Industrial Fluids
Any fluid or solution, which may be chemically, biologically or otherwise contaminated or polluted in a form or concentration, which would constitute a hazard if introduced into an approved water supply.

1.38 Industrial Piping System
Any system used for transmission of or to confine or store any fluid, solid or gaseous substance other than an approved water supply. Such a system would include all pipes, conduits, tanks, receptacles, fixtures, equipment and appurtenances used to produce, convey or store substances which are or may be polluted or contaminated.

1.39 Internal Protection
The appropriate type or method of backflow prevention within the consumer's potable water system at the point of use, commensurate with the degree of hazard.

1.40 Isolation
See Internal Protection, 1.39.

1.41 Manifold Assembly
An assembly comprised of backflow prevention assemblies (DC or RP) of the same manufacturer, model and size. Manifold adaptor fittings on both the inlet and outlet of the manifold assembly are considered integral components. The size of the manifold assembly is determined by the inlet and outlet connections of the manifold adaptor fittings.

1.42 Negative Pressure
Any pressure below atmospheric pressure.

1.43 Non-health Hazard
See Pollution, 1.47.

1.44 Parallel installation
Two or more backflow prevention assemblies of the same type installed in parallel, that is having a common inlet, outlet and direction of flow.

1.45 Plumbing Hazard
An internal or plumbing type cross-connection in a consumer's potable water system with either a pollutant or contaminant.

1.46 Point of Delivery
See Service Connection. 1.62

1.47 Pollution
An impairment of the quality of the water to a degree which does not create a hazard to the public health but which does adversely and unreasonably affect the aesthetic qualities of such waters for domestic use.

1.48 Potable Water
Water from any source which has been investigated by the health agency having jurisdiction, and which has been approved for human consumption.

1.49 Pressure
A uniform force applied over a surface, measured as a force per unit area. Typically water pressure is measured in pounds per square inch or psi.

1.50 Pressure Fluctuation
The changes of pressure within a system.

1.51 Pressure Gradient
A description of the direction and rate of change of pressure over time.

1.52 Pressure Vacuum Breaker Backsiphonage Prevention Assembly (PVB)
An assembly containing an independently operating internally loaded check valve and an independently operating loaded air inlet valve located on the discharge side of the check valve. The assembly is to be equipped with properly located resilient seated test cocks and tightly closing resilient seated shutoff valves attached at each end of the assembly. (See Chapter 10, for additional details.) This assembly is designed to protect against a non-health hazard (i.e., pollutant) or a health hazard (i.e., contaminant) under a backsiphonage condition only. See figure 1.6.

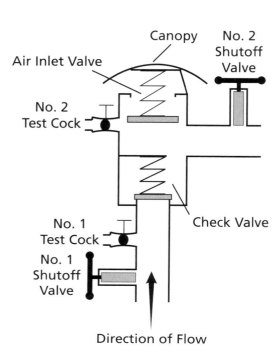

Figure 1.6
Pressure Vacuum Breaker Backsiphonage Prevention Assembly (PVB)

1.53 Public Potable Water System
Any publicly or privately owned water system operated as a public utility under a valid health permit to supply water for domestic purposes. This system will include all sources, facilities and appurtenances between the source and the point of delivery such as valves, pumps, pipes, conduits, tanks, receptacles, fixtures, equipment and appurtenances used to produce, convey, treat or store potable water for public consumption or use.

1.54 Readily Accessible
Capable of being reached for testing and/or maintenance, without the need of removing any access panel, door, or similar obstruction.

1.55 Reclaimed Water
Water which, as a result of treatment of wastewater, is suitable for a direct beneficial use or a controlled use that would not otherwise occur. Reclaimed water is not safe for human consumption.

1.56 Recycled Water
See Reclaimed Water

1.57 Reduced Pressure Principle Backflow Prevention Assembly (RP)
An assembly containing two independently acting approved check valves together with a hydraulically operating, mechanically independent pressure differential relief valve located between the check valves and at the same time below the first check valve. The unit shall include properly located resilient seated test cocks and tightly closing resilient seated shutoff valves at each end of the assembly. (See Chapter 10, for additional details.) This assembly is designed to protect against a non-health hazard (i.e., pollutant) or a health hazard (i.e., contaminant). This assembly shall not be used for backflow protection of sewage or reclaimed water. (Note: Check with local administrative authority for acceptable uses.) See figure 1.7.

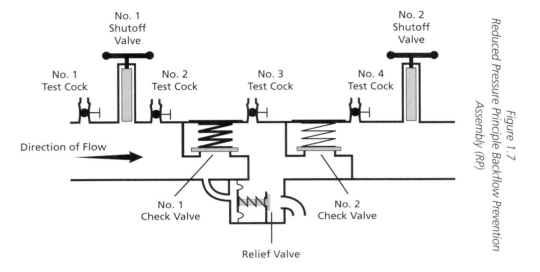

Figure 1.7
Reduced Pressure Principle Backflow Prevention Assembly (RP)

1.58 Reduced Pressure Principle Detector Backflow Prevention Assembly (RPDA)

A specially designed assembly composed of a line-size approved reduced pressure principle backflow prevention assembly with a specific bypass containing a specific water meter and an approved reduced pressure principle backflow prevention assembly. The meter shall register accurately for rates of flow up to 2 gpm and shall show a registration for all rates of flow. (See Chapter 10, for additional details.) This assembly shall be used to protect against a non-health hazard (i.e., pollutant) or a health hazard (i.e., contaminant). The RPDA is primarily used on fire sprinkler systems. See figure 1.8.

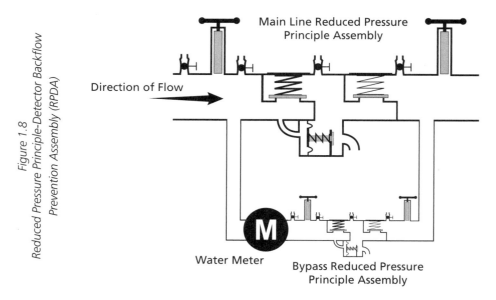

Figure 1.8
Reduced Pressure Principle-Detector Backflow Prevention Assembly (RPDA)

1.59 Reduced Pressure Principle-Detector Backflow Prevention Assembly-Type II (RPDA-II)

A specially designed assembly composed of a line-size approved reduced pressure principle backflow prevention assembly with a specific bypass around the second check valve containing a specific water meter and an approved check valve. The meter shall register accurately for rates of flow up to 2 gpm and shall show a registration for all rates of flow. (See Chapter 10, for additional details.) This assembly shall be used to protect against a non-health hazard (i.e., pollutant) or a health hazard (i.e., contaminant). The RPDA-II is primarily used on fire sprinkler systems. See figure 1.9.

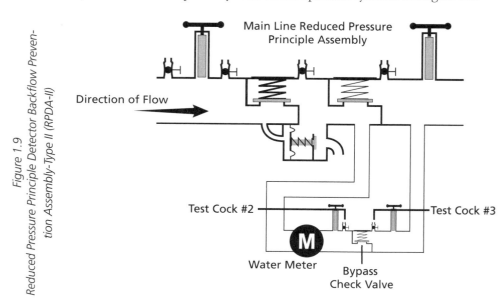

Figure 1.9
Reduced Pressure Principle Detector Backflow Prevention Assembly-Type II (RPDA-II)

1.60 Reused Water
See Reclaimed Water

1.61 Sanitary Sewer
The pipe that carries sewage.

1.62 Service Connection
The terminal end of a service connection from the public potable water system, (i.e., where the water supplier may lose jurisdiction and sanitary control of the water at its point of delivery to the consumer's water system). If a water meter is installed at the end of the service connection, then the service connection shall mean the downstream end of the water meter.

1.63 Service Protection
The appropriate type or method of backflow protection at the service connection, commensurate with the degree of hazard of the consumer's potable water system.

1.64 Spill-Resistant Pressure Vacuum Breaker Backsiphonage Prevention Assembly (SVB)
An assembly containing an independently operating internally loaded check valve and independently operating loaded air inlet valve located on the discharge side of the check valve. The assembly is to be equipped with a properly located resilient seated test cock, a properly located bleed/vent port, and tightly closing resilient seated shutoff valves attached at each end of the assembly. (See Chapter 10, for additional details.) This assembly is designed to protect against a non-health hazard (i.e., pollutant) or a health hazard (i.e., contaminant) under a backsiphonage condition only. See figure 1.10.

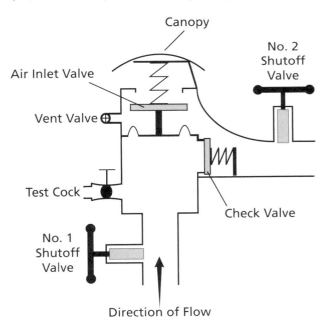

Figure 1.10
Spill-Resistant Pressure Vacuum Breaker Backsiphonage Prevention Assembly (SVB)

1.65 Static Pressure
The water pressure in any system under non-flowing conditions.

1.66 System Hazard
An actual or potential threat of severe danger to the physical properties of the public or the consumer's potable water system or of a pollution or contamination, which would have a protracted effect on the quality of the potable water in the system.

1.67 Thermal Expansion
The resulting effect when water in a closed system, such as a piping system downstream of a backflow preventer heats up. In effect, the heat causes the water volume to expand, but since the system is closed, the pressure increases.

1.68 Used Water
Any water supplied by a water purveyor from a public potable water system to a consumer's water system after it has passed through the service connection and is no longer under the control of the water purveyor.

1.69 Venturi
A piping apparatus with a constricted region designed to increase the velocity and thus decrease the pressure of an incompressible fluid in the constricted region.

1.70 Venturi Effect
When an incompressible fluid's velocity increases as a result of flowing through a constricted area of piping, the pressure will decrease.

1.71 Water Supervisor
The consumer or a person on the premises appointed by the consumer charged with the responsibility of maintaining the consumer's water system(s) on the property free from unprotected cross-connections and other sanitary defects, as required by regulations and laws.

1.72 Water Supplier
The public or private owner or operator of the potable water system supplying an approved water supply to the public.

NOTES:

Definitions

Chapter 2 — *History*

History

Inside this Chapter

History of Cross-Connection Control 20
    Ancient Civilizations 20
    The Modern Age 20
    Backflow Preventers 21
    Development of Regulations 24
    University of Southern California 25
    The State of the Industry 32

2 History

History of Cross-Connection Control

Ancient Civilizations

Civilizations have grown where there has been an abundance of drinking water, starting from prehistoric times to the present. When populations could not live directly adjacent to the drinking water supply, such as rivers and lakes; it became necessary to transport this essential water to the people for drinking purposes and irrigation of crops. Early forms of water transportation started with clay jars, animal skins and, where the geography would allow it, the development of simple ditches and canals to carry the water over long distances. The Romans developed an intricate network of aqueducts to carry water to their cities for drinking, bathing, decorative fountains and finally into the sewer system to carry away their waste. Perhaps the close proximity of the drinking water system and the waste system could have been interconnected inappropriately, constituting one of the earliest forms of a cross-connection. Even before Rome it is believed that complicated irrigation systems were used to raise water in elevation to irrigate the Hanging Gardens of Babylon, one of the Seven Wonders of the Ancient World.

The Modern Age

Many of the early water systems in cities in the United States were developed primarily for the suppression of fires in industrial areas. The first municipal water utility in the United States was established in Boston, Massachusetts in 1652 to provide domestic water and fire protection[1]. Pipes were fabricated out of hollowed logs. When needed to fight a fire the firefighters would drill a hole in the wooden water mains and plug them when finished. This was known as the "fireplug."

Figure 2.1
Early pipes were made of hollowed out logs

In many of the large industrial factories and mills along rivers, a system of pumps and piping from the nearby river dedicated for fighting fires in their facility would be installed. To ensure that water would be available, from either source, the water distribution system or the river, the two would be directly tied together. The fire underwriters that carried the insurance of the factory insisted that, in buildings where fire sprinklers were installed, a secondary water supply be provided to supplement the water system. A single check or gate valve was typically the only protection to prevent the river water from entering the water system. During some fire fighting efforts, it was found that not all of the water being pumped from the river went to suppressing the fire. In fact, river water was found to be pumped back into the water distribution system. The paper "Secondary Water Supplies, Their

---

[1]   Hanke, S.H. 1972. Pricing Urban Water. Pp.283-306 In: Public Prices for Public Products. S. Mushkin (ed). Washington, DC: The Urban Institute.

Danger and Values" presented at a 1910 New England Water Works Association meeting discussed the many cases where secondary supplies contaminated the water distribution system. The typical type of check valve at that time was an iron body with cast iron swing checks or clappers, producing a metal to metal seating surface. Tubercular rust growth between the iron components and infrequent inspection, if any, lead to very limited backflow protection. To improve the effectiveness of these metal to metal checks, it was reported that multiple check valves were installed in series, in some cases as many four or five were bolted together. In addition, some installation standards required a length of straight pipe installed between the check valves, long enough so that a rag or pair of overalls stuck inside couldn't foul multiple check valves.

Backflow Preventers

Check Valves

Figure 2.2 shows a sample of a check valve design dated 1903 which uses a linkage between the check valves so that when one check valve opens, the other one opens. In addition, when there is a flow of water through check valve "A", a separate linkage will close drain valve "C". Once the flow of water stops and the check valves close, the drain valve will open. All of the components are mechanically linked together.

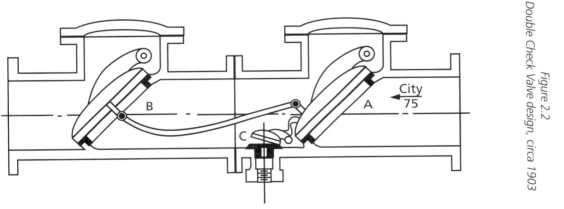

Figure 2.2
Double Check Valve design, circa 1903

As water distribution systems became more prevalent, bringing water to more and more customers, there was an increasing number of water using equipment, fittings and fixtures being developed. The convenience of running water changed how water was used on a customer's property. The advent of indoor plumbing had designers and inventors coming up with new ways to use water. The increased number of water users also meant that there was a greater risk of interconnecting the potable water supply to some non-potable or hazardous source.

Vacuum Breakers

Early United State patents were granted for devices that would prevent the return of contaminated sewage into the potable water supply. As shown in figure 2.3, in 1909 patent No. 926,968, Device for Preventing The Return of Waste Water into Water Supply Pipes, was issued which might be considered one of the first atmospheric vacuum breakers.

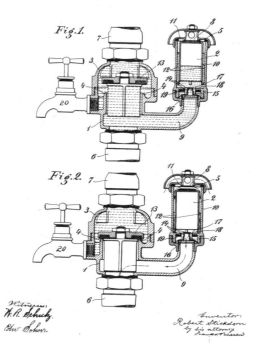

Figure 2.3
A patent applicaton from 1909 for a "Device for Preventing the Return of Water into Water Supply Pipes."

The common operating characteristic of the atmospheric vacuum breaker is that an opening will admit air into the piping arrangement to "break" the vacuum. Hence, the name vacuum breaker.

An Early Reduced Pressure Principle Device

In a neighborhood just west of downtown Los Angeles, drillers looking for oil struck 104 degree sodium rich water. This discovery turned into the Bimini Water Company, and this water would supply the local residents until the city water mains were installed in 1915. The thermal springs were fed to large public swimming pools in 1903, and local residents flocked to the soothing waters. The Bimini Baths (named after the Bahamian island where legend says the Spanish explorer Ponce de Leon searched for the Fountain of Youth) continued to expand their operation and a complex eventually covered fourteen acres.

Figure 2.4
Bimini Baths in Los Angeles, circa 1910

The owners of Bimini Baths asked the local water supplier Los Angeles Department of Water and Power (LADWP) for a potable water service connection in 1937. Mr. Norman Slane, LADWP's first cross-connection control inspector, made an inspection of the Bimini Baths and required the installation of a double check valve assembly on the service line to the Baths. The Operating Engineer for Bimini Baths was Mr. Orien Kersey Entriken, known to most people as O.K. Entriken. The President of the Bimini Baths was Mr. Frank Carlton, and during a visit to the Baths Mr. Entriken explained to Mr. Carlton the requirement from LADWP for a backflow prevention valve, and showed him a device that he thought would meet the needs. Mr. Entriken had been working on various designs of check valve mechanisms for several years, and had filed for a patent in 1933 for a backflow preventer so that "… there can be no contamination of the public supply from any particular service line." Mr. Carlton recognized the potential value of this valve arrangement, and formed a company to produce this type of product. The E.C. Valve Company (Entriken and Carlton) was formed, and produced various designs called E.C. #1, E.C. #2, etc. The Model E.C. #3 used a mechanical linkage from the inlet shutoff valve to open a relief valve when the inlet shutoff valve was closed.

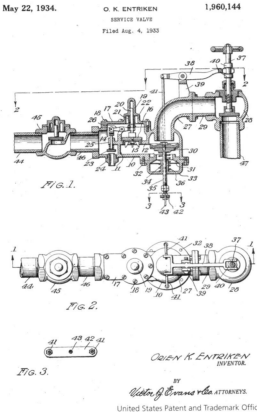

Figure 2.5
An early design by O.K. Entriken

Around 1941 Mr. Leonard Snyder was hired as an engineer by the EC Valve Company and the model EC #5 was developed which also contained an air inlet valve. In 1945 Mr. Snyder applied for a patent for a backflow preventer which used the reduced pressure principle. The patent stated: "This invention relates to backflow preventers and more particularly to a backflow preventer adapted to be positioned between a service or induction line and a consumer's or eduction line to prevent backflow from the consumer's line to the service line."

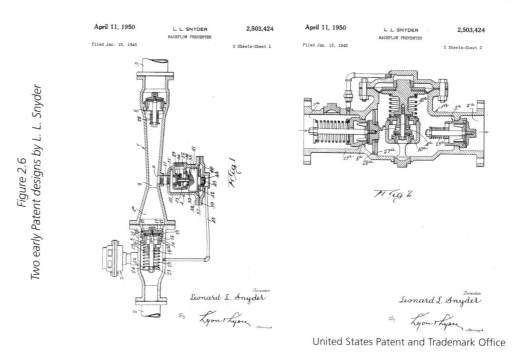

Figure 2.6
Two early Patent designs by L. L. Snyder

United States Patent and Trademark Office

In August 1948 Mr. Snyder asked Mr. David Guinn to join a newly formed company known as the Backflow Engineering and Equipment Company which later was shortened to BEECO. It was at this time when the BEECO model 6 was developed, the first reduced pressure principle assembly.

Development of Regulations

In 1912, the United States Congress passed the Public Health Service Act which authorized surveys and studies for water pollution, especially as it affects the public health. One of the initial water related regulations adopted was prohibiting the use of the "common-cup" on trains, ships, etc. Prior to this, passengers used a common cup to share water while on board. The first standards under the Public Health Service Act became law in 1914 which introduced the concept of maximum contaminant levels (MCLs) for drinking water served on interstate transportation, intending to protect the traveling public. Many state and local governments adopted these standards as guidelines.

Unfortunately, waterborne disease outbreaks were a fairly common occurrence, with one of the notable outbreaks taking place during the 1933 World's Fair in Chicago. An improperly designed plumbing system caused a backflow incident to occur, with 1,409 people coming down with amoebic dysentery and 98 deaths.

Backflow incidents such as this made it clear that the protection of the potable water in both the distribution system and internal plumbing system was critical and that all parties involved must participate to provide safe drinking water to the public. Cross-connection control programs were implemented in different areas of North America with different administrative authorities taking the lead in their respective regions. In some locations the primary force came from those administrative authorities directly involved with the oversight of the internal plumbing systems. Plumbing and listing laboratories were established in some of the larger cities (i.e., Chicago, New York, Los Angeles, etc.) to research, test and list plumbing products for use in their respective areas.

In the City of New York in 1938 a vacuum test bench was constructed to simulate the type of siphoning action which could take place in a typical plumbing system. Ormonde J. Burke, Chief Inspector at the New York Bureau of Water Supply described the test bench as being located on the fourth floor of a department repair shop, to permit the installation of a simulated riser of pipe over 33.9 feet tall. This was done in part to convince the public and code officials that vacuum conditions could be created in the plumbing system, sufficient enough to cause backflow from plumbing fixtures.

For a vacuum breaker to be acceptable to the administrative authority, tests had to be completed on this test bench by the Department of Water Supply, Gas and Electricity. Similar research and testing of plumbing arrangements and fixtures were conducted in other cities.

In other regions the water supplier developed as the lead agency overseeing the protection of the water distribution system and, in some cases, the internal plumbing system. In the 1930's some water agencies implemented cross-connection control programs where they would not only provide backflow protection at the service connection but they also attempted to oversee the protection of internal plumbing systems. However, as the number of water customers would grow in an area, it may have been too burdensome to continue overseeing both system and internal backflow protection programs.

As an example: In the 1930's, the LADWP operated a cross-connection control program in which they intended to inspect the internal plumbing systems of all their major water users and require backflow protection on all internal water applications. At one particular large complex, the LADWP inspectors went from building to building recording all the hazards and unprotected cross-connections. The intent was to work with the customer to have proper backflow protection installed on all internal uses of potable water. The inspectors had completed about one-half of the on-site buildings when they discovered a piece of water using equipment that they had not seen before. Wanting to make sure that they didn't overlook this on one of the previously completed buildings, the inspectors revisited one of the completed buildings. Upon reentering the building, it was found that the internal plumbing system had already been changed and new cross-connections created. It was at this point that the inspection team realized that internal changes were inevitable and that they did not have sufficient staff to re-inspect all facilities on a regular basis. The focus of their cross-connection control program was revised so that the LADWP would perform on site hazard assessments but require backflow protection at the service connection only.

In December 1941, a Cross-Connection Control Advisory Committee was formed through a joint resolution passed by the City of Los Angeles Department of Building and Safety, the Department of Health, the Board of Mechanical Engineers and the Department of Water and Power. It's chief purpose was to formulate and promote uniform policies and procedures for the elimination of cross-connections in all water piping systems throughout the city and in all buildings and structures containing water, sewer and plumbing fixtures, for the protection of public health and to prevent overlapping inspection by the various departments.

Under the authority of the Advisory Committee, inspectors from the various departments working on cross-connection elimination were deputized by each of the other departments so that each inspector carried the authority of all the departments represented by the Committee and could act in any individual instance from any of the departments. The scope of the work carried out by inspectors working under the authority of the Advisory Committee was much broader than that of an inspector working for one department only. Whereas the Water Department was interested only in protecting the quality of the water in its water mains and was satisfied to accept protection at the service connection, the Advisory Committee was interested in the protection of the water piping system on the consumer's premises where the health of persons working or living there were endangered. The Advisory Committee wanted to see the elimination of cross-connections on the consumer's property in addition to, or in lieu of, protection placed at the service connection.

The University of Southern California
In the early 1940's the USC Viterbi School of Engineering conducted training courses for those involved in the protection of the potable water supplies, including inspectors from the local water agencies and building and safety departments. So that the proper operation of plumbing components could be demonstrated to the students, a demonstration laboratory was constructed on the first floor of Biegler Hall. This laboratory contained operating cutaways of all types of mechanical devices and plumbing fixtures and the students could observe the water flowing through them. This training was

invaluable for the inspectors that had to understand and recognize these items in the field during their routine inspections.

Two of the primary instructors for this demonstration laboratory were Mr. Roy Van Meter and Mr. Harry Hayes. Each of them had been instrumental in the early development of the LADWP cross-connection control program and their many years of field experience conducting cross-connection control inspections made them excellent instructors.

In 1943, during World War II, a supply ship took on supplies in the Los Angeles/San Pedro Harbor. In their haste to return to sea, wrong valves were turned during the filling of the potable water tanks. Instead of potable water, the tanks were pumped full of salt water out of the harbor. Before the ship could get past the breakwater of the harbor, many of the people onboard ship had become ill. The ship had to return to port and an investigation revealed that this was caused by a cross-connection between the city water supply and the harbor water.

Because of this and other similar backflow incidents occurring in the area, a group of concerned individuals, believing that the unbiased efforts of an educational institution would serve the ultimate aim of protecting the potable water supplies best, approached the University of Southern California asking that research be done in the area. After several conferences this group worked out an agreement with the University and, in September 1944, the USC Board of Trustees established the Foundation for Cross-Connection Control Research (Foundation). An anonymous donation to the University was made to establish a laboratory and hire a team of researchers. Dr. Everett Cloran was named the first Director of the Foundation. He was joined by Mr. E. D. Alterton and Mr. William Tibbetts as the initial laboratory technicians.

The Foundation
A hydraulic laboratory was constructed on the campus of the University, housing a large vacuum system (11 feet in diameter by 30 feet long) for simulating backsiphonage conditions.

Figure 2.7 Original Foundation Laboratory on the University Park Campus of USC, circa 1945

Included in the early objectives of the Foundation were:
- To evaluate and supplement existing information on mechanical backflow prevention devices operating under line pressure
- To develop standard techniques for testing and approving mechanical backflow prevention devices
- To prepare specifications establishing minimum performance of mechanical backflow prevention equipment
- To prepare maintenance and test schedules and report forms

- To recommend materials, clearances, flow loss design, and similar related matters as they apply to backflow prevention equipment
- To investigate the effect of pipe sizes, vacuum conditions, industrial and domestic water pollution problems, and variations of pressures in water distribution systems
- To compare the relative value of the various methods now used as a means of eliminating or controlling dangerous cross-connections, with particular reference being given to the air gap system, the parallel piping method, double check valve installations, and model mechanical backflow prevention devices

Figure 2.8
North View of Original Laboratory, circa 1945

Some testing of various check valve designs was conducted in the laboratory. However, much of the initial work conducted by the laboratory technicians was field testing a variety of existing backflow preventers and gathering data regarding their effectiveness in preventing backflow. This data was compiled as part of an initial document by the Foundation and published by the School of Engineering as Paper No. 5. Papers No. 1 through 4 produced by other departments of the School of Engineering were on subjects unrelated to cross-connection control. Paper No. 5 reviewed the existing products used for backflow prevention, and contained a detailed set of design, material and operating specifications for double check valve devices and reduced pressure principle backflow prevention devices. For a product to comply with these specifications required that they not only successfully complete a set of laboratory tests (i.e., hydrostatic, flow rate versus pressure loss, etc.), but also exhibit acceptable performance under actual field conditions. The Field Evaluation was established to ensure that these health protecting products worked under "real-world" conditions. The Field Evaluation program had a two part process. After successfully completing the Laboratory Evaluation, three devices would have to be installed in acceptable field sites. The Foundation's laboratory technicians would field test these devices for a period of six months and, if they provided acceptable performance, the devices were granted "provisional" approval by the Foundation. Then three additional devices had to be installed, and these devices were evaluated in the field for a period of three years. At the successfully completion of the three year Field Evaluation the devices were granted "full approval."

In 1947 the original funding for the Foundation ran out and the original research team found work elsewhere. But, fortunately for the Foundation, the School of Engineering had just secured the services of Dr. Kenneth C. Reynolds as Professor of City Engineering. Dr. Reynolds, just arrived from the Massachusetts Institute of Technology, was asked to assume the directorship of the Foundation as part of his teaching load. However, during this period of time there was no regular funding for the operation of the Foundation. Then in 1956 a small study contract was entered into for the re-evaluation of the Paper No. 5. A much more expanded set of specifications were developed and published in 1959 as University of Southern California Engineering Center (USCEC) Report 48-101.

This work was accomplished under the direction of Dr. Reynolds with the assistance of Captain Walter H. Alback (U.S. Navy, retired) and Professor E. Kent Springer.

In 1959 the Foundation reviewed their evaluation process of the backflow preventers with representatives of the Oak Ridge National Laboratory (ORNL). The ORNL representatives were concerned about the use of backflow preventers in nuclear and radioactive service applications. Tests were conducted at ORNL in 1962 on a reduced pressure principle assembly simulating an installation to a radioactive process. A non-radioactive tracer solution used in their testing showed no detectable backflow.

The Southern California Water Utilities Association (SCWUA), a group formed in 1932 to promote awareness of topics appropriate to the water industry, established a special Specifications and Manual of Cross-Connection Control Practice Committee. They expanded USCEC 48-101 to include general definitions as well as recommended cross-connection control practices. This committee work was then edited by Dr. Reynolds and published by the Foundation as the *Manual of Cross-Connection Control-Recommended Practice* in 1960.

At the time, almost concurrently with the retirement of Dr. Reynolds in 1965, the University was enjoying a major building program on the University Park Campus. The site of the Foundation's laboratory was located in the middle of the Viterbi School of Engineering's new building complex, and the laboratory had to be demolished. During this period of time the evaluation work of the Foundation was carried out almost entirely by leasing time in industrial laboratories where the required capacity and availability could be found. The directorship of the Foundation was taken over by Professor E. Kent Springer in June 1965 as a part of his teaching load. A variety of new hydraulic research projects conducted by the Foundation lead to the expanded name Foundation for Cross-Connection Control and Hydraulic Research.

In 1966, the SCWUA reorganized itself by incorporation so that they could provide more support for the work of the Foundation through a membership program of water suppliers. In March of 1967 the SCWUA presented a financial gift to USC as the first of the Foundation Membership fees. The presentation to President Norman Topping of USC from Walter Weight, President of the SCWUA is shown in figure 2.9. Shown in the photo from left to right are: Glade Cookus, General Manager, Palos Verdes MWD; Carlton Peterson, General Manager/Chief Engineer, Diamond Bar MWD; E. Kent Springer, Director of the Foundation; Alfred Jorgensen, VP Lowry Associates; Walter Weight, President of SCWUA; William Whiteside, General Manager/Chief Engineer, California Domestic Water Company; Dr. Norman Topping, President of USC; George Sopp, Water Service Superintendant, LADWP; Gus Lenain Superintendant, City of Anaheim Water Department; Duncan Blackburn, General Manager, City of Pasadena Department of Water and Power; Dr. Alfred Ingersoll, Dean, USC Viterbi School of Engineering.

Figure 2.9
Presentation of check to USC President Norman Topping from the SCWUA

In the early 1960's an abandoned pumping station of the LADWP was found to be available and negotiations were initiated for the use of the space as the Laboratory of the Foundation. In April 1965, LADWP provided the laboratory new equipment that the Foundation could use for its continued operation and testing of devices as well as doing other hydraulic research. The LADWP also maintained part of the laboratory facility for their own testing requirements on large water meters and recording equipment. This was called the Riverside Laboratory.

Figure 2.10
Riverside Laboratory after renovation in 1968

Figure 2.11
Shortly after the renovation of the Riverside Laboratory in 1968.

Due to the Foundation's extensive background in the operation of the backflow prevention assemblies, the Foundation was called upon to train field personnel in the intricacies of field testing of the assemblies. Training seminars were conducted in the late 1960's throughout the United States, and in 1970 the Foundation conducted their first five-day *Course for the Training of Backflow Prevention Assembly Testers*. Once there was a need for training of administrators to run cross-connection control programs, the Foundation developed the *Course for the Training of Cross-Connection Control Program Specialists* in 1988.

Timelines

Directors
1944-1947: Dr. Everett Cloran
1947-1965: Dr. Kenneth C. Reynolds
June 1965-Sept 1984: Professor E. Kent Springer, P.E.
Sept 1984-present: Professor J. J. Lee, Ph.D., P.E.

Figure 2.12
Professor E. Kent Springer and Dr. J. J. Lee at the Foundation's 50th Anniversary Celebration

Publications
Paper No. 5 - April 1948
- Classified backflow preventers
- Double check valve assembly
- Superior pressure principle assembly
- Reduced pressure principle assembly
- Pressure vacuum breaker
- Non-pressure type vacuum breaker
- Established Laboratory and Field Evaluation Program for RP and DC
- Report of field investigations
- Established field test procedures

USCEC 48-101-January 30, 1959
- Established maximum allowable pressure losses for the RP and DC. Paper No. 5 did not contain a maximum allowable pressure losses.
- Established maximum working water pressure of 125 psi

Manual of Cross-Connection Control, 1st Edition-1960
- Field test procedures for RP (w/mercury manometer) and DC (w/duplex gage)
- Standards for RP and DC, sizes ¾"-10"

Manual of Cross-Connection Control, 2nd Edition-1965

Manual of Cross-Connection Control, 3rd Edition-1966

Manual of Cross-Connection Control, 4th Edition-1969
- Requirement for inline repairability
- Standards for RP and DC expanded to sizes ¼"-16"
- Maximum working water pressure raised from 125 psi to 150 psi
- Field Evaluation changed to twelve months, previously had been separate six month and three year field evaluations

Manual of Cross-Connection Control, 5th Edition-1974
- New Standard for PVB
- Field test procedures added for PVB
- Field test procedures for RP used differential pressure gage

Manual of Cross-Connection Control, 6th Edition-1979
- New standard for double check detector assembly (DCDA)

Manual of Cross-Connection Control, 6th Edition Revised-1982

Manual of Cross-Connection Control, 7th Edition-1985
- The term *assembly* was used throughout the Manual in place of *device* to reinforce the backflow preventer's inclusion of shutoff valves and test cocks as integral components
- Requirement for resilient seated shutoff valves and test cocks on all new assemblies.
- Standards required assemblies to be evaluated in the laboratory at elevated temperatures.
- New standard for AVB
- Field test procedures added preliminary steps (Notify, Identify, Inspect, Observe), and recommend yearly accuracy verification of gage equipment

Manual of Cross-Connection Control, 8th Edition-1988
- New Standard for Reduced Pressure Principle Detector Assembly (RPDA)

Manual of Cross-Connection Control, 9th Edition-1993
- Field test procedures contain illustrations for each step
- Field test procedure for DC changed from backpressure test to direction of flow test.
- New standard for spill resistant pressure vacuum breaker (SVB)
- Standards for all assemblies require a life-cycle test
- Standards for all assemblies require elevated temperature test at MWWP and MWWT.
- Requirement of replacement seats for all assemblies. Prior standards only required replacement seats for ferrous bodied assemblies.

State of the Industry

The Safe Drinking Water Act (SDWA) was originally passed by Congress in 1974 to protect public health by regulating the nation's public drinking water supplies. Initially, the SDWA focused primarily on the treatment processes as the means of providing safe drinking water. The 1986 and 1996 amendments (Public Law 104-182) greatly enhanced the existing law by recognizing source water protection, operator training, funding for water system improvements and public information as important components of a safe drinking water program. The SDWA authorizes the United States Environmental Protection Agency (US EPA) to set national health-based standards for drinking water. Under the SDWA, states may be granted primacy, which means the state must have regulations that are no less stringent than the regulations promulgated by the US EPA. The US EPA, states and water purveyors must then work together to make sure that these standards are met. To keep the drinking water safe, the SDWA sets up multiple barriers against hazards entering the drinking water supply. These barriers include: source water protection, treatment, distribution system integrity and public information. The SDWA applies to every public water system in the United States. A public water system is defined as having at least 15 service connections or serving at least 25 people per day for 60 days of the year.

Public water systems are divided into different categories:

- Community water system (there are approximately 54,000 in the United States): A public water system that serves the same people year-round. Most residences including homes, apartments and condominiums in cities, small towns and mobile home parks are served by Community Water Systems.
- Non-community water system: A public water system that serves the public but does not serve the same people year-round. There are two types of non-community systems:
  - Non-transient non-community water system (there are approximately 20,000): A non-community water system that serves the same people more than six months per year, but not year-round, for example, a school with its own water supply is considered a non-transient system.
  - Transient non-community water system (there are approximately 89,000): A non-community water system that serves the public but not the same individuals for more than six months, for example, a rest area or campground may be considered a transient water system.

There is no specific language in the SDWA addressing cross-connection control; however, most states recognize cross-connection control as an integral part of their water quality program. Under the SDWA, water suppliers are only responsible for the quality of the water delivered to the service connection, or water meter. There are other federal rules, such as the Lead and Copper Rule (LCR), which require the water suppliers to monitor the quality of water at the water user's tap. This testing on the water user's side of the service connection (i.e., water user's internal plumbing system) for the LCR requirements has created some confusion as it relates to cross-connection control programs. Since the water supplier is monitoring water quality at the water user's tap, some mistakenly interpret this to mean that the SDWA is also responsible for correcting/protecting cross-connections in the water user's internal plumbing system. However, the water supplier is not responsible for contaminants added to the water by circumstances under the control of the water user. Under the SDWA, the water supplier is only responsible for the quality of water delivered (i.e., service connection or water meter) to the water user.

Due to the lack of specific requirements for cross-connection control in the SDWA, states have developed their own set of rules/regulations regarding cross-connection control. The various state rules/regulations vary greatly, from those simply outlawing cross-connections, to those with very detailed and comprehensive cross-connection control programs. The US EPA has recognized the need for more detailed guidance and uniformity for cross-connection control programs throughout the United States, and there is consideration for the development of federal cross-connection control rules.

History

> Under the SDWA, states may be granted primacy, which means the state must have regulations that are no less stringent than the regulations promulgated by the US EPA

In conformance with the Federal Advisory Committee Act (FACA) the Microbial/Disinfection By-Products Rule Federal (M/DBP) Advisory Committee produced an Agreement in Principle which stated that the FAC recognized:

> ...that cross connections and backflow in distribution systems represent a significant public health risk...as part of the 6-year review of the Total Coliform Rule, EPA should evaluate available data and research on aspects of distribution systems that may create risks to public health and, working with stakeholders, initiate a process for addressing cross connection control and backflow prevention requirements and consider additional distribution system requirements related to significant health risks.

Subsequent workshops in 2000 and 2002 developed nine issue papers covering issues related to water distribution systems, with cross-connection control considered as one of the highest priorities. As of the publication date of this manual, the EPA was in the process of revising the Total Coliform Rule (TCR) to provide a more comprehensive approach for addressing water quality in the distribution system. To that end, the National Academies' Committee on Public Water Supply Distribution Systems: Assessing and Reducing Risks (Water Science and Technology Board) conducted meetings, with one of their tasks to identify and prioritize issues of greatest concern for distribution systems based on review of published material. Their first interim report in March 2005 identified cross-connections and backflow as one of the highest priorities based on the associated potential health risks. In September 2006 the final report, "Drinking Water Distribution Systems: Assessing and Reducing Risks" was published.

This report made several regulatory recommendations including:

- The EPA should work closely with representatives from states, water systems and local jurisdictions to establish the elements that constitute an acceptable cross-connection control program.
- For utilities that desire to operate beyond regulatory requirements, adoption of G200 or an equivalent program is recommended to help utilities develop distribution system management plans.
- More attention should be paid to having adequate facilities, instructors and apprentice programs to train utility operators, inspectors, foremen and managers.

At the time of printing, it is not known what new Federal Regulations may be developed regarding cross-connection control, but it is anticipated that this will be at least a basic framework of program elements (see the ABC's of a Cross-Connection Control Program in Chapter 3). This will support those states with existing programs more fully and help those states with limited programs to make improvements.

More information on the Total Coliform Rule and potential revisions and distribution system requirements can be found on the EPA's web site, www.epa.gov. This web site describes the TCR: purpose and requirements, the FAC recognition of the public health risks of cross-connections and backflow in distribution systems, the EPA's review process of the TCR as well as nine white papers that review information and research regarding potential public health risks associated with distribution systems. One of the nine issue papers of particular interest is "Potential Contamination Due to Cross-Connections and Backflow and the Associated Risks.[2]" The goal of this document is to review what is known regarding: (1) causes of contamination through cross-connections; (2) the magnitude of risk associated with cross-connections and backflow; (3) costs of backflow contamination incidents; (4) other problems associated with backflow incidents; (5) suitable measures for preventing and correcting problems caused by cross-connections and backflow; (6) possible indicators of a backflow incident; and, (7) research opportunities.

---

2   Environmental Protection Agency, *Potential Contamination Due to Cross-Connections and Backflow and the Associated Risks,* http://www.epa.gov/ogwdw/disinfection/tcr/regulation_revisions.html#whitepapers.

History

Chapter 3: Hydraulics

Hydraulics

Inside this Chapter

Hydraulics 38
    Water... 38

Backflow 43
    Backsiphonage 43
    Backpressure 46

Cross-Connections 46

Degree of Hazard 47

The Backflow Incident 48

How to Prevent Backflow 48
    Check Valve 49
    Double Check Valve Assembly (DC) 50
    Reduced Pressure Principle Assembly (RP) 52
    Atmospheric Vacuum Breaker (AVB) 54
    Pressure Vacuum Breaker (PVB) 55
    Spill-Resistant Pressure Vacuum Breaker (SVB) 56

Three Questions 57

Hydraulics of the Backflow Prevention Assemblies 59
    Double Check Valve Assembly 59
    Reduced Pressure Principle Assembly 65
    Pressure Vacuum Breaker Backsiphonage Prevention Assembly 74
    Spill-Resistant Pressure Vacuum Breaker
        Backsiphonage Prevention Assembly 78

3 Hydraulics

Hydraulics

To understand cross-connection control, it is essential for one to have a good understanding of the properties of water. A basic knowledge of hydraulics is, therefore, necessary to comprehend the concepts involved in the backflow of fluids.

Water

In general, people think water is something simple, but actually it is quite a complex substance. Water is the only known substance that can be a solid, a liquid and a gas under normal conditions of temperature and pressure. For example, a glass of water (liquid) with ice (solid) in it is refreshing on a hot summer day when the humidity, or water vapor (gas) in the air, is high.

Ice is another example of how water is unique in its properties. All other liquids become more dense as the temperature drops, but water becomes lighter as it becomes ice, and ice occupies more volume than the water it comes from. This is why ice floats in a glass of water.

If one takes a garden hose and fills a box that is exactly 12 inches by 12 inches by 12 inches, filling it completely with water, the result will be a cubic foot of water. One cubic foot of water weighs about 62.4 pounds. See figure 3.1.

1 cubic foot of water = 62.4 pounds

Figure 3.2 — A Cubic foot of water weighs 62.4 pounds

The "footprint" (or square at the bottom) of this cubic foot of water is 12 inches by 12 inches or 144 square inches. Taking the 62.4 pounds of water and dividing it by 144 gives 0.433 pounds of water exerted on each square inch of surface area of the footprint of the cubic foot. This is 0.433 pounds for every square inch in that footprint, or 0.433 pounds per square inch (lbs./in^2) or psi. See figure 3.2.

A height of twelve inches of water generates a pressure of 0.433 psi

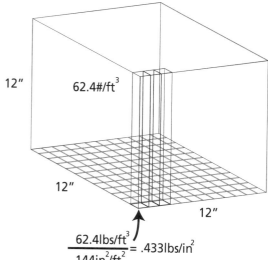

Figure 3.2

$$\frac{62.4 \text{lbs/ft}^3}{144 \text{in}^2/\text{ft}^2} = .433 \text{lbs/in}^2$$

psi is a unit of measurement used to measure pressure. Pressure is a force (pounds) exerted over a unit area (square inch). Water pressure in an open (or non-pressurized) system, such as an open tank, is generated by the elevation or height of water. As demonstrated above, a height of twelve inches of water generates a pressure of 0.433 psi at the bottom. If one wants to create a pressure of 1.0 psi, one would have to raise the height of water to 2.31 feet. See figure 3.3.

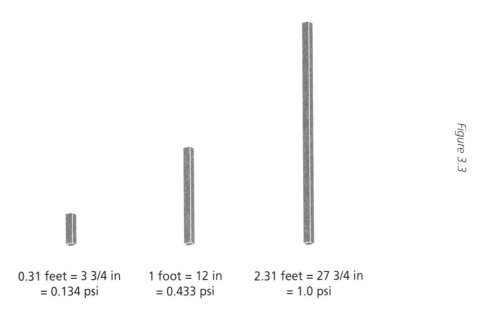

Figure 3.3

0.31 feet = 3 3/4 in = 0.134 psi

1 foot = 12 in = 0.433 psi

2.31 feet = 27 3/4 in = 1.0 psi

A water column 27 ¾ inches or 2.31 feet high will generate a pressure of 1.0 psi at the bottom. The diameter of the water column will not affect the pressure at the bottom of the water column. Even through there is more water in the larger diameter water column, this larger weight of water is spread out over a larger area. As shown in figure 3.4, if a container contains one pound of water and has a surface area at the bottom of the container of one square inch, the pressure at the bottom of the container is one pound per square inch, or 1 psi. If the container contains ten pounds of water, but the bottom of the container has a area of ten square inches, the pressure is still one pound per square inch. The same holds true with a container holding 100 pound of water with an area of 100 square inches at the bottom. The pressure would still be one pound per square inch.

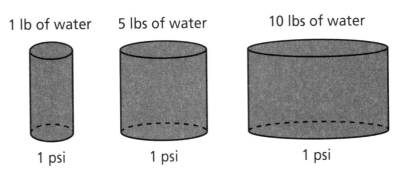

Figure 3.4

One psi is generated by an elevation of 27 3/4 inches of water. The diameter of the pipe or vessel is irrelevant. The pressure is generated by the elevation of the water.

In discussing pressure we must distinguish between *gage pressure* and *absolute pressure*. Gage pressure is the pressure above atmospheric pressure. Absolute pressure is the total pressure, that is, the sum of gage pressure and atmospheric pressure. See figure 3.5.

$$P_{ABSOLUTE} = P_{GAGE} + P_{ATMOSPHERIC}$$
$$P_{GAGE} = P_{ABSOLUTE} - P_{ATMOSPHERIC}$$

Figure 3.5

Atmospheric pressure is the pressure excerpted on the earth by the atmosphere above it. The weight of air is pushing down on the surface of the earth.

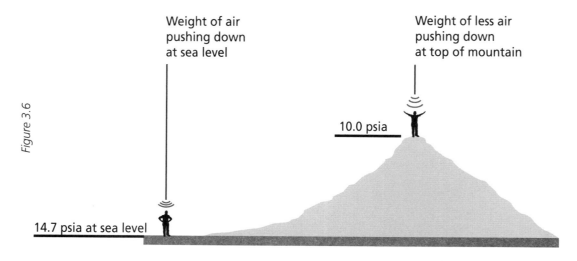

Figure 3.6

Atmospheric pressure, at sea level, is a nominal 14.7 psia (pounds per square inch, absolute). See figure 3.6. If there was no air pushing down on the surface of the earth, there would be no pressure, or the pressure would be 0.0 psia. This is known as a perfect vacuum. When pressure drops below atmospheric pressure, it is called negative pressure. Negative pressure is, in reality, a pressure which is below atmospheric pressure (i.e., nominal 14.7 psia, or 0.0 psig). Negative pressure is also called sub-atmospheric pressure.

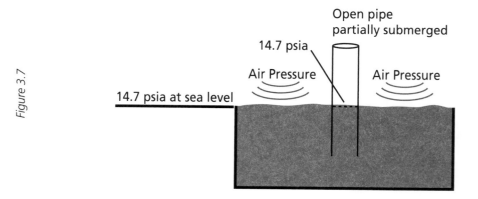

Figure 3.7

Pressure always seeks to equalize. As an example, if an open pipe is partially submerged into water, as shown in figure 3.7, pressure acts equally on the surface of water both outside and inside the pipe.

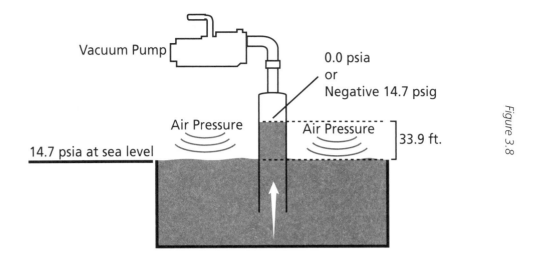

Figure 3.8

Now if this pipe is closed on the top and a vacuum is applied to the closed end of the pipe, the water level inside the pipe will rise due to the negative pressure caused by the vacuum pump. Theoretically, the highest level a perfect vacuum can pull water up in a tube at sea level is 33.9 feet (Fig. 3.8).

If a hose is filled with water and each end of the hose is submerged in a bucket filled with water, the pressures inside the hose will tend to equalize. An equilibrium condition will exist because the pressure inside the hose at point 'A' will be the same as at point 'B' (Fig. 3.9).

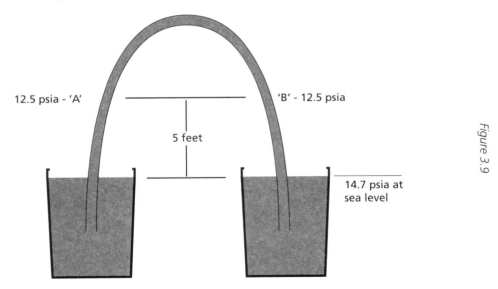

Figure 3.9

If one of the buckets is lowered, then the pressure inside the hose will change. Momentarily, the static pressures inside the two legs of the hose are different as shown in figure 3.10. The pressure at 'A' is now 12.5 psia, and at 'B' it is 10.4 psia. Water tends to move from a higher pressure to a lower pressure, so the water will begin to move from the higher bucket on the left, flow over the loop of the hose, into the lower bucket on the right.

3 Hydraulics

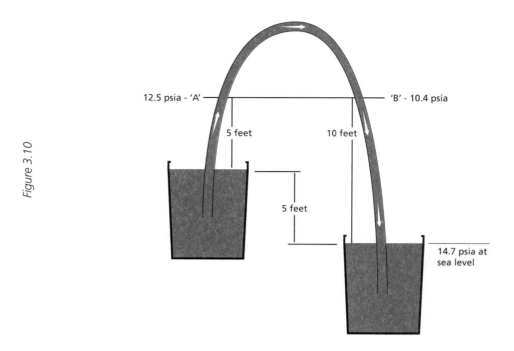

Figure 3.10

Once the water starts flowing through the hose, the dynamic pressures inside the hose will change slightly. This siphoning action will continue until the upper bucket is emptied or air enters the upper end of the hose.

This same siphoning principle is what one might use to empty a fish tank for cleaning or fill a gas can from the gas tank of a car.

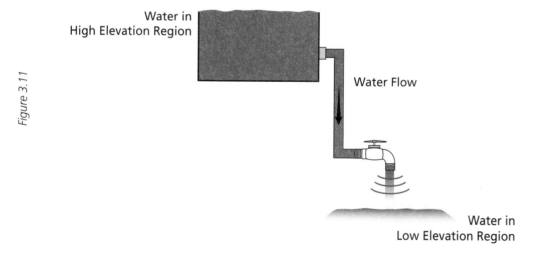

Figure 3.11

Backflow as defined in Chapter 1 is the "undesirable reversal of flow of water or other substances..."

Water at a higher elevation will seek to go to a lower elevation as shown in figure 3.11. The discharge end of a garden hose, for example, is at atmospheric pressure or 0.0 psig. When the hose bibb is opened, water enters the hose. The pressure will force the water to flow through the hose to the discharge end. In this same way, water flows through the pipes of a distribution system. Water flowing through the pipe tends to lose pressure due to friction between the inside of the pipe and the flowing water. Pressure must be generated from a pump or elevated storage, as shown in figure 3.12, to distribute or push the water throughout the water distribution system.

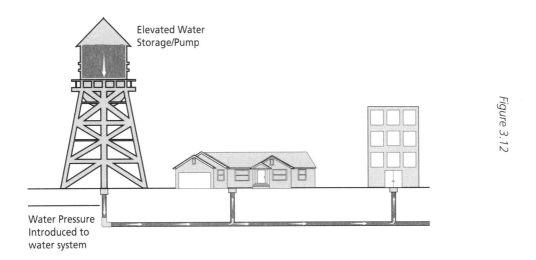

Figure 3.12

When the hydraulics of the system experience anomalies, the normal flow of the water distribution system is interrupted and problems may occur.

Backflow

Water within a pipe is free to travel in either direction. Depending upon the forces acting upon the water in the pipe, water tends to flow from areas of higher pressure to areas of lower pressure.

As water travels in a water distribution system to its intended end use, there are hydraulic conditions which may alter the normal direction of flow.

Backflow as defined in Chapter 1 is the "undesirable reversal of flow of water or other substances..." When a water pipe breaks, for example, the water in the pipe is not operating under normal operating conditions. The water in the pipe will flow to the region of lowest pressure which, in this case, would be the location of the pipe break which is open to atmosphere (0.0 psig). This may cause water in the pipe to flow in the other-than-intended direction (Fig. 3.13).

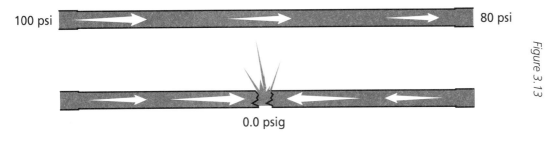

Figure 3.13

Backsiphonage

Hydraulic conditions may be such that the water pressure may be lowered to a sub-atmospheric condition. The resulting reversal in flow may not only bring water to the low pressure region but anything to which the water is connected as well. If there is a connection to a non-potable substance, (e.g., a hose in a bucket of insecticide) the sub-atmospheric pressure could draw the substance into the potable water pipe. This is called backsiphonage.

When pressure drops below atmospheric pressure, it is called negative pressure. Negative pressure is, in reality, a pressure, which is below atmospheric pressure.

3 Hydraulics

Several hydraulic conditions may cause backsiphonage. A water main break, as an example, could easily cause backsiphonage to occur. As illustrated in figure 3.14, a fire hydrant is knocked off opening the distribution system to atmosphere.

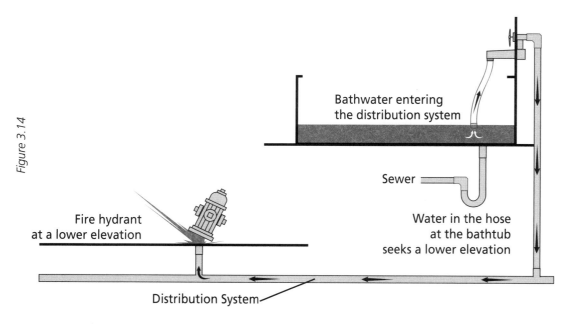

Figure 3.14

If this fire hydrant is located at a lower elevation than other parts of the nearby piping, backsiphonage may occur. A sucking of the water located in the bathtub occurs, drawing the bath water into the distribution system. The water in the hose at the bathtub seeks a lower elevation in order to equalize the pressure in the system.

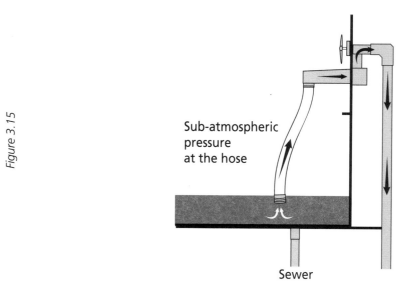

Figure 3.15

A drop in system pressure may create a sub-atmospheric pressure at the hose, sucking water from the bathtub into the drinking water system.

The pressure at the end of the hose is less than atmospheric pressure because the water between the hydrant and the hose is pulling downward toward the hydrant. This pulling creates a sub-atmospheric pressure at the end of the hose, sucking water from the bathtub into the drinking water system, as illustrated in figure 3.15. This is called *siphoning* and is a common hydraulic phenomenon.

The *aspirator effect* may also cause backsiphonage to occur. As water flows through a pipe, the pressure against the wall of the pipe decreases as the velocity of the water increases. Backsiphonage could occur if a large quantity of water is flowing through a pipe, with another pipe attached to it. The water flow jetting across the opening of the attached pipe may create an area of low pressure. Water could be siphoned from the attached pipe into the flowing pipe because of the low pressure at the connection between the two pipes. See figure 3.16.

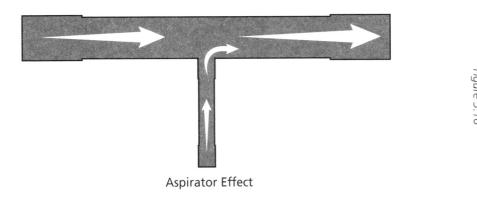

Aspirator Effect

Figure 3.16

A *venturi,* illustrated in figure 3.17, is a piece of equipment that is specifically designed to create a low-pressure region. It is essentially a piece of pipe with a smaller diameter in the center section of the pipe than at both ends. This causes the velocity of the water to increase through this narrow section. As this occurs the pressure decreases, possibly to sub-atmospheric pressure. A hose or pipe is connected to this low-pressure region in the center in order to use the vacuum as a siphon.

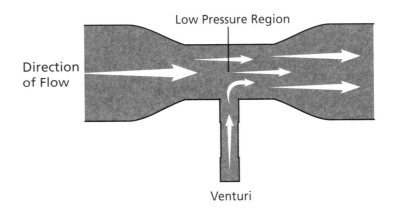

Venturi

Figure 3.17

Venturis are used commonly in laboratories to create a suction, which may be needed for various tasks. These are also commonly used in chemical or fertilizer applicators for home use. The garden hose is attached to one end of the applicator with a small siphon tube attached to the center of a venturi within the applicator. The small siphon tube is dropped into the chemicals or fertilizer, which is held in a the container of the applicator. As the water flows through the venturi, the chemicals are siphoned up from the container and distributed with the water onto the lawn, or plants. This is illustrated in figure 3.18.

Venturis are used quite commonly in laboratories to create a suction, which may be needed for various tasks

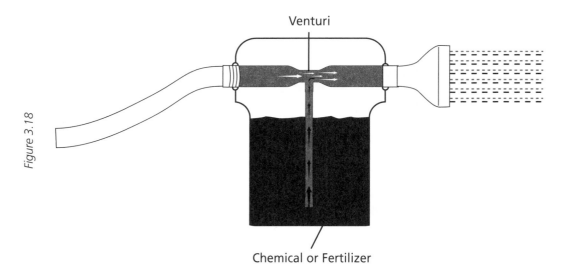

Figure 3.18

Backpressure

If the water pressure in the downstream system is higher than the supply pressure, (such as might occur when a pressurized re-circulating system is attached to the potable water supply through a water make-up line) this pressure may overcome the potable water supply causing water (or other substances) to be forced back into the potable water supply, as shown in figure 3.19. This is called *backpressure*.

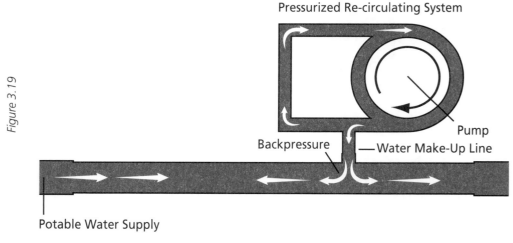

Figure 3.19

Cross-Connections

The passages through which backflow may occur are called cross-connections. A cross-connection is an actual or potential connection between a potable water supply and any non-potable substance or source. There are two types of cross-connections: indirect cross-connections and direct cross-connections. An *indirect cross-connection* is a cross-connection subject to backsiphonage only. Figure 3.20 shows an example of this: a garden hose with the end in a bucket with chemicals or some other hazardous material.

An indirect cross-connection is subject to backsiphonage only, while a direct cross-connection is subject to backpressure and backsiphonage

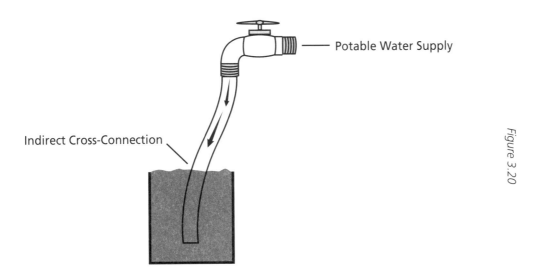

Figure 3.20

Should the water pressure drop to a sub-atmospheric condition, the hazardous material would be backsiphoned into the potable water supply. There is no way to push water back through an indirect cross-connection, since any addition to the bucket would simply overflow onto the surrounding ground. So, by definition, any indirect cross-connection cannot be subject to backpressure.

A *direct cross-connection* is a cross-connection which is subject to backpressure and backsiphonage. An example of a direct cross-connection may be a water make-up line attached to a boiler system, which has chemically treated water re-circulating through the boiler. If the pressure in the boiler system increases above the supply pressure, the treated boiler water would be pushed, or backpressured, into the potable water supply (Fig. 3.21).

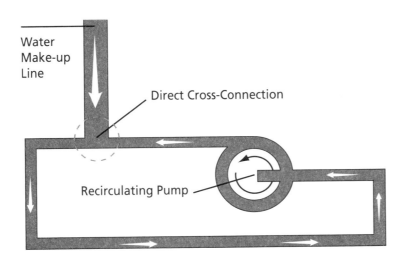

Pressurized Re-circulating System

Figure 3.21

A pollutant, is aesthetically objectionable, however, it will not cause illness or death.

A contaminant is a substance, which could cause illness or death.

Degree of Hazard

A *pollutant*, or non-health hazard, is considered to be a substance which is aesthetically objectionable. However, it will not cause illness or death. The substance may have an undesirable taste, smell, odor or look, but poses no health hazard. A *contaminant* (or health hazard), on the other hand, is a substance which could cause illness or death. The term *lethal hazard* is used specifically for radioactive material and raw sewage. It is important to know the degree of hazard, as there are different means of protecting against different degrees of hazard.

The Backflow Incident

In order for a backflow incident to occur three conditions must be met. First of all there must be a cross-connection. In other words, a passage must exist between the potable water system and another source. Second, a hazard must exist in this other source to which the potable water is connected. This may be a contaminant, a pollutant or a lethal hazard. Third, the hydraulic condition of either backsiphonage or backpressure must occur. When these three conditions are present, then a backflow incident occurs.

How to Prevent Backflow

Depending upon the type of cross-connection and the degree of hazard involved, there are different acceptable ways to prevent backflow from occurring.

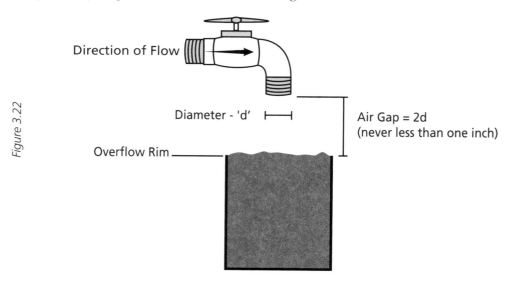

Figure 3.22

An *air gap* (Fig. 3.22) is a physical separation between the water supply line and the flood-level rim of the receiving vessel. The flood-level rim of the receiving vessel is the point at which water would flow over to the surrounding ground. This separation must be at least twice the diameter of the supply line but never less than one inch. An example of an air gap would be the water flowing from a pipe into a tank. If the pipe has a diameter of three-quarters of an inch and is at least one and one-half inches above the flood-level rim of the tank, then a proper air gap exists. Since the air gap actually creates a separation between the supply water and the substance in the receiving vessel, it is extremely effective for preventing backflow. It may be used to protect against a pollutant or a contaminant and it may be used to protect against a direct or indirect cross-connection. An air gap is the preferred method of protection against a lethal hazard.

A direct cross-connection may occur downstream of the receiving vessel. Should backpressure occur the resulting backflow would only flow over the rim of the receiving vessel but would never be introduced into the supply line. It should also be noted that an air gap should not be used in an area with a hazardous atmosphere, since backsiphonage could draw the hazardous atmosphere into the supply line.

Although the air gap is a very effective means of preventing backflow; it is not practical in every case. One must realize that once potable water passes through an air gap into an open vessel, two things occur. Sanitary control of the water is lost, and the supply pressure is lost. (See 1.4.)

> *In order for a backflow incident to occur three conditions must be met: There must be a cross-connection. A hazard must exist and the hydraulic condition of either backsiphonage or backpressure must occur*

Check Valve

Although a single check valve is not an acceptable means of preventing backflow, it is often an important component of backflow prevention assemblies. A check valve is a one-way valve, which will allow water to pass in only one direction (Fig. 3.23). This may be accomplished in many different configurations and designs but typically in the backflow prevention assembly the check valves will be one of two general types: an axially moving poppet or hinged swing check.

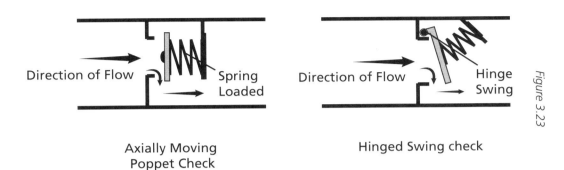

Figure 3.23

In either case, check valves used in backflow preventers must be biased to the closed position. This means that there is some force holding the check valve closed. This force is typically generated in the form of an internal spring loading. Internal spring loading means that the spring or springs are not able to be affected unless the check valve is disassembled. Many older assemblies used a weight for loading the check valve. However, the weight-loaded assembly has been phased out for the most part. Weights were predominantly composed of lead. Due to the dangers associated with lead in the drinking water, many of these have been removed from service and none are currently being produced using lead as the weight.

In order to ensure that water can only pass through check valves in one direction, check valves in backflow prevention assemblies are designed to have a drip tight seal. This cannot be accomplished effectively with two hard surfaces (e.g., metal to metal), so the check valves typically use a soft sealing surface or elastomer (i.e., rubber) disc to seal against a seat. See figure 3.24.

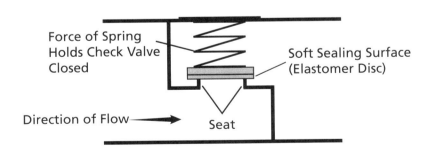

Figure 3.24

A check valve is a one-way valve, which will allow water to pass in only one direction.

The elastomer is forced to close against the seat because of a loading such as a spring. Opening the check valve requires that the water pressure in front of the check valve must be greater than the force holding it closed. When the inlet (upstream) pressure increases sufficiently the water pressure will overcome the force of the spring and open the check valve (Fig. 3.25). This allows water to flow from the inlet, through the check valve to the outlet (downstream). When the flow of water ceases, the check valve will close.

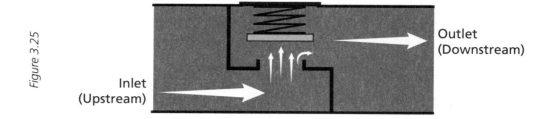

Figure 3.25

If the upstream pressure drops, the check valve closes because of the force of the spring. If backpressure occurs trying to force the water from the downstream side of the check valve to the upstream side, the force of the spring and the backpressure closes the check valve. Thus, water only flows in one direction through a properly operating check valve. A single check valve is not considered protection against backflow. A single check valve can become fouled by debris in the water line and, therefore, not provide adequate protection against backflow.

Double Check Valve Assembly (DC)

The *double check valve assembly* (Fig. 3.26) consists of two internally loaded check valves in series, two resilient seated shutoff valves and four resilient seated test cocks for field-testing purposes. (When a shutoff valve or test cock is resilient seated, this means that the seating surface or sealing surface, is not a metal to metal seal.) If one of the check valves would fail to seal properly, because of debris in the water or any other reason, the other check valve acts as a backup and should prevent backflow from occurring.

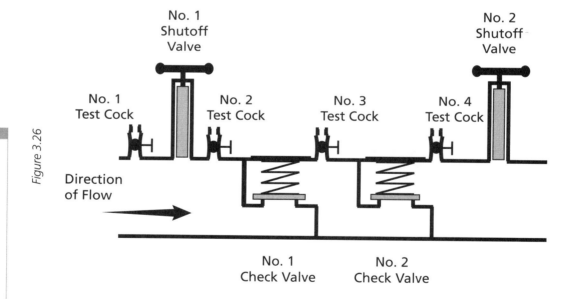

Figure 3.26

The DC may be used to protect against a direct or indirect cross-connection, and may be used to protect against a pollutant (non-health hazard) only.

Two assemblies similar to a double check valve assembly are the double check detector assembly (DCDA) and the double check detector assembly-type II (DCDA-II).

The DCDAs and DCDA-IIs are usually used on water lines that are static, but need the ability to detect water use, such as a fire sprinkler system. Since fire sprinkler systems are not normally metered, the DCDA and DCDA-II provide a means to detect any unauthorized use of water, or any leaks in the system.

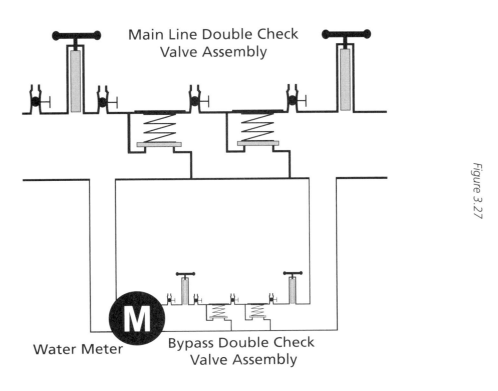

Figure 3.27

The DCDA (Fig. 3.27) includes a mainline double check valve assembly and a bypass arrangement. The bypass arrangement attaches to the mainline assembly from upstream of the No. 1 check valve to downstream of the No. 2 check valve. The bypass arrangement includes a specific double check valve assembly along with a specific water meter. These assemblies are designed to require all flows of 2 gallons per minute (gpm) or less to flow exclusively through the bypass arrangement and register accurately on the water meter. At greater flows water will flow through the mainline assembly as well as the bypass arrangement. The meter on the bypass arrangement will continue to register, however, registration will not be accurate, as the majority of the flow will be flowing through the mainline assembly, which is not metered.

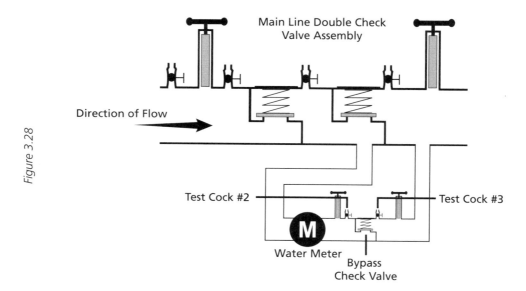

Figure 3.28

The DCDA-II (Fig. 3.28) is similar to the DCDA. However, the bypass arrangement is attached to the mainline assembly from downstream of the No. 1 check valve to downstream of the No. 2 check valve and contains a single check valve and a water meter. The DC, DCDA and the DCDA-II provide the same level of backflow protection.

Reduced Pressure Principle Assembly (RP)

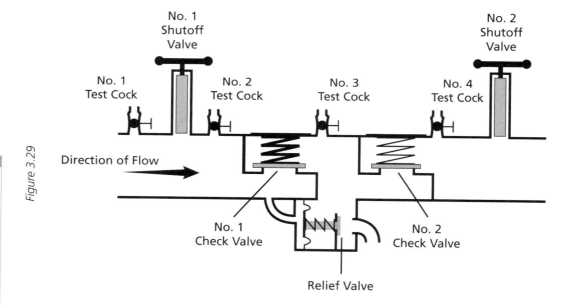

Figure 3.29

Hydraulics

The relief valve of an RP is mechanically independent of the check valves, yet hydraulically dependent upon the differential across the No. 1 check valve.

The *reduced pressure principle assembly* is a mechanical backflow prevention assembly. See figure 3.29. The RP consists of two internally loaded check valves in series and a mechanically independent differential pressure relief valve located between the two check valves. This assembly also includes two resilient seated shutoff valves and four resilient seated test cocks for field-testing purposes.

The differential pressure relief valve, although mechanically independent of the rest of the assembly, is hydraulically dependent upon the difference in pressure across the first check valve. In essence, the relief valve senses the difference in pressure between upstream of the No. 1 check valve and downstream of the No. 1 check valve. The relief valve is designed to discharge the water between the two check valves when the differential pressure across the first check drops to a certain point. This point must be at least 2.0 psid. This keeps the pressure downstream of the No. 1 check valve at least 2.0 psi less than the pressure upstream of the No. 1 check valve. This ensures that water will only flow in the proper direction, since water generally flows from a higher pressure to a lower pressure. Essentially, this means the relief valve can sense if backflow is beginning to occur or if one of the check valves is failing and release pressure from the region between the check valves. This will either correct the hydraulic condition or it will continue to discharge water from between the check valves in order to keep contaminated water from moving upstream into the potable water supply.

Because the RP is designed to prevent backflow even if one or both of the check valves fail, it is a very reliable means of backflow prevention. The reduced pressure principle assembly may be used to protect against a pollutant or a contaminant and may be used to protect against a direct or indirect cross-connection.

Two assemblies similar to a reduced pressure principle assembly are the reduced pressure principle detector assembly (RPDA) and the reduced pressure principle detector assembly-type II (RPDA-II). The RPDA and RPDA-II are designed for use on fire sprinkler systems. Since fire sprinkler systems are not typically metered, this provides a means to detect any unauthorized use of water or any leaks in the system.

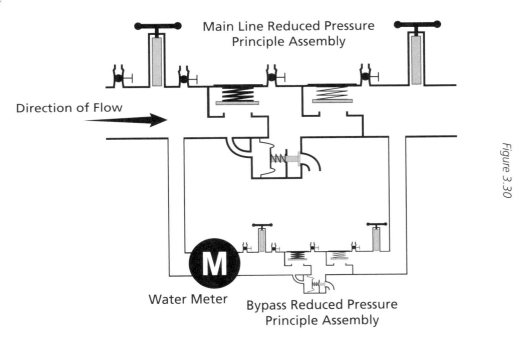

Figure 3.30

The RPDA (Fig. 3.30) includes a mainline reduced pressure principle assembly and a bypass arrangement. The bypass arrangement is attached to the mainline assembly upstream of the No. 1 check valve and downstream of the No. 2 check valve. The bypass arrangement includes a specific reduced pressure principle assembly along with a specific water meter. These assemblies are designed to require all flows of 2 gpm or less to flow exclusively through the bypass arrangement and register accurately on the meter. At larger flows water will flow through the mainline assembly as well as the bypass arrangement. The meter on the bypass arrangement will continue to register. However, registration will not be accurate, as the majority of the flow will be flowing through the mainline assembly, which is not metered.

A shutoff valve is a tightly closing valve used as an integral part of many backflow preventers.

A test cock is a tightly closing valve located at specific locations on backflow preventers for testing purposes.

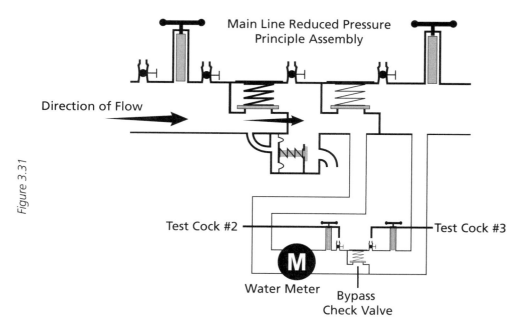

Figure 3.31

The RPDA-II (Fig. 3.31) is similar to the RPDA. However, the bypass arrangement is attached from downstream of the No. 1 check valve to downstream of the No. 2 check valve and contains a single check valve and a water meter. The RP, RPDA, and the RPDA-II provide the same level of backflow protection.

Atmospheric Vacuum Breaker (AVB)

The *atmospheric vacuum breaker* (Fig. 3.32) consists of an air inlet valve, a check seat and an air inlet port(s). Water flow into the atmospheric vacuum breaker causes the air inlet valve to close. This allows water to flow through the assembly. When the water flow stops, the air inlet valve drops down to the check valve seat. Additionally, as the air inlet opens it allows air into the body of the assembly through the air inlet port, which compensates for any vacuum and separates the downstream non-potable water from the upstream potable water. This assembly is effective when used in the proper application.

The AVB is used to protect against either a pollutant or a contaminant under backsiphonage conditions only. The AVB may only be under pressure for twelve out of any twenty-four hour period.

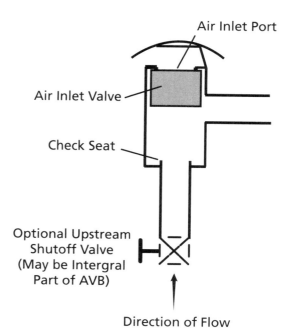

Figure 3.32

The AVB may only be used to prevent backsiphonage but it may be used to protect against either a pollutant or a contaminant. It is important to note that the atmospheric vacuum breaker must be installed at least six inches above all downstream piping and outlets. Additionally, this assembly may not be subjected to continuous pressure. It may only be in use for twelve hours out of any twenty-four hour period and may have no shutoff valves or control valves downstream.

Another type of atmospheric vacuum breaker is commonly found on hose bibbs. These atmospheric vacuum breakers (also known as hose bibb vacuum breakers) are designed to be attached on the discharge side of a hose bibb, hydrant or faucet which has hose threads. The hose bibb vacuum breaker consists of a check valve loaded to the closed position, and an air inlet valve loaded to the open position. Water flow into the hose bibb vacuum breaker causes the check valve to open and the air inlet valve to close. When the water flow stops and the device is not under pressure, the air inlet valve opens allowing air to enter. There are different classifications of hose bibb devices. Some of these may contain two check valves, or be freeze resistant and have automatic draining features for cold weather applications. The hose bibb is not to be subjected to more than twelve hours out of any twenty four hour period of continuous water pressure and may be used to protect against either a pollutant or contaminant. They are designed to be used to protect against backsiphonage and low head backpressure created from a hose raised no more than ten feet above the hose bibb.

Other specialized types of vacuum breakers or anti-siphon valves, may be incorporated into ballcocks and flushometer valves used on the potable water supply to toilets and urinals. See Chapter 7.

Pressure Vacuum Breaker (PVB)

The *pressure vacuum breaker* (Fig. 3.33) consists of an internally loaded check valve, a loaded air inlet valve, two resilient seated shutoff valves and two resilient seated test cocks for field-testing and maintenance.

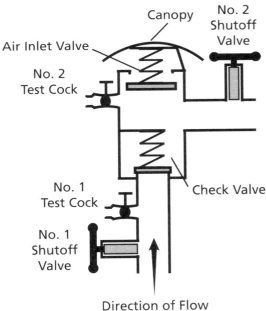

Figure 3.33

The PVB may be used to protect against either a pollutant or a contaminant under backsiphonage conditions only. It may be used under continuous pressure.

The PVB is designed so that, when the assembly is pressurized, the check valve opens and the air inlet valve closes allowing water to flow through the assembly. The loaded check valve prevents water from flowing in the reverse direction of flow. The air inlet valve opens when the body pressure is reduced allowing air into the body of the assembly to separate the upstream potable water from the non-potable water downstream. The air inlet valve is loaded to ensure the air inlet will open when the water pressure drops in the body, even if the assembly has been subjected to continuous pressure. The pressure vacuum breaker may be used to protect against either a pollutant or a contaminant but it may only be used to protect against backsiphonage. This assembly must be installed at least twelve inches above all downstream piping and outlets; it may be used under continuous pressure.

Spill-Resistant Pressure Vacuum Breaker (SVB)

The *spill-resistant pressure vacuum breaker* (Fig. 3.34) is essentially the same as a pressure vacuum breaker, in application. However, it is designed to minimize the discharge of water from the air inlet port. This enables the end user to use the SVB in situations where it would be undesirable to have water discharged into the surrounding areas.

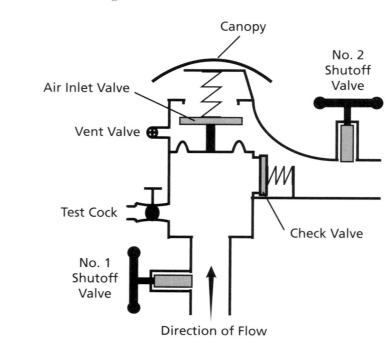

Figure 3.34

The SVB has an internally loaded check valve and a loaded air inlet valve, two resilient seated shutoff valves, one test cock and one vent valve. The SVB is designed so that, when the assembly is pressurized, the air inlet valve will close before the check valve opens. This will keep water from discharging through the air inlet valve when the assembly is pressurized. The SVB may be used to protect against a pollutant, or contaminant, but may only be used to protect against backsiphonage. The SVB may be used under continuous pressure and it must be installed at least twelve inches above all downstream piping and outlets.

The SVB is a pressure vacuum breaker designed to be resistant to spilling.

Three Questions

By answering three questions about any cross-connection; one is able to determine which type, if any, of backflow prevention assembly is needed.

1. Is the connection a direct or indirect cross-connection?
2. Is the cross-connection to a pollutant (non-health hazard) or a contaminant (health hazard)?
3. Is the connection under continuous use or pressure (used for more than twelve of any twenty-four hours)?

With the answers to the above questions, one may turn to the flow chart in figure 3.35 or the table in figure 3.36 and determine which backflow prevention assembly would be acceptable. For example, should there be a contaminant under continuous pressure through a direct cross-connection, one would look at the table or flowchart and determine that an air-gap or a reduced pressure principle assembly would be acceptable.

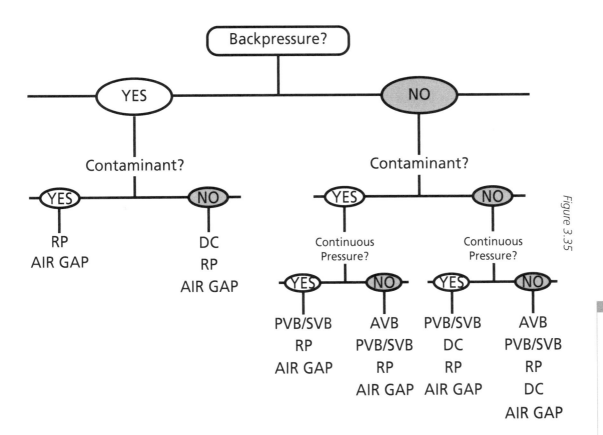

Figure 3.35

TYPES OF BACKFLOW PROTECTION

	INDIRECT Backsiphonage Only		DIRECT Backsiphonage & Backpressure
	Continous Use	Non- Continous Use	
Health Hazard (Contaminent)	PVB/SVB RP AIR GAP	AVB PVB/SVB RP AIR GAP	RP AIR GAP
Non-Health Hazard (Pollutant)	PVB/SVB RP DC AIR GAP	AVB PVB/SVB DC RP AIR GAP	DC RP AIR GAP

Figure 3.36

Hydraulics of the Backflow Prevention Assemblies

Double Check Valve Assembly

Normal Flow

The operating mechanics of a double check valve backflow prevention assembly follow some basic principles. To understand its operation further, assume an assembly has been installed in a piping system and is being filled and pressurized with water. As the inlet, or No. 1 shutoff valve is being opened, the incoming water fills the chamber between the No. 1 shutoff valve and No. 1 check valve (Fig. 3.37). Once this chamber is filled, the pressure increases. The amount of water pressure necessary to cause the check valve to move or open will depend upon the amount of force or loading trying to keep the check valve in a closed position. This force must be at least 1.0 psi.

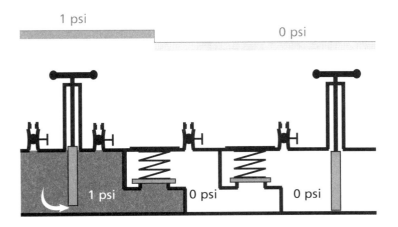

Figure 3.37
The Assembly is Pressurized as the No. 1 Shutoff Valve is Opened.

The water pressure will increase to the point that it overcomes the force keeping the No. 1 check valve closed. The No. 1 check valve opens and water fills the chamber between the No. 1 check valve and the No. 2 check valve (Fig. 3.38).

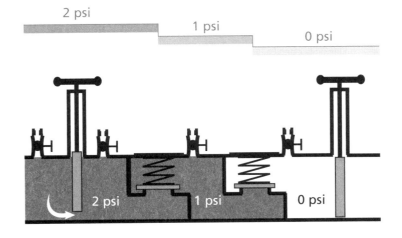

Figure 3.38
Water fills the Chamber between the Check Valves

Once the chamber between the check valves fills, the water pressure increases to the point that it overcomes the force keeping the No. 2 check valve closed. This force is at least 1.0 psi. The No. 2 check valve opens and water fills the chamber between the No. 2 check valve and the No. 2 shutoff valve (Fig. 3.39). Once this final chamber fills, the water pressure continues to increase until the assembly is fully pressurized. Once the assembly is fully pressurized and there is no further flow into the assembly, both the No. 1 and No. 2 check valves close.

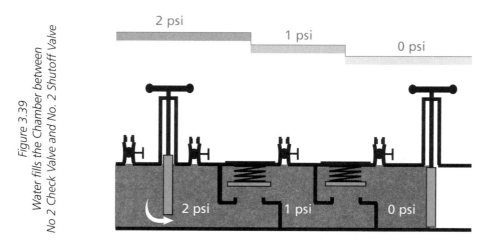

Figure 3.39
Water fills the Chamber between No 2 Check Valve and No. 2 Shutoff Valve

After the No. 1 shutoff valve is fully opened, the No. 2 shutoff valve is fully opened allowing flow through the assembly (Fig. 3.40).

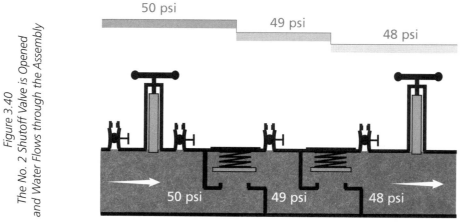

Figure 3.40
The No. 2 Shutoff Valve is Opened and Water Flows through the Assembly

The amount the check valves open depends on the amount of water flowing through the assembly. If there is no water use downstream, the water is in a static (non-flowing) condition and the No. 1 and No. 2 check valves are closed. However, if there is water use downstream, the water is in a dynamic (flowing) condition and the No. 1 and No. 2 check valves open. The more water usage, the more water that flows through the assembly and the more fully the check valves will open.

The amount the check valve opens depends on the flow rate (Fig. 3.41).

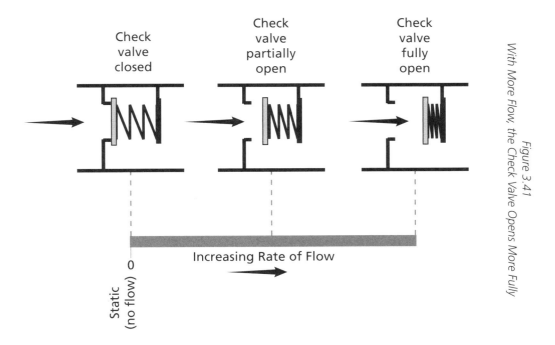

Figure 3.41
With More Flow, the Check Valve Opens More Fully

Backpressure Condition

Under normal operating conditions, the check valves open and close depending on the amount of water flowing through them.

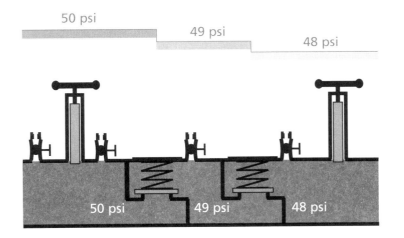

Figure 3.42
With the Flow at a Static Condition, The Check Valves are Closed.

As the flow rate decreases to a static condition, the check valves close (Fig. 3.42). When the downstream pressure increases (backpressure), the No. 2 check valve remains closed. This prevents the downstream water from flowing to the upstream side of the No. 2 check valve. The No. 1 check valve is also closed (Fig. 3.43).

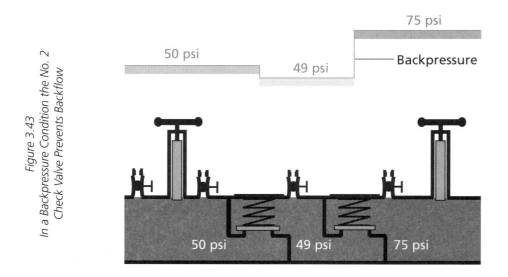

*Figure 3.43
In a Backpressure Condition the No. 2
Check Valve Prevents Backflow*

If there is a backpressure condition and the No. 2 check valve is leaking so that water can flow backwards through the No. 2 check valve, the No. 1 check valve remains closed (Fig. 3.44). This prevents the downstream water from flowing to the upstream side of the No. 1 check valve, even though the No. 2 check valve is leaking.

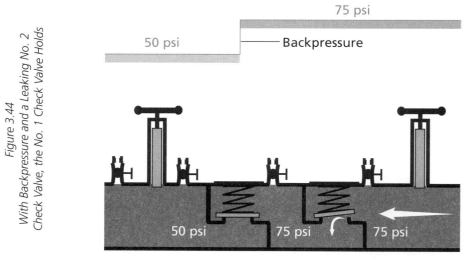

*Figure 3.44
With Backpressure and a Leaking No. 2
Check Valve, the No. 1 Check Valve Holds*

If there is a backpressure condition with both check valves leaking, the downstream water will flow to the upstream side of the assembly. Under this condition, backflow occurs through the assembly (Fig. 3.45). This is one of the reasons the double check valve backflow prevention assembly may only be used to protect against a pollutant.

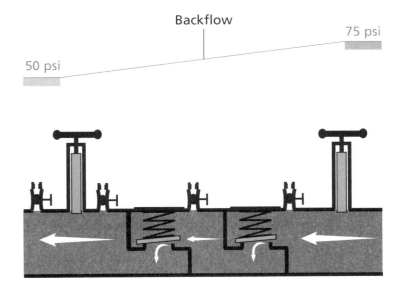

Figure 3.45
If Both Check Valves are Fouled the Assembly will Leak in a Backpressure Condition.

The amount of backflow will depend upon the size of the leak through the check valves. If the check valves are fouled due to small debris in the line or the sealing surfaces of the check valves are slightly damaged or worn, then the leak may be small. If the check valves are grossly fouled or damaged, the amount of backflow is greater.

Backsiphonage Condition

Under normal operating conditions, the check valves open and close depending on the amount of water flowing through them. As the flow rate decreases to a static condition, the check valves close. Should there be a decrease in the upstream pressure so that it is below atmospheric pressure (backsiphonage), the No. 1 check valve remains closed (Fig. 3.46). This prevents the water between the two check valves from being siphoned to the upstream side of the No. 1 check valve. The No. 2 check valve also remains closed.

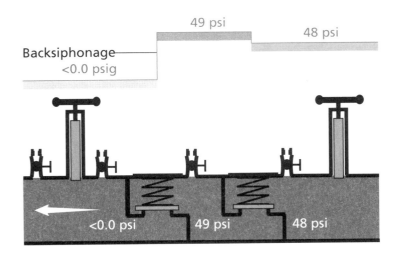

Figure 3.46
In a Backsiphonage Condition the No. 1 Check Valve Holds

If there is a backsiphonage condition and the No. 1 check valve is leaking, the pressure between the two check valves also drops to a sub-atmospheric pressure. However, the water downstream of the No. 2 check valve is not able to flow backwards into the chamber between the two check valves because the No. 2 check valve remains in the closed position. See figure 3.47.

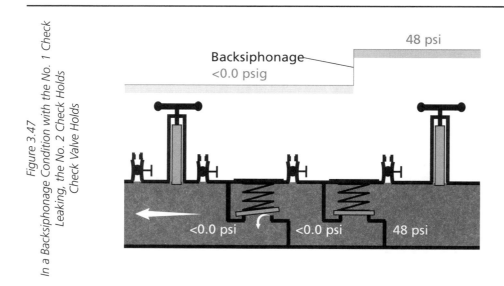

Figure 3.47
In a Backsiphonage Condition with the No. 1 Check Leaking, the No. 2 Check Holds Check Valve Holds

Under a backsiphonage condition with both the check valves leaking, the downstream water is siphoned through the check valves to the upstream side of the assembly. See figure 3.48. Again, this is one of the reasons the double check valve backflow prevention assembly may only be used to protect against a pollutant.

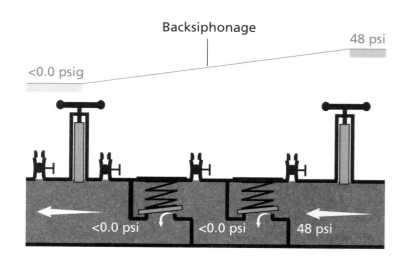

Figure 3.48
If Both Check Valves are Fouled the Assembly will Leak in a Backsiphonage Condition

Reduced Pressure Principle Assembly

The operating mechanics of a reduced pressure principle backflow prevention assembly follow some basic principles. First it is important to be familiar with the components of the relief valve as shown in figure 3.49.

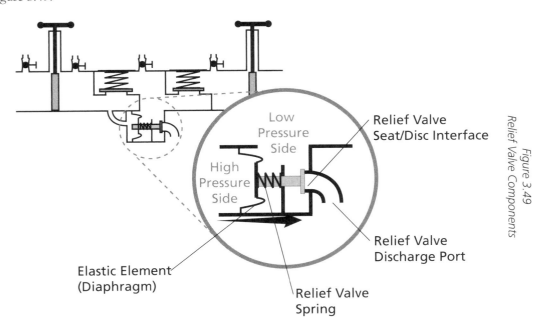

Figure 3.49 Relief Valve Components

The relief valve is mechanically independent and hydraulically dependent upon the differential pressure across the No. 1 check valve. It is designed to keep pressure between the check valves at least 2.0 psi less than the upstream pressure.

To understand its operation further, assume an assembly has been installed in a piping system and is ready to be filled and pressurized with water. With no pressure in the assembly the No. 1 check valve is closed and the relief valve is open. See figure 3.50

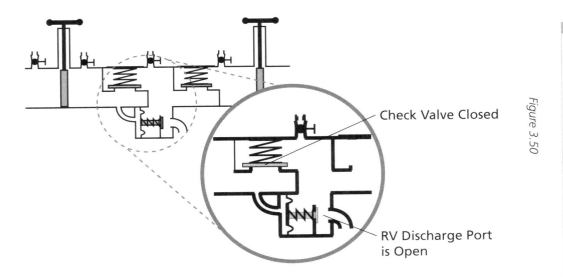

Figure 3.50

As the inlet or No. 1 shutoff valve is opened, the incoming water fills the chamber between the No. 1 shutoff valve and No. 1 check valve. This also fills the region on the upstream side of the relief valve as shown in figure 3.51.

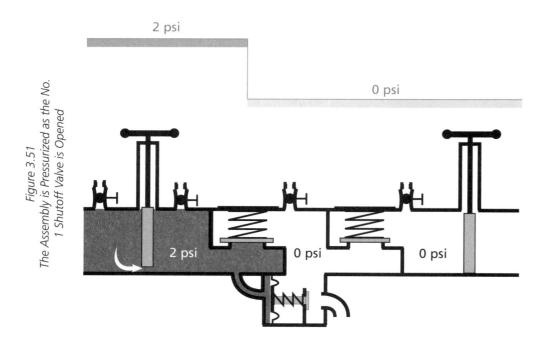

*Figure 3.51
The Assembly is Pressurized as the No. 1 Shutoff Valve is Opened*

Once this chamber, including the upstream side of the relief valve, is filled, the pressure will increase to the point that the relief valve closes. See figure 3.52.

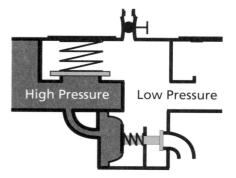

*Figure 3.52
The Relief Valve Closes*

Then, this pressure overcomes the force keeping the No. 1 check valve closed, which must be at least five psi. The relief valve closes before the No. 1 check valve opens because the forces keeping the relief valve open are less than the forces keeping the No. 1 check valve closed.

After the relief valve closes, the No. 1 check valve opens and water fills the chamber between the No. 1 check valve and the No. 2 check valve. See figure 3.53.

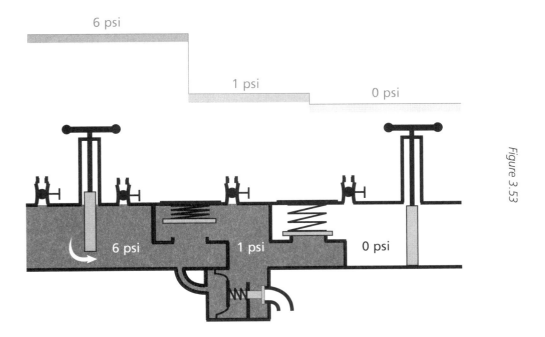

Figure 3.53

Once the chamber between the check valves fills, the water pressure increases to the point that it overcomes the force keeping the No. 2 check valve closed. This force must be at least 1.0 psi.

The No. 2 check valve opens and water fills the chamber between the No. 2 check valve and the No. 2 shutoff valve. Once this final chamber is filled, the water pressure increases until the assembly is fully pressurized. See figure 3.54. When the assembly is fully pressurized and there is no further flow into the assembly, both the No. 1 and No. 2 check valves close.

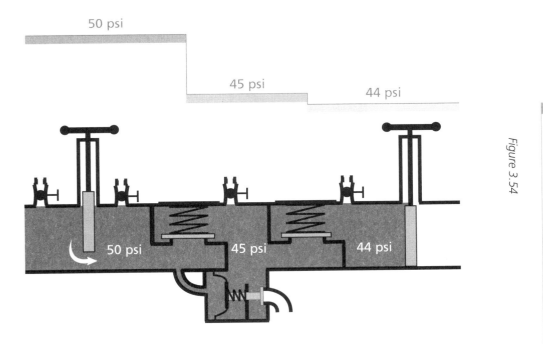

Figure 3.54

3 Hydraulics

After the No. 1 shutoff valve is fully opened, the No. 2 shutoff valve is fully opened as shown in figure 3.55.

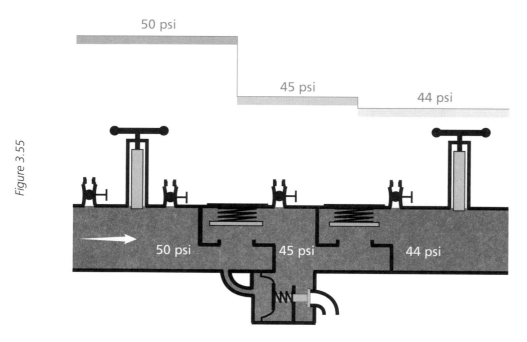

The amount the check valves open is dependent upon the rate of flow through the assembly. If there is no water use downstream, the water will be in a static (non-flowing) condition and the No. 1 and No. 2 check valves along with the relief valve will be closed. However, if there is water use downstream, the water is in a dynamic (flowing) condition. The No. 1 and No. 2 check valves are open while the relief valve remains closed. The greater the water usage, the more water flows through the assembly and the more fully the check valves open.

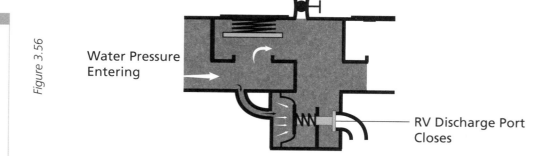

The relief valve closes when the force acting on the high pressure side of the elastic element, such as a rubber diaphragm, are greater than the forces acting on the low pressure side. See figure 3.56. The force acting on the high pressure side is the water pressure from the upstream side of the assembly. The forces acting on the low pressure side is the water pressure from the chamber between the two check valves plus the force of the relief valve spring. See figure 3.57

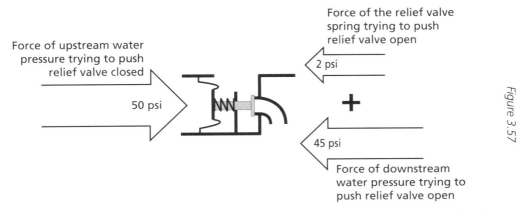

Figure 3.57

The high pressure side of the diaphragm is pressurized with the water from the upstream side of the No. 1 check valve. The low pressure side of the diaphragm is pressurized with the water from the downstream side of the No. 1 check valve (chamber between the check valves). This is why the operation of the relief valve is hydraulically dependent upon the differential pressure across the No. 1 check valve.

The relief valve will open when the forces acting on the low pressure side of the diaphragm are greater than the forces acting on the high pressure side, discharging water from the chamber between the check valves to keep the pressure in this chamber lower than the pressure upstream of the No. 1 check valve (at least 2.0 psi lower). The force from the relief valve spring is acting to push the relief valve open.

A Leaking No. 1 Check Valve with no Flow through the Assembly

If the No. 1 check valve leaks, the pressures on either side of the check valve will attempt to equalize. Before they equalize, however, the relief valve will open, discharging water from the chamber between the two check valves. See figure 3.58. The relief valve is designed to keep the pressure between the check valves at least 2.0 psi less than the upstream pressure. The volume of water discharging from the relief valve is typically equivalent to the leak across the No. 1 check valve.

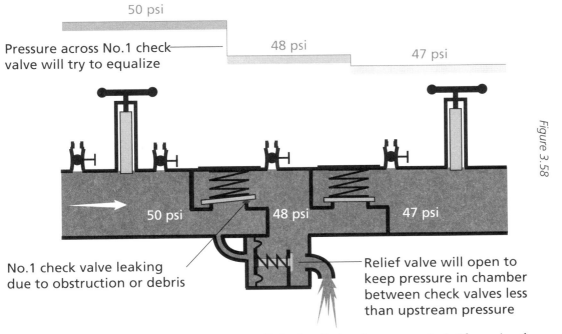

Figure 3.58

FIELD NOTE: A common occurrence is the relief valve discharging upon arrival. If opening the No. 3 test cock or the No. 4 test cock causes relief valve to stop discharging, this is a good indication that the No. 1 check valve is leaking.

A Leaking No. 1 Check Valve with Flow through the Assembly

If the No. 1 check valve leaks, the pressure on either side of the check valve will attempt to equalize with the pressure on the other side. If there is a flow of water through the assembly that is larger than the leak through the No. 1 check valve, the relief valve remains closed. This flow will tend to maintain the normal pressure gradient across the No. 1 check valve. As long as there is a flow passing through the assembly, the differential pressure across the No. 1 check valve does not equalize. Therefore, the relief valve remains closed. Once the flow of water through the assembly stops, the pressure on either side of the leaking No. 1 check valve tries to equalize again, causing the relief valve to open and discharge.

Backpressure Condition

Under normal operating conditions, the check valves open and close depending upon the amount of water flowing through them. As the flow rate decreases to a static condition, the check valves close. With an increase in the downstream pressure (backpressure), the No. 2 check valve remains closed. This prevents the downstream water from flowing to the upstream side of the No. 2 check valve. Since the pressure upstream of the No. 2 check valve has not been affected; the No. 1 check valve and the relief valve remain closed. See figure 3.59.

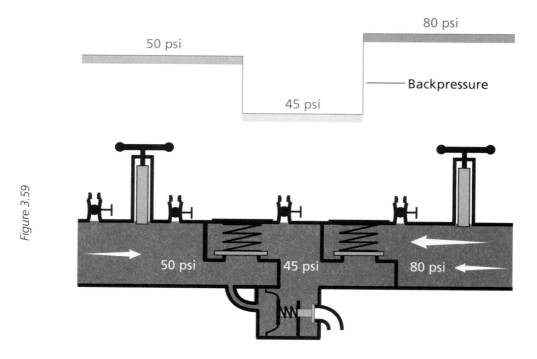

Figure 3.59

If there is a backpressure condition and the No. 2 check valve is leaking so that water can flow backwards through the No. 2 check valve, the water that is flowing backwards into the chamber between the check valves causes the pressure in this chamber to increase. This increased pressure together with the force of the relief valve spring overcomes the force of the pressure on the upstream side of the diaphragm, causing the relief valve to open. The volume of water discharging from the relief valve is equivalent to the leak in the No. 2 check valve. See figure 3.60. The pressure in the chamber between the check valves remains at least 2.0 psi less than the upstream pressure and the No. 1 check valve remains closed.

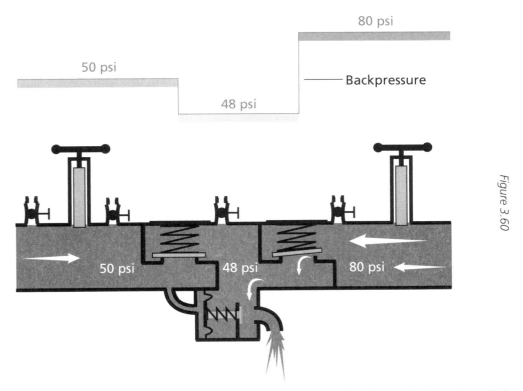

Figure 3.60

If there is a backpressure condition with both the No. 1 and No. 2 check valves are leaking, the relief valve opens and discharges water from the chamber between the two check valves. This is due to both the leak in the No. 1 check valve and the leak in the No. 2 check valve with backpressure. See figure 3.61.

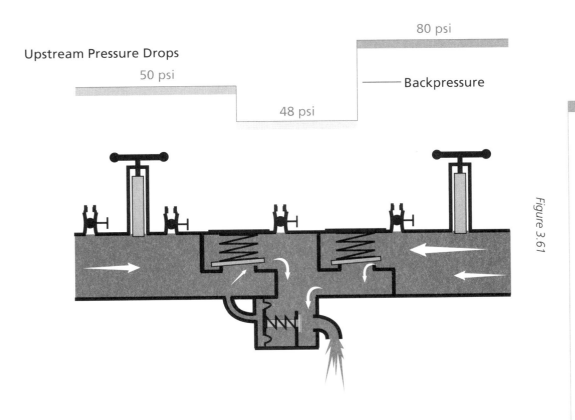

Figure 3.61

3 Hydraulics

71

Backsiphonage Condition

Under normal operating conditions, the check valves open and close depending upon the amount of water flowing through them. As the flow rate decreases to a static condition, the check valves close. Should there be a decrease in the upstream pressure, the relief valve will discharge before the upstream pressure drops to within 2.0 psi of the pressure between the two check valves. The discharging water ensures that the pressure in the chamber between the two check valves remains at least 2.0 psi less than the upstream pressure.

When the upstream pressure continue to drop, the relief valve continues to discharge ensuring that the chamber between the two check valves remains at least 2.0 psi less than the upstream pressure. When the upstream pressure continues to drop to atmospheric or below (backsiphonage), the relief valve fully opens discharging all of the water from the chamber between the two check valves. During this backsiphonage, the check valves remain closed preventing any downstream water from backflowing through the assembly. See figure 3.62.

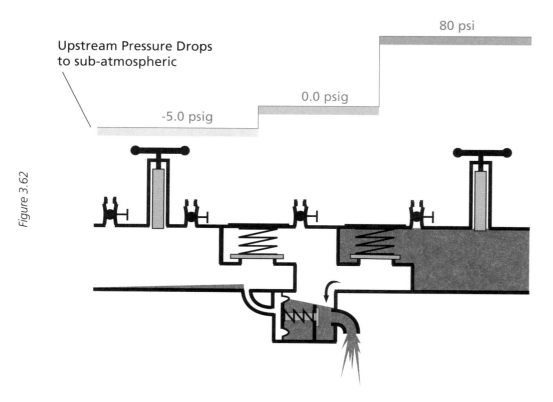

Figure 3.62

Should the upstream pressure drop to sub-atmospheric (backsiphonage), the relief valve will fully open discharging all of the water from the chamber between the two check valves.

During this backsiphonage, should the No. 1 check valve leak, air is drawn into the relief valve and back through the leaking No. 1 check valve as shown in figure 3.63.

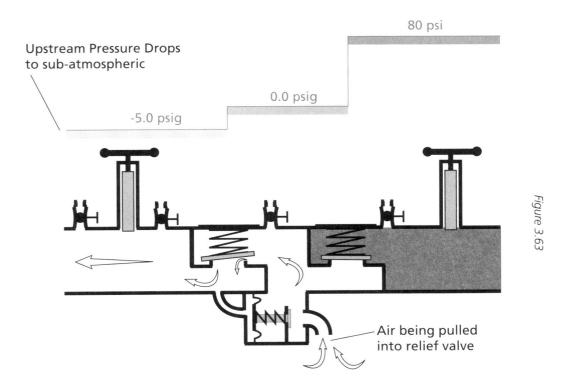

Figure 3.63

The No. 2 check valve remains closed, preventing the water downstream of the No. 2 check valve from flowing back into the supply side.

When the upstream pressure drops to sub-atmospheric (backsiphonage), the relief valve opens fully discharging all of the water from the chamber between the two check valves. During this backsiphonage, if both check valves leak; air is drawn into the relief valve and back through the leaking No. 1 check valve. Additionally, any water that may be backflowing through the No. 2 check valve (under a backpressure condition), flows out through the relief valve. See figure 3.64

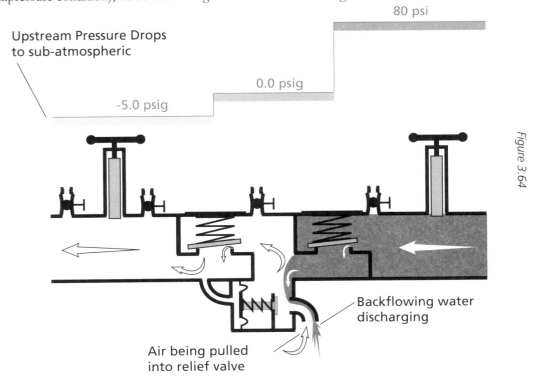

Figure 3.64

3 Hydraulics

Pressure Vacuum Breaker Backsiphonage Prevention Assembly

The operating mechanics of a pressure vacuum breaker backsiphonage prevention assembly follow some basic principles. To understand its operation further, assume that the assembly has just been installed in a piping system and is ready to be filled and pressurized with water. As the inlet or No. 1 shutoff valve is being opened, the incoming water fills the chamber between the No. 1 shutoff valve and the check valve. See figure 3.65.

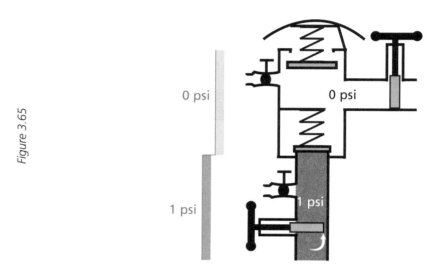

Figure 3.65

Once full, the pressure in the chamber increases. The amount of water pressure necessary to cause the check valve to open will depend upon the amount of force, or loading, trying to keep the check valve closed. This force must be at least 1.0 psi.

The water pressure increases to the point that it overcomes the force keeping the check valve closed. The check valve gradually opens and water fills the chamber between the check valve and the No. 2 shutoff valve. See figure 3.66.

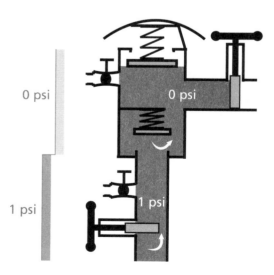

Figure 3.66

Once the chamber between the check valve and the No. 2 shutoff valve is filled, the water pressure increases to the point that it overcomes the force of the air inlet valve loading, closing the air inlet valve. This force must be at least 1.0 psi. See figure 3.67

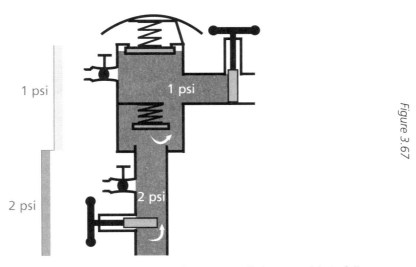

Figure 3.67

Once this chamber is filled, the water pressure continues to increase until the assembly is fully pressurized. With the assembly fully pressurized and no further flow into the assembly, both the check valve and the air inlet valve are closed. After the No. 1 shutoff valve is fully opened, the No. 2 shutoff valve is fully opened allowing flow through the assembly. See figure 3.68.

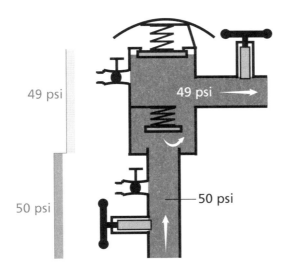

Figure 3.68

The amount the check valve is opened depends upon the rate of flow through the assembly. If there is no water use downstream, the water is in a static (non-flowing) condition and the check valve is closed. However, if there is water use downstream, the water is in a dynamic (flowing) condition and the check valve is open. The amount the check valve opens depends on the amount of water flowing. The higher the flow rate through the assembly, the more fully the check valve opens.

The air inlet valve closes when the force acting on the inside of the air inlet valve is greater than the force acting on the outside of the air inlet valve. The force acting on the inside of the air inlet valve is the water pressure in the chamber between the check valve and the No. 2 shutoff valve. See figure 3.69.

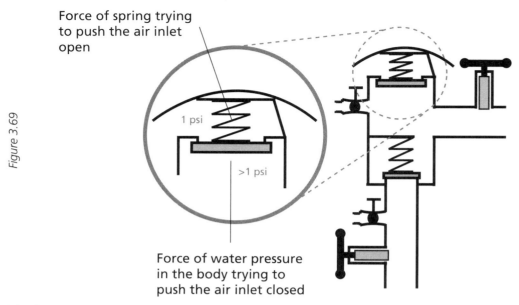

Figure 3.69

The force acting on the outside of the air inlet valve is the air inlet valve spring. This is a minimum of 1.0 psi. The air inlet valve opens when the force acting on the outside of the air inlet valve is greater than the forces acting on the inside of the air inlet valve. See figure 3.70.

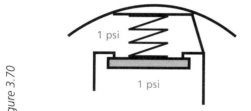 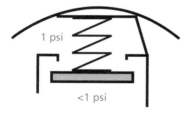

Figure 3.70

Air inlet closed - Force keeping air inlet closed is equal to force trying to open

Air inlet open - Force pushing air inlet open is greater than force trying to close

Backsiphonage Condition

Under normal operating conditions, the check valve opens and closes depending upon the rate of flow through it. As the flow rate decreases to a static condition, the check valve closes. When the upstream pressure drops below atmospheric pressure (backsiphonage), the check valve remains closed. This prevents the water between the check valve and the No. 2 shutoff valve from being siphoned to the upstream side of the check valve. The air inlet valve also remains closed.

In a backsiphonage condition with the check valve leaking, the pressure between the check valve and the No. 2 shutoff valve drops. When this pressure drops to a certain point, the air inlet opens. When this happens the pressure between the check valve and No. 2 shutoff valve is at least 1.0 psi. Once the pressure between the check valve and the No. 2 shutoff valve drops below atmospheric pressure, air is drawn through the air inlet valve and the leaking check valve. This entering air breaks the vacuum so no downstream water is siphoned back through the assembly. See figure 3.71.

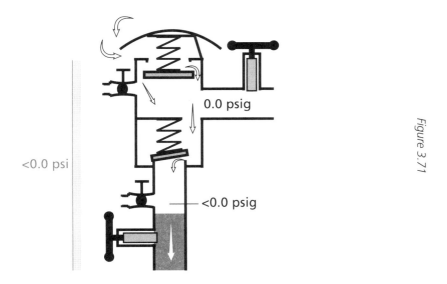

Figure 3.71

Backpressure Condition

NOTE: The pressure vacuum breaker is not designed to be used in a backpressure condition. This is shown, only to demonstrate why the assembly will not work in a backpressure condition.

This assembly is designed to provide protection against backsiphonage only and will not provide protection against backpressure. If, however, the assembly is installed improperly such that a backpressure condition exists, and the check valve is leaking, the water downstream of the assembly is forced backward through the check valve. In this case, the air inlet valve stays in the closed position since the pressure in the chamber between the check valve and the No. 2 shutoff valve does not drop. This allows the water downstream of the No. 2 shutoff valve to backflow through the assembly as shown in figure 3.72.

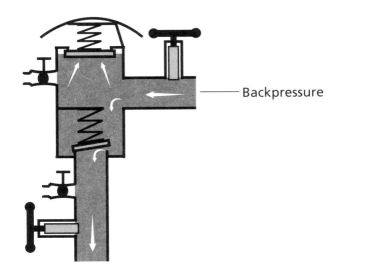

Figure 3.72

The pressure vacuum breaker is not designed to be used in a backpressure condition

3 Hydraulics

Spill-Resistant Pressure Vacuum Breaker Backsiphonage Prevention Assembly

The operating mechanics of the spill resistant pressure vacuum breaker backsiphonage prevention assembly follow some basic principles. To understand its operation further, assume that the assembly has just been installed in a piping system and is ready to be filled and pressurized with water. As the inlet, or No. 1 shutoff valve is opened, the incoming water fills the chamber immediately downstream of the No. 1 shutoff valve. See figure 3.73.

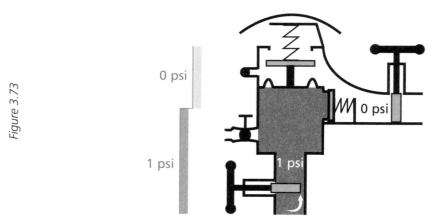

Figure 3.73

Once this chamber is filled, the water pressure increases to the point that it overcomes the force of the air inlet valve loading, closing the air inlet valve. The air inlet valve closes before the check valve opens because the forces keeping the check valve closed are greater than the forces keeping the air inlet valve open. The air inlet valve loading force is at least 1.0 psi. See figure 3.74.

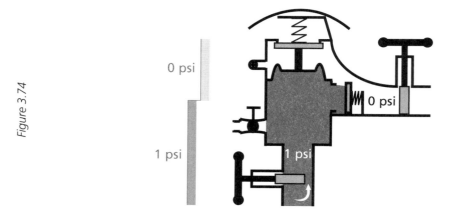

Figure 3.74

Once the air inlet is closed, the water pressure increases to the point that it overcomes the force keeping the check valve closed. This force is at least 1.0 psi. The check valve gradually opens and water fills the chamber between the check valve and the No. 2 shutoff valve. See figure 3.75.

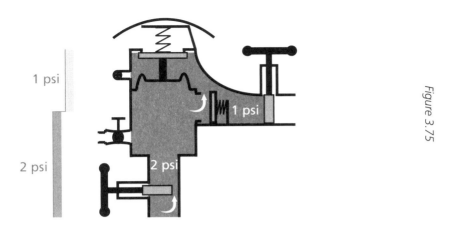

Figure 3.75

The amount of water pressure necessary to cause the check valve to open will depend upon the amount of force, or loading, trying to keep the check valve in a closed position. This force must be at least 1.0 psi. Once the chamber between the check valve and the No. 2 shutoff valve is filled, the water pressure continues to increase until the assembly is fully pressurized. When the assembly is fully pressurized and there is no further flow into the assembly, both the check valve and the air inlet valve close. After the No. 1 shutoff valve is fully opened, the No. 2 shutoff valve is fully opened allowing flow through the assembly. See figure 3.76.

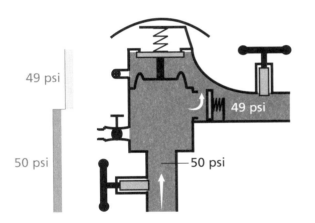

Figure 3.76

The amount the check valve opens depends upon the amount of flow going through the assembly. With no water use downstream, the water is in a static (non-flowing) condition and the check valve is in the closed position. However, if there is water use downstream, the water is in a dynamic (flowing) condition and the check valve opens. The amount the check valve opens depends upon the rate of water flow. As the flow rate increases through the assembly, the check valve opens more fully.

Under normal operating conditions, the pressure within the assembly is greater than the force of the air inlet valve spring, thus keeping the air inlet valve closed.

Backsiphonage Condition

Under normal operating conditions, the check valve opens and closes depending upon the amount of water flowing through it. As the flow rate decreases to a static condition, the check valve closes. When there is a decrease in the upstream pressure so that it is below atmospheric pressure (backsiphonage), the check valve remains closed. This prevents the water between the check valve and the No. 2 shutoff valve from being siphoned to the upstream side of the check valve. In this condition, the air inlet valve also remains closed. See the close-up in figure 3.77.

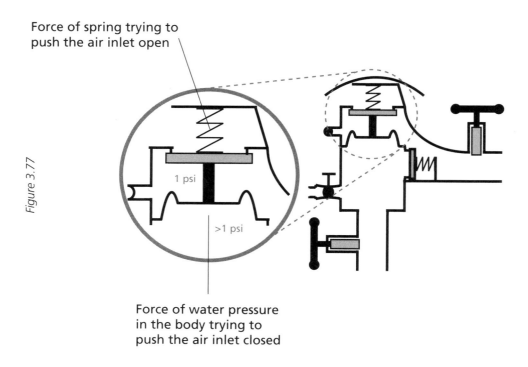

Figure 3.77

Force of spring trying to push the air inlet open

Force of water pressure in the body trying to push the air inlet closed

If there is a backsiphonage condition and the check valve leaks, the pressure between the check valve and the No. 2 shutoff valve drops. When this pressure drops to within 1.0 psi of the atmospheric pressure, the air inlet valve opens. See figure 3.78.

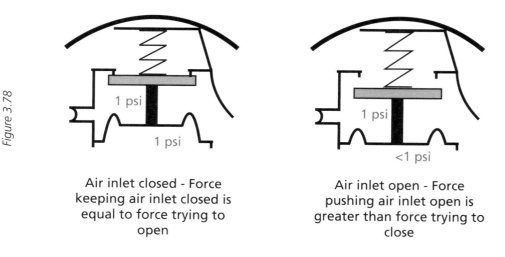

Figure 3.78

Air inlet closed - Force keeping air inlet closed is equal to force trying to open

Air inlet open - Force pushing air inlet open is greater than force trying to close

Once the pressure between the check valve and the No. 2 shutoff valve drops below atmospheric pressure, air is drawn through the air inlet valve and the leaking check valve. This entering air breaks the vacuum so that no downstream water is siphoned back through the assembly as shown in figure 3.79.

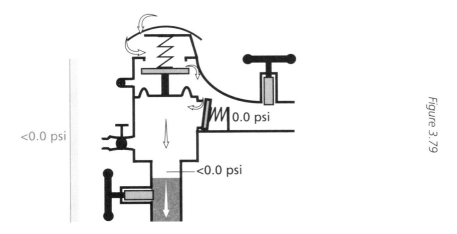

Figure 3.79

Backpressure Condition

NOTE: The spill-resistant pressure vacuum breaker is not designed to be used in a backpressure condition. This is shown, only to demonstrate why the assembly will not work in a backpressure condition.

This assembly is designed to provide protection against backsiphonage only and will not provide protection against backpressure. If however, the assembly is installed improperly such that a backpressure condition exists, and the check valve is leaking, the water downstream of the assembly forces its way back through the check valve. The air inlet valve stays closed since the pressure in the chamber between the check valve and the No. 2 shutoff valve does not drop. This allows the water downstream of the No. 2 shutoff valve to backflow through the assembly as shown in figure 3.80.

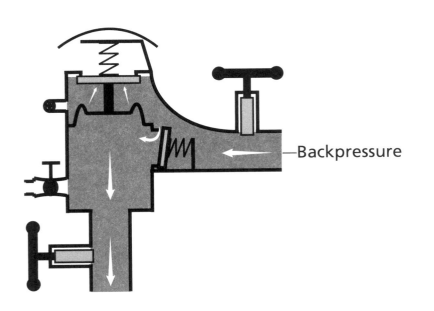

Figure 3.80

3 Hydraulics

Hydraulics

Chapter 4 — *Elements of a Program*

Elements of a Program

Inside this Chapter

Jurisdiction	86
Responsibilities	86
Responsibility of the Health Agency	86
Responsibility of the Water Supplier	87
Responsibility of the Plumbing Official	87
Responsibility of the Consumer	87
Responsibility of the Certified Backflow Prevention Assembly Tester	88
Responsibility of the Repair and Maintenance Technician	88
The ABC's of a Cross-Connection Control Program	88
Authority	89
Backflow Preventers	89
Certified Testers and Specialists	92
Defensible and Detailed Records	93
Education and Training	94
Policies & Procedures	95
Administration of the Program	96
Authority	96
Auxiliary Water Systems	96
Certified Backflow Prevention Assembly Testers	97
Change of Occupancy or Use	97
Combined Services	98
Critical Services	98
Dual Services	98
Equipment	98
Existing Assemblies	98
Fire Sprinkler Systems	99
Incident Response	99
Irrigation Systems	100
Low Water Pressure	102
Multiple Services	103
Non-Compliance	103
Plumbing Code	103
Recycled Water Systems	103
Restricted or Classified Services	104
Sump and Lift Stations	104
Single and Multiple Family Dwellings	104
Service Across Political or Water Agency Boundaries	105
Typical High Hazard Services	105
Summary	105

4 Elements of a Program

Elements of a Program

Jurisdiction

In general, the water supplier only has the responsibility for water quality up to the service connection. Typically the water supplier loses sanitary control of the water at its point of delivery to the consumer's water system. Beyond this point the piping is considered part of the plumbing system, which is governed by the local plumbing code. The plumbing code enforcement agencies (i.e., building, plumbing and health agencies, etc.) have jurisdiction over the internal plumbing system, issuing permits at the time of construction or major revision. The plumbing codes typically do not require retroactive changes of a water users plumbing system. The Uniform Plumbing Code states: "Existing Construction: No provision of this code shall be deemed to require a change in any portion of a plumbing or drainage system or any other work regulated by this code in or on an existing building or lot when such work was installed and is maintained in accordance with law in effect prior to the effective date of this code..."[1] The exception is when any such plumbing or drainage system or other work regulated by the administrative authority is determined to be unsafe dangerous, unsanitary or a menace to life, health or property.

In a similar manner, the International Plumbing Code states: "Existing installations: Plumbing systems lawfully in existence at the time of adoption of this code shall be permitted to have their use and maintenance continued..."[2] The exception would be the same as stated above.

Responsibilities

The implementation of regulations for the effective control of cross-connections requires the full cooperation of the water supplier, the health agency, the plumbing official and the consumer. Each has its responsibilities and each must carry out its phase of a coordinated cross-connection control program in order to prevent pollution or contamination of the potable water supplies. The responsibilities of each are outlined hereafter.

Responsibility of the Health Agency

General
The health agency has the responsibility for promulgating and enforcing laws, rules, regulations and policies to be followed in carrying out an effective cross-connection control program.

Public Potable Water System
The health agency has the primary responsibility of ensuring that the water supplier operates the public potable water system free of actual or potential sanitary hazards, including unprotected cross-connections. The health agency also has the responsibility of ensuring that the water supplier operates the public water system to meet Federal and State drinking water standards. Since cross-connections are a serious health hazard and can result in the contamination of the public water supply, the health agency should require the water agencies to have a comprehensive cross-connection control program. This agency has the further responsibility of ensuring that the water supplier provides an approved water supply at the point of delivery to the consumer's water system and, further, that he requires the consumer to install, test and properly maintain an approved backflow prevention assembly(s) on the service connection(s) when required.

Consumer's Water System(s)
The health agency has the primary responsibility of ensuring that the consumer's potable water system is provided with an approved water supply and that its potable water system(s) is maintained free of sanitary hazards, including unprotected cross-connections. Generally, on site cross-connection

---

1   International Association of Plumbing and Mechanical Officials, 2006, *Uniform Plumbing Code*, Section 101.4.1.3

2   International Code Council, 2006, International Plumbing Code, Section 102.2

control and backflow prevention requirements will be addressed in State and Local Plumbing Codes and will be enforced by the local plumbing and building officials. In some states, local health agencies have the authority to regulate on site cross-connection control requirements.

Responsibility of the Water Supplier

The water supplier has the responsibility to maintain their public water system in compliance with all Federal and State drinking water standards. Since cross-connections can cause contamination of the public water supply, water suppliers should have a cross-connection control program in place. The program should include an ordinance or rules of service to give the water supplier the authority to implement the cross-connection control program. Under the cross-connection control regulations or rules of most states and territories the water supplier has primary responsibility to prevent unapproved water sources, or any other substance, from entering the public water supply system. The water supplier is prohibited by these regulations or rules from installing or maintaining a water service connection to a consumer's water system within its jurisdiction where a health, system, plumbing or pollutional hazard exists, or will probably exist, unless the public potable water supply is protected against backflow by an approved backflow prevention assembly(s) installed at the service connection(s), i.e., point of delivery.

The water supplier's responsibility begins at the source and includes all of the public water distribution system, including the service connection and ends at the point of delivery to the consumer's water system(s). In addition, the water supplier must exercise reasonable vigilance to ensure that the consumer has taken the proper steps to protect the public potable water system. To ensure that the proper precautions are taken the water supplier is required to determine the degree of hazard to the public potable water system. When it is determined that a backflow prevention assembly is required for the protection of the public system the water supplier shall require the consumer, at the consumer's expense, to install an approved backflow prevention assembly at each service connection, to test immediately upon installation, relocation and annually or more often, to properly repair and maintain such assembly or assemblies and to keep adequate records of each field test and subsequent maintenance and repair, including materials or replacement parts.

Responsibility of the Plumbing Official

The building and safety department of the appropriate political jurisdiction has the responsibility to not only review building plans and inspect plumbing as it is installed; but, it has the explicit responsibility of preventing cross-connections from being designed and built into the structures within its jurisdiction. Where the review of building plans detects the potential for cross-connections being made as an integral part of the plumbing system the plumbing official has the responsibility under most building codes for requiring that such cross-connection practices be either eliminated or provided with approved backflow prevention equipment.

The plumbing official's responsibility begins at the point of service (i.e., the downstream side of the water meter or service connection) and carries throughout the consumer's water system. The plan inspector should enquire about the intended use of water at any point where it is suspected that a cross-connection might be made or where one is actually called for by the plans. When such is discovered, a suitable approved backflow prevention assembly shall be required and be properly installed.

Responsibility of the Consumer

The consumer has the responsibility of preventing pollutants and contaminants from entering his/her potable water system(s) or the public potable water system. The consumer's responsibility starts at the point of delivery from the public potable water system and includes all of his/her water systems. The consumer, at his/her own expense, shall install, operate, test and maintain approved backflow prevention assemblies as directed by the authority having jurisdiction. The consumer shall maintain accurate records of tests and repairs made to backflow prevention assemblies and provide the

4 Elements of a Program

In general, the water supplier only has the responsibility for water quality up to the service connection. Typically the water supplier loses sanitary control of the water at its point of delivery to the consumer's water system.

administrative authority having jurisdiction with copies of such records. The records should be on forms approved by the administrative authority having jurisdiction and include the list of materials or replacement parts used. Following any repair, overhaul, re-piping or relocation of an assembly the consumer must have it tested to ensure that it is in good operating condition and will prevent backflow. Tests, maintenance and repairs of backflow prevention assemblies must be made by a certified backflow prevention assembly tester and/or a repair technician.

When requested by the water supplier or health agency, the consumer must appoint a water supervisor who shall be responsible for conformance with all applicable laws, rules and regulations pertaining to cross-connection control; for the installation, operation and use of all water piping systems, backflow prevention assemblies and water-using equipment on the premises; and for the avoidance of unprotected cross-connections. The water supervisor should be the consumer or any full-time employee appointed by the consumer who has a thorough knowledge of the installation, operation and maintenance of all water systems and backflow prevention assemblies on the consumer's premises. In the event of pollution or contamination of the public or the consumer's potable water system due to backflow on or from the consumer's premises, the owner or water supervisor shall promptly take steps to confine further spread of the pollution or contamination within the system and should notify the local health officer and the water supplier of the condition. The water supervisor shall take appropriate measures to free the water system(s) of any pollutants or contaminants.

Responsibility of the Certified Backflow Prevention Assembly Tester

When directed to perform a field test, a certified backflow prevention assembly tester will have the following responsibilities:

The tester shall be responsible for performing accurate field tests and for making reports of such field tests to the consumer and responsible authorities on forms approved by the administrative authority having jurisdiction. The tester shall be equipped with and be capable of using all the necessary tools, gages and other equipment to properly field test backflow prevention assemblies. A certified tester shall perform and be responsible for the accuracy of all tests and reports.

Responsibility of the Repair and Maintenance Technician

The repair technician shall be responsible for installing, repairing, overhauling and maintaining backflow prevention assemblies and making reports of such repairs to the consumer and the administrative authority having jurisdiction. The report shall include a list of all materials or replacement parts used. The technician shall be equipped with and be capable of using all the tools and other equipment necessary to repair, maintain and overhaul backflow prevention assemblies. It will be the technician's responsibility to ensure that original manufactured replacement parts are used in the repair or maintenance of the backflow prevention assembly. It will be the technician's further responsibility not to change the design, material or operational characteristics of an assembly during repair or maintenance. A repair technician shall perform the work and be responsible for the accuracy of reporting such work. The technician shall have all state and local licenses and permits needed to repair, maintain and overhaul backflow prevention assemblies.

The ABC's of a Cross-Connection Control Program

There are several elements necessary for a comprehensive cross-connection control program. These elements are summarized under the following headings.

A. Authority
B. Backflow Preventers
C. Certified Testers and Specialists
D. Defensible and Detailed Records
E. Education and Training

Elements of a Program

There are several elements necessary for a comprehensive cross-connection control program: Authority, Backflow Preventers, Certified Testers and Specialists, Defensible and Detailed Records, Education and Training

Authority

Authority is essential to carry out a cross-connection control program. The administrative authority must have the legal authority in place to implement policies, conduct site surveys and require backflow protection. This authority is usually in the form of a local ordinance or law. Of course, this will depend on the administrative authority and the type of regulations under which it falls. For example, a county health organization may have a different type of legal document than would a municipal water supplier.

The legal document should address the following issues:

- Acceptable backflow prevention assemblies
- Field test procedures
- Annual field test requirement
- Maintenance
- Installation
- Certified testers
- Certified specialists
- Site surveys
- Record keeping
- Incident documentation
- Backflow Incident response
- Public information
- Non-compliance penalties

The legal basis for a cross-connection control program must be specific in several ways and yet flexible in others. For example, the administrative authority must have the specific right to enforce the program through fines, termination of water service, etc. The adoption of reference sources, however, may need to be more flexible. For example, if an administrative authority decides to make the requirement that backflow preventers installed in their system are approved by a specific approval or listing agency, the administrative authority may specify that such assemblies must appear on the approval agency's List of Approved Backflow Prevention Assemblies. This can be done in two ways. One administrative authority may site the List of Approved Backflow Prevention Assemblies dated 1 January 2009. This would be fine, except that there may be additions to the List on 2 January 2009. Another possibility is to state assemblies must appear on "the most current List of Approved Backflow Prevention Assemblies" as published by the specific approving or listing agency. This would ensure that assemblies are approved or listed at the time they are installed and also allow for additions to the List without having to go through the entire legal process of changing the ordinance or law. Some agencies may want to include a statement, which allows them to make more specific requirements. Such a statement may state, "assemblies approved by this department must be listed on the most current List of Approved Backflow Prevention Assemblies, and meet any additional requirements deemed necessary by this department." Whatever type of statement is used, it is necessary for the administrative authority to have all the authority they need to effectively carry out their cross-connection control program, so that all reasonable and prudent actions are taken to prevent backflow.

A model ordinance is available in Chapter 8 of this Manual as a basis for an administrative authority to create an ordinance.

Backflow Preventers

For any cross-connection control program to be effective the administrative authority needs to have the appropriate means to prevent backflow, which include mechanical backflow preventers. There are several listing and approval agencies, which may be used. One consideration is that the administrative authority must have a non-biased means of determining which backflow preventers may be used under their jurisdiction. There needs to be clear guidelines so that each backflow preventer has the same listing/approval requirements placed on them if they are to be used in the jurisdiction of the administrative authority.

Additionally, it is very important that any additional requirements placed on the backflow preventers are reasonable. For example, the color of the assembly is not a technically valid reason for not accepting a backflow preventer. This is why it is best to rely on the listing/approval of a backflow preventer by an entity that specifically lists/approves backflow preventers.

Data show that a field evaluation is a very effective means of determining the in situ operational characteristics of an assembly. Over one-third of the assemblies that passed the Laboratory phase of the Approval Program at the Foundation for Cross-Connection Control and Hydraulic Research at the University of Southern California, did not pass the Field Evaluation phase the first time. This is one of the reasons the field evaluation is considered to be invaluable.

There are several approving and/or listing agencies with standards for backflow prevention assemblies.

Approvals/Listings

The responsible officer or employee of the administrative authority must adopt a list of acceptable backflow prevention assemblies. This list may apply to assemblies used as system protection, internal protection or both depending upon the jurisdiction of the administrative authority.

There are several agencies that have standards or specifications for backflow prevention assemblies; so, it is essential that the administrative authorities have an understanding of the various listing agencies.

AWWA

The American Water Works Association (AWWA) is an international non-profit educational and scientific organization. It was founded in 1881 and is now the world's largest organization of water supply professionals.

The American Water Works Association (AWWA) has established Standards C510 and C511. These standards cover the double check valve assembly and the reduced pressure principle backflow prevention assembly respectively and replaced the original Standard C506 in 1989. The AWWA does not approve any backflow prevention assembly. It is the manufacturers who make the claim that the assembly meets the requirements of the AWWA Standards.

ASSE

The American Society of Sanitary Engineering (ASSE) is comprised of members from all segments of the plumbing industry. The ASSE is a unique organization because its membership is a cross-section of the industry, including contractors, engineers, inspectors, journeymen, apprentices and others who are involved in various segments of the industry.

The Society develops standards for plumbing products, including backflow prevention products. The standards are used by inspectors, contractors and engineers when they specify, install and inspect plumbing installations.

The ASSE Standards for backflow prevention require a series of laboratory evaluations. Testing is conducted by ASSE recognized laboratories.

IAPMO

The International Association of Plumbing and Mechanical Officials (IAPMO) writes standards and codes. IAPMO maintains standards for some plumbing products. For other products they recognize standards of other organizations. IAPMO has their own laboratories, and also recognizes other laboratories to perform testing.

Factory Mutual (FM)
FM Global is a commercial and industrial property insurance and risk management organization. FM Global maintains standards for double check valve assemblies and reduced pressure backflow prevention assemblies (Class No. 1221).

FM requires assemblies to conform to the standards of one of the following:

- ASSE
- AWWA
- USC

Additionally, the assemblies must meet FM performance evaluation requirements which include body strength and friction loss tests.

Underwriters Laboratories (UL)
UL is an independent, non-profit, product-safety testing organization. UL standard 1469 covers, "backflow special check valves," consisting of reduced pressure principle assemblies, double check valve assemblies, as well as detector assemblies. This standard tests body strength and pressure loss only.

The Foundation for Cross-Connection Control and Hydraulic Research at the University of Southern California (USC)
USC publishes standards for backflow prevention assemblies in Chapter 10 of this manual. In order for a backflow prevention assembly to be approved by USC each size and model of assembly must meet the requirements of Chapter 10. The Foundation has its own laboratory where all of the laboratory evaluations of backflow prevention assemblies are conducted. Additionally, the successful completion of the one-year field evaluation is required for the assembly to be approved. The field evaluation requires that a minimum of three of every size and model of assembly submitted must be installed in different water systems. Each of these assemblies will be field tested on a nominal thirty-day schedule. At the end of the twelve-month field evaluation period, the assembly is disassembled and inspected. Three of each size and model must successfully complete twelve months of simultaneous trouble-free service and pass the final inspection.

Regional and Local Approvals/Listings
There are some regional agencies that require local approval of water using equipment. As an example the City of Los Angeles requires that any backflow prevention assemblies used within the City be listed by the City's Mechanical Testing Laboratory, which is a division of the Department of Building and Safety. Also, the City of New York requires that any backflow prevention assembly used in their jurisdiction be listed on the Materials and Equipment Acceptance Division of the Department of Buildings of the City of New York.

List of Approved/Listed Backflow Prevention Assemblies
A list of approved or listed backflow prevention assemblies may be adopted from one of the approving or listing agencies mentioned previously. This is the most common means used by administrative authorities, as opposed to creating a list of their own. Many administrative authorities do, however, make additions, deletions or modifications to the list based on specific parameters. For example, one administrative authority may require assemblies to be approved after a specific date, or it may require assemblies to have a specific characteristic, such as replaceable seats.

Whatever method is used, the administrative authority must make sure to have a standard impartial means of determining what backflow preventers are acceptable in their system.

Installation Guidelines

When a backflow prevention assembly is required, it is necessary to provide installation guidelines to the customer to ensure the assembly is installed correctly. Guidelines for backflow prevention assembly installations may be found in Chapter 8.

Certified Testers and Specialists

Certified Backflow Prevention Assembly Testers

Once backflow prevention assemblies are properly installed, they must be periodically field-tested and properly maintained. Who will field-test these assemblies? That may be determined by a state regulation or a local administrative authority. There are some very important issues that need to be considered when determining who will be permitted to field-test the backflow preventers.. First of all, what method of field-testing is to be used? Some may assume if the assembly is not visibly leaking, then it is working properly. This is not likely to be the type of field-testing any administrative authority would want to see in their jurisdiction. Chapter 9 of this manual contains the field-test procedures for the double check valve assemblies, the double check detector assemblies, the double check detector assemblies-type II, the reduced pressure principle assemblies, the reduced pressure principle detector assemblies, the reduced pressure principle detector assemblies-type II, the pressure vacuum breaker assemblies and the spill-resistant vacuum breaker assemblies. These procedures are widely accepted by many administrative authorities and certifying bodies. AwwaRF report 90928 reports that 74% of the respondents indicated that the field-test procedures used in their jurisdiction are from either the 8th or 9th Edition of the *Manual of Cross-Connection Control*.[3] These procedures enable the tester to accurately determine the condition of the backflow preventers, even with minor shutoff valve leaks. But, there are other field-test procedures available and the administrative authority must make the determination as to what field-test procedure is to be used in their jurisdiction.

Once the field test procedures are adopted there must be some means of determining if a tester is qualified to test using these procedures. Training and certification are necessary to ensure qualified personnel field test the backflow preventers. While many training organizations offer training courses to train individuals on the intricacies of testing and diagnosing problems on the various backflow prevention assemblies; passing these training courses does not necessarily certify the attendee. The attendees may be issued a diploma or certificate of completion, which states that the individual did successfully complete the course. But certification is usually a process instituted in which individuals are recognized for their demonstrated proficiency in field testing backflow preventers. The certification of backflow prevention assembly testers typically requires a written and performance examination with a recertification every two or three years, to ensure the tester maintains their proficiency with the current field test procedures. The recertification should, like the original certification, include a written and performance exam. It also may require the certified tester to attend an update seminar that includes a review of field test procedures prior to recertification.

Many administrative authorities conduct their own certification program. Others adopt the certification program of a third party organization or another administrative authority.

One of the most important aspects of a certification program is recertification, as mentioned in the previous paragraph. Recertification is necessary to ensure that certified testers maintain their proficiency in performing the field test. For example, if an administrative authority does not have a certification program (including recertification), but only requires training, a tester could come in to the administrative authority and show a certificate of completion from ten, twenty, or even thirty years prior. The tester meets the qualifications, if those qualifications only state that the tester must have attended an acceptable training course. It is possible that the tester hasn't tested a backflow preventer in thirty years. If he has, he has probably continued to use outdated methods. This is why recertification is so important. Of course, recertification may take on many forms. For some types

---

Elements of a Program

> A certified cross-connection control program specialist is trained to administer a cross-connection control program. The specialist must be very familiar with all of the essentials of any program.

[3] Lee, J. J., et al, 2003, *Impacts of Cross-Connections in North American Water Supplies,* Denver, AwwaRF

of certification only a renewal fee is necessary. However, for backflow prevention assembly testing it is essential that an examination be required, not only a written examination, but also a performance examination. There is no way to tell if a tester can actually test and ascertain the condition of an assembly if they do not demonstrate this fact to an impartial proctor. Whether the administrative authority conducts their own certification program, or adopts one already in operation it is imperative that the program requires recertification with both a written and performance examination, and recertification no more than every three years. See guidelines for a certification program in Chapter 8.

Certified Cross-Connection Control Program Specialists

The Administrative Authority has the responsibility of requiring site surveys to determine if backflow protection is needed at the site. This is one of the tasks of the cross-connection control program specialist. A certified cross-connection control program specialist is trained to administer a cross-connection control program. The specialist must be very familiar with all of the elements of any program. One of the most critical tasks is conducting site surveys, where the program specialist determines what type of backflow protection is needed.

How the specialist goes about this will depend on the type of protection for which the administrative authority is responsible—system protection or internal protection. If the administrative authority has a containment or system protection program, it is the responsibility of the administrative authority to determine if there are unprotected cross-connections on the premises, which may pose a threat to the quality of the water in the distribution system. If there is a hazard, the administrative authority must require the appropriate type of backflow preventer at the service connection in order to protect the potable water distribution system.

If, however, the administrative authority is responsible for internal protection, the assessment of which type of backflow protection is necessary, if any, must be made for each point of water use. A program specialist will need to conduct a site survey to make the determination. The program specialist will need to have a thorough knowledge of the plumbing code as well as an understanding of the various hydraulic conditions, which allow backflow to occur. Additionally, it will be necessary to have a good understanding of the degree of hazard. Knowledge in these areas is necessary in order to make a determination as to which type of backflow protection, if any, would be needed at each point of use.

In a similar fashion to the tester certification, there are third-party certification programs as well as agency certification programs for the program specialist. These certification programs typically require training and/or experience and a certification exam.

Defensible and Detailed Records

Another important element of a cross-connection control program is record keeping. Records should be retained for a specific period of time. Some areas have specific requirements for this. Other areas go by the statute of limitations on any action, which may be brought against the administrative authority for failing to administer a viable cross-connection control program. But retaining the records for a certain period of time is only one aspect of record keeping. Records must be detailed and defensible. The records must contain specific information, which supports the actions of the administrative authority.

The records must be "defensible." This means the records should be sufficient to show that the administrative authority is meeting all of the requirements necessary to carry out their cross-connection control program. Should, for example, there be a backflow incident in the jurisdiction, the administrative authority needs to be able to show through their records, that they have done all they could to survey sites, require approved backflow preventers and require regular field testing of these assemblies (show due diligence). If the records are insufficient to do this, it could leave the administrative authority in non-defensible position.

The certification of backflow prevention assembly testers typically requires a written and performance examination with a recertification every two or three years, to ensure the tester maintains their proficiency.

Education and Training

Education

Education is important to every aspect of the cross-connection control program. First of all the staff of the administrative authority must be educated in the specifics of cross-connection control. Specific training for personnel will be discussed in the following text, but a general cross-connection control education of all personnel is very important. The concepts of backflow, what a cross-connection is, how water from a customer's premise can get into the potable water supply are important concepts to introduce to all personnel. This isn't so that all personnel can become experts and handle all cross-connection issues, but it certainly helps when personnel are familiar with basic concepts and know to whom people should be referred when certain situations arise.

As an example: personnel in a particular water supplier are introduced to some of the concepts of cross-connection control. The meter readers, repair personnel, and anyone having a field job are requested to inform the cross-connection control program specialist of any potential cross-connections they observe while conducting their normal duties at facilities that may warrant a cross-connection control field survey. An agency employee in the field notices that a facility seems to be changing and large pieces of equipment are being brought into the facility. When the program specialist makes inquiries, it is determined that the facility has completely changed the type of work being conducted, and will now be using contaminants under pressure for various processes and now needs to have a reduced pressure principle assembly installed. This is just one example, and hopefully, this type of change would be picked up through permitting processes for any remodeling or modification to the plumbing system. However, one must be aware that permits are not always sought on every change that is made, even if it is a sizable change.

Education is also a very important public relations tool. As those who have been involved in cross-connection control for any length of time will testify, public relations is a big part of the job. Whenever a customer is going to be required to comply with rules or regulations, even if it is to protect the public health, public relations can help the process go smoothly. When requiring a customer to install backflow protection, it may not be easy to convince the customer of this need. Some administrative authorities have taken the position of forcing compliance without explaining any of the concepts of cross-connection control to the customer. Providing informational brochures to customers will help them understand the hazards which exist, and why backflow protection is necessary.

Education, will certainly help convince the customer of their need to protect the potable water system. As the customer has a better understanding of what cross-connection control is all about, they can become an ally in preventing further cross-connections from being created at their facility and also in helping other customers in the jurisdiction accept the program.

AwwaRF Report 90928[4] points out that public education in the area of cross-connection control is a major area of concern. Uninformed public and uninformed agency personnel were noted as major hindrances to the administration of a cross-connection control program. However, the same survey indicated that the agencies considered a public education program a minimal part of the cross-connection control effort, at most. This explains why the public and agency personnel may not be well informed. There are several educational tools for the administrative authority to use in providing the public and their own personnel a general education on cross-connection control.

Training

Although training and education sound very similar, there is a distinct difference. Training is used to teach personnel how to carry out specific tasks. Those individuals that are required to be certified to field test backflow preventers must first receive specific training. This is not just to "pass the exam," but also to learn how to field test and diagnose the condition of each of the backflow preventers. Field testing requires proper manipulation of the field test kit, so hands-on training is critical to the learning experience. It is important that the individuals have the opportunity to learn the actual

---

[4] Ibid.

Records should be sufficient to show that the administrative authority is meeting all of the requirements necessary to carry out their cross-connection control program.

field testing procedure with a backflow preventer in front of them. They also need the opportunity to practice the field test procedures with actual backflow preventers—both properly operating and malfunctioning assemblies.

The training for a cross-connection control program specialist should include information and instruction on such items as rules and regulations, policies and procedures, record keeping, public relations, and site surveys. Actual site surveys as a training exercise will help personnel understand the significance of a site survey. Tracing water lines, locating water using equipment and identifying unprotected cross-connections are some of the goals of a site survey.

To summarize, the five key elements of an effective cross-connection control program are:

A. Authority
B. Backflow Preventers
C. Certified Testers and Specialists
D. Defensible and Detailed Records
E. Education and Training

Policies & Procedures

The legal authority mentioned in previous pages gives the administrative authority the legal authority needed to carry out the cross-connection control program. Policies and procedures, however, are needed to guide the administrative authority in the details of carrying out the program. The administrative authority should have specific policies and standard operating procedures in place so that consistency in administering the program is achieved.

The legal document of the administrative authority discussed earlier gives the administrative authority the authority to implement and operate a cross-connection control program. The legal document should only go so far as to implement the program—not to go into all the operational details of the program. That detail should be explained in the Policies and Procedures document created by the administrative authority.

Before going into the details of specific issues covered by the Policies and Procedures it would be best to discuss the several options that an administrative authority has for the implementation of their cross-connection control program. These may include:

- The water supplier's program will be limited to system protection (containment) with all internal protection (isolation) problems handled by the local health agency and/or plumbing or building department..
- The water supplier will contract with the local health agency for operation of its system protection program. (The health agency is still responsible for the internal protection.)
- The water supplier will contract with a private enterprise for the operation of its system protection program. (The health agency is still responsible for the internal protection.)
- The local health agency (County or City) can contract with a water supplier for the operation of the internal protection program along with the water supplier's system protection program.
- The State Health agency may have responsibility over the entire system and internal protection program.
- There might be other combinations of responsibility unique to an area or jurisdiction.

The following are some, but not necessarily all, of the specific issues that should be covered by an administrative authority's Policies and Procedures document. Also, keep in mind that an administrative authority's Policies and Procedures should be framed in such a way that additions and deletions may be made easily—i.e., not having to go back through the entire legal process required for the adoption of a new legal document.

4 Elements of a Program

The five key elements of an effective cross-connection control program are:

Authority
Backflow Preventers
Certified Testers and Specialists
Defensible and Detailed Records
Education and Training

Administration of the Program

In both the legal document that established the cross-connection control program and in the Policies and Procedures document the administration of the program should be specifically assigned to an office within the administrative authority (not by a person's name). Which office is given this responsibility depends upon the size and administration of the agency. The responsibility for the program is placed in the hands of the General Manager, Chief Engineer or other office as explained in the legal document. For some small agencies it is possible that employees have multiple job responsibilities; while in the larger agencies, one or more employees may be responsible for the cross-connection control program only.

A mechanism for handling appeals should be established. (Appeals referenced here may come from a customer's objection to being required to install a backflow prevention assembly.) It is important that appeals be handled in an equitable manner. To facilitate handling any appeal it is important that a fully documented record be kept from the initial site survey to the final acceptance of the installation of the protection. Records should also be kept of all field testing and maintenance that occur after the initial installation.

As detailed earlier in the chapter responsibility for the protection of the potable water distribution system lies with the water supplier. This does not preclude some or all of the cross-connection control program being contracted to a third party. But, the ultimate responsibility for the protection of the potable water distribution system still lies with the water supplier. So, in the Policies and Procedures statement on Administration it will be necessary to explain exactly how the various aspects of the program are to function and to affirm where the ultimate responsibility lies.

Authority

Even though there is a legal statement, ordinance, rule, resolution or other legal document, it is best to also cite the legal authority that the agency has for the establishment of the cross-connection control program. The general line of authority starts with the Federal Safe Drinking Water Act Amendments of 1996. Then, based on this is the State enactment that establishes the State's primacy in the control of the water within each State. Then the county may enact some form of cross-connection control program; and, finally the local water supplier with its legal statement. It is not often that a full recitation of all the legal history is necessary to convince a customer that the water supplier does, in fact, have the authority to enact and enforce controls concerning the water within its distribution system, but it is prudent to have the information in the Policies & Procedures. Additionally, within the statement of Authority there should be a reference to the ability of the water supplier to terminate water service to the customer if the customer fails to provide the required backflow prevention equipment, its maintenance and its periodic testing.

Auxiliary Water Systems

The definition of an auxiliary water system is given in Chapter 1. There are many documented cases where a customer's well, or pond, or even a stream through the property has been the source of a cross-connection that resulted in illness. It is important to have in place a policy that covers auxiliary water supplies. There are many types of auxiliary water supplies. As an example in San Francisco there is not only the Bay that constitutes a source of auxiliary water; but, within the City there is a system of cisterns specifically for fire fighting purposes. Also, in another city along the Bay there are large diameter concrete pipe sections sunk with the axis vertical and the open top covered by a grating that have been provided as a source of water for fire fighting.

Another type of auxiliary water source is that of an adjacent water supplier. The water is considered by the Health Department as potable water; but neither water supplier wants water from the other's distribution system unknowingly introduced into its system. An industrial complex could be located right at the political boundary of the two water agencies and, for insurance purposes, the fire sprinkler system must be supplied by not only two service connections from the host agency but also by a service connection from the neighboring agency. These are just some examples. An auxiliary water supply is just that, any source of water on the property in question.

A policy must be in place that allows the water supplier to protect the potable water distribution system from auxiliary water supplies.

Certified Backflow Prevention Assembly Testers

The Policies and Procedures manual of each agency needs to specifically identify what basis will be used for the establishment of a List of Certified Backflow Prevention Assembly Testers for that agency. This basis may be mandated by a state or other regulation. Only persons who have demonstrated their ability in field testing these assemblies to the satisfaction of the water supplier or health agency should be permitted to field test backflow preventers. For the purpose of administrating the cross-connection control program of a water or health agency it is necessary to have available a list of people who have demonstrated to the satisfaction of the administrative authority that they do, in fact, know how to properly field test any type of assembly used in the administrative authority's jurisdiction. In many instances a company will want to be listed; but the administrator of a program needs to know that the specific person who performs the field test and signs the report is certified. The nature of this work, the protection of the public health, is such that a water or health agency should not accept a delegation of responsibility that is implied when the owner (manager or supervisor) of a company signs off for an employee who actually did the testing. Furthermore, the customer who is responsible for installing, maintaining and field testing the assembly should be given the opportunity to select a tester from among a number of people available. The Administrative Authority may conduct their own certification program or adopt the certification program of a third party organization or another administrative authority.

Change of Occupancy or Use

The on site water systems of some facilities change regularly. But, it is a troublesome task to survey the same facility repeatedly. One method of keeping up with changes that may affect the quality of the water is to interface with the building or plumbing enforcement agencies. This contact plus field personnel will help to alert the administrative authority of changes of occupancy, apparent use, or noticeable changes in water usage. Then the administrative authority can make a follow-up survey to determine what changes have occurred in the property and what, if any, modifications are in order with respect to cross-connection control.

The occupancy may change without a change of ownership, or a change of property management. Therefore, if the occupancy changes and the water use alters, the only thing that may alert the water supplier to this may be a new sign over the door of the building. This is why field personnel should be trained to be aware of any variations from the norm that they may observe in the field. Education through seminars or informal meetings is important to help field personnel understand their role in the cross-connection control program.

Elements of a Program

Combined Services

A combined service is a water service that provides water for both domestic uses and other non potable uses (e.g., industrial systems, irrigation systems, fire sprinkler systems) through the same service connection. Although the domestic portion of the water line may not require backflow protection, the assessment of the degree of hazard for system protection should be made based on the entire system.

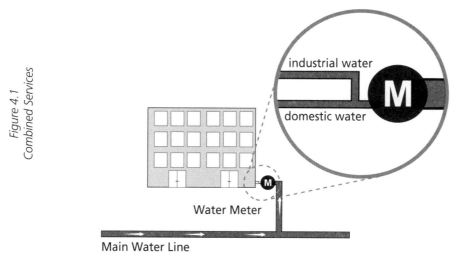

*Figure 4.1
Combined Services*

Critical Services

Critical services are water services where the water cannot be shut off—even for a few moments at any time. Typically critical services are found at hospitals, emergency care centers, film-processing laboratories, industrial plants where the water is in continuous use. When the need for a critical service is made known, the water supplier usually will need to do one of two things. The water supplier could require multiple services. In this case, all services would need to be protected by the same level of backflow protection. (See Multiple Services.) If only a single water service is available to the property, parallel backflow protection may be required. In this case a single service connection is provided with two or more backflow prevention assemblies. One may be shut off for field testing or maintenance while the other(s) provides water.

Dual Services

See *Multiple Services*.

Equipment

Specific pieces of water using equipment will require backflow protection due to the operating characteristics of the equipment. Chapter 7 discusses a wide range of equipment explaining what it is, how it works and where it's used.

Existing Assemblies

How the administrative authority deals with existing backflow prevention assemblies depends on whether the agency has had or has not had a cross-connection control program in place.

If existing assemblies are approved backflow prevention assemblies then they may be left in service and the make, model, size and serial number are entered into the agency's records. If there has been a cross-connection control program, it is not as simple. If the assemblies are currently approved then they remain in service. If, however, some of the assemblies in service are no longer approved, but were approved when they were installed, they may be left in service as long as they continue to function properly. In this case more frequent field testing may be required. If the assemblies were never recognized as approved assemblies they should be removed and replaced with currently approved assemblies. It is very important that records of each of these actions be maintained in detail.

> A combined service is a water service that provides water for both domestic uses and other non potable uses

Fire Sprinkler Systems

Industrial fire protection systems may consist of sprinklers, fire hose connections and hydrants. Sprinkler systems may be dry or wet, open or closed. A fire sprinkler system operates automatically when heat activated components discharge water over a specific area. It is standard practice to equip automatic fire sprinkler systems with fire department connections, through which the fire department can pump supplemental water into the sprinkler system.

A fire sprinkler system is connected to the domestic water system to supply water to the sprinkler heads to reduce the effect of a fire. Sprinkler heads will vary according to their general design and performance characteristics. There are various types of fire sprinkler systems and their use depends on the intended application. The common types of fire sprinkler systems include the wet pipe system, dry pipe system, deluge system and pre-action system.

Generally, fire protection systems should require backflow protection commensurate with the degree of hazard. Wet pipe systems typically use black iron pipe that is not approved for potable water systems, so the quality of the water, which sits stagnant in these pipes, will degrade and should be protected with a minimum of a double check valve assembly. A wet pipe system which contains chemicals (i.e., anti-freeze, fire extinguishing foams, etc.) is a contamination hazard and should be protected with a reduced pressure principle assembly.

In situations where there is a need to provide backflow protection and a means to detect unauthorized use of water the Double Check Detector Assembly (DCDA) or the Double Check Detector Assembly-Type II (DCDA-II) could be used for non-health hazards whereas the Reduced Pressure Principle Detector Assembly (RPDA) or the Reduced Pressure Principle Detector Assembly-Type II (RPDA-II) could be used for health hazard conditions. These are specifically matched assemblies in which very low water usage will be accurately recorded by the detector water meter.

The retrofitting of a backflow prevention assembly into a fire sprinkler system may require a recalculation of the hydraulic operating conditions to ensure that the fire system continues to perform satisfactorily. (See Chapter 7, p. 144.)

Incident Response

Water suppliers and other administrative authorities may pursue a comprehensive cross-connection control program and still backflow incidents may occur. If an incident does occur, the administrative authority should have a plan describing the details of how to respond to such an incident.

The Source of Backflow

A water quality complaint may be the first indication of a backflow incident occurring. Water quality complaints should be followed up as soon as possible to determine if they are the result of a cross-connection. Trained personnel should determine the cause of the water quality complaint. When the source of the quality complaint is determined to be due to a backflow incident, the backflow incident response plan should be put into effect.

Isolate Source of Backflow

The source of the backflow should be isolated as soon as it is discovered. The extent of the contamination needs to be investigated to determine if the contamination is restricted to one area of the customer's facility, the entire facility or the water supplier's distribution system. Multiple samples may be necessary to identify the affected areas. Samples should be taken before any system or on-site flushing is performed.

Determine Extent of the Incident

It will be important to determine if the problem is restricted to the customer's water system, or if it is coming from the distribution system. If the source of the problem can be traced to the customer's system, then the service connection to their facility may need to be turned off to prevent any contamination from getting back into the distribution system. If the problem is with the customer's internal plumbing system, then the customer needs to be informed so that proper changes to the plumbing system can be made.

If it is found that the contamination has backflowed into the distribution system or the source of the on-site contamination is coming from the distribution system; the investigation must move to additional sampling in the distribution system. Identifying that the contamination has reached the distribution system means that some or all customers on that distribution system may be affected. Isolate that portion of the distribution system, which may have been affected by the contamination.

Notification
If the backflow incident is limited to one facility, the occupants or employees should be notified not to use the water until the system is decontaminated. If it is discovered that the backflow incident has affected the distribution system, the public must be notified in order to take appropriate precautions.

Decontamination
The affected area, either in the distribution system or on the customer's premises must be decontaminated. Adequate flushing will be necessary as well as following proper protocols for decontamination of the system. Different administrative authorities may have different jurisdictional responsibilities. For example, the water supplier may decontaminate the distribution system while the local health agency may be responsible for ensuring that the internal plumbing system is decontaminated properly.

Backflow Protection
If it is discovered that a specific cross-connection on the customer's property was the source of the backflow incident, either the cross-connection must be eliminated or appropriate backflow protection should be required to prevent a repeat of the incident. The water supplier may require backflow protection at the service connection as well.

Documentation
Documentation of backflow incidents may or may not be a regulatory requirement. However, it is important to document backflow incidents in order to help prevent similar incidents in the future. Documentation gathered as quickly possible and detailed in a backflow incident report. See sample Backflow Incident Report, 8.18.

Irrigation Systems
Irrigation systems are not limited to commercial irrigation; but, this section of the Policies and Procedures covers any use of water taken from the potable water distribution system that is for lawns, gardens, landscaping, et al., whether it be residential or commercial. Under this heading it is wise to include separate paragraphs relating to the various types of irrigation such as: gravity flow for orchards or row crops, set sprinklers with impact heads, drip irrigation and other forms (all of these with or without chemical additives). Under each type of irrigation system the type of backflow prevention assembly should be clearly spelled out (i.e., RP, PVB, or AVB).

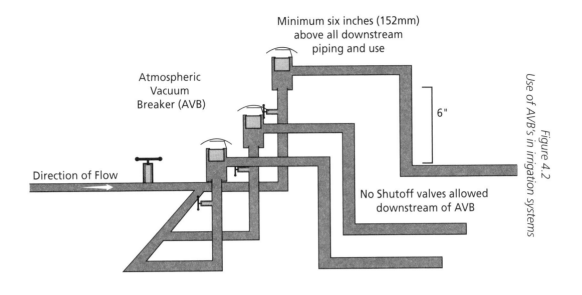

Figure 4.2
Use of AVB's in irrigation systems

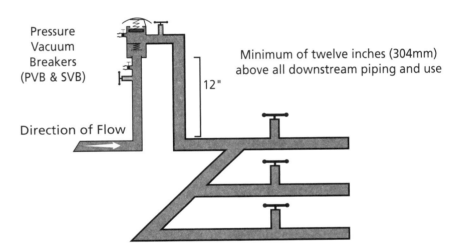

Figure 4.3
Proper Installation of PVB's and SVB's

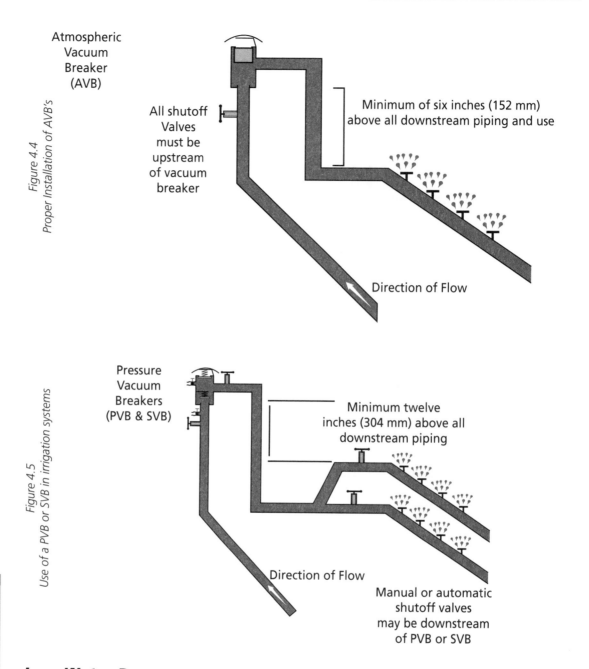

Figure 4.4
Proper Installation of AVB's

Figure 4.5
Use of a PVB or SVB in irrigation systems

Low Water Pressure

Plumbing codes typically specify a minimum pressure at the point of use of 15 psi. With a low distribution pressure the customer may need to use a booster pump within the property in order to adequately serve all the needs. Hence, if the customer does install a booster pump it will also be required, as a minimum, to install a double check backflow prevention assembly so that the water within the customer's system cannot be backpressured into the water supplier's system.

Multiple Services

A multiple service is defined as being a condition whereby two or more water services are provided to a single site. These may be interconnected on site so there is a continuous water supply where the water cannot be shut off (see Critical Service.) or they may be for different on site uses, such as domestic and irrigation. Such a set of conditions could be for a house, condominium or even a commercial or an industrial water user. It is normally recommended that the same level of protection be provided to all service connections. This may or may not apply to fire systems.

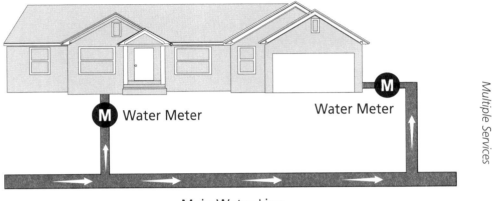

Figure 4.6 Multiple Services

Non-Compliance

While the administrative authority's legal document may give the agency the right to refuse water service to any premise where there is an unprotected cross-connection it is also appropriate to have a statement in the Policies and Procedures setting forth the right of the water supplier to refuse to serve water or to terminate water service whenever a survey of the premises shows that a required backflow prevention assembly has either not been installed properly; or, if installed, has not been field tested or maintained as required. Standard operating procedures for termination of water service may entail turning off and locking the service connection (e.g., curb stop) or physically removing the water meter. Once the service connection is turned off or disconnected the hazard to the distribution system has been eliminated. Before the service connection is turned back on, the customer would have to show compliance with all of the initial requirements, such as installing a backflow preventer and having it tested. Also, this section should discuss the certification of testers and the right of the agency to rescind certification when a tester is found to have falsified records, improperly field tested assemblies, etc. Agencies should spell out the appeal process for either the customer who does not wish to comply or for the tester whose certification is rescinded.

Plumbing Code

The plumbing codes cover the water using equipment and applications of water use within a property. This document, in its latest form, should be adopted, by reference, as a part of the Policies and Procedures.

Recycled Water Systems

When an administrative authority allows the use of recycled water, specific precautions must be made in order to ensure the continued separation of the recycled system and the domestic water system. The two systems must be completely separate without any interconnection. On premises where there are both potable and recycled water supplies, it is necessary to periodically check to confirm the systems have remained separate. This cross-connection test may be performed by injecting dye into the recycled water or lowering the pressure in one or both of the water systems. Evidence that there is an interconnection (cross-connection) between the potable and recycled water systems will require a detailed inspection to locate and eliminate the cross-connection(s). In some cases, backflow prevention assemblies may be used on the recycled system service connection to prevent backflow from any on-site use getting into the recycled water distribution system. The backflow prevention assemblies installed in the recycled water systems must not be field tested with the same test equipment used to field test the assemblies on the potable system.

Restricted or Classified Services

Whenever the administrative authority is refused admission to a property for the purpose of performing a cross-connection control survey the water supplier has two options. The agency could either refuse to serve water to the property or require maximum backflow protection at the service connection. This maximum protection can be in the form of an air gap separation or a reduced pressure principle backflow prevention assembly.

Sump and Lift Stations

Whenever a property is located so that the drainage water or the sanitary discharge must be collected in a sump and pumped up to the storm drain or the sanitary sewer respectively, the water supply to this property must contain a reduced pressure principle assembly. If there is a hose bib located nearby for clean-out purposes, a hose attached to such a hose bib constitutes at least a potential (if not an actual) cross-connection.

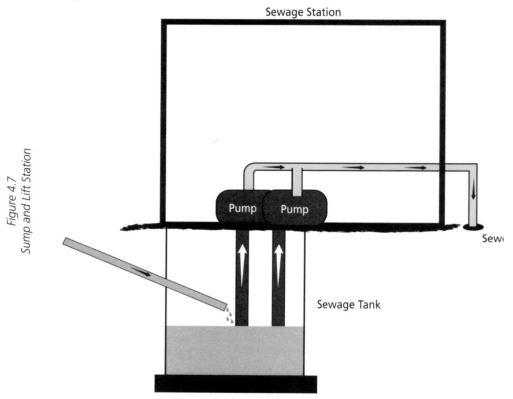

Figure 4.7
Sump and Lift Station

Single and Multiple Family Dwellings

This policy mainly concerns housing on single lots or a single parcel of land. In some respects this ties in with Multiple Services. As a matter of policy water agencies do not typically provide more than one service connection to a property. The reason is that, if two service connections existed, the owner or occupant could easily create cross-connections within the property that would then permit water from one service connection to flow through the property and back into the agency's distribution system at the other service connection. The solution for this multiple use condition is to have a single service connection which then branches out to the various points of use within the property.

Service Across Political or Water Supplier Boundaries

Often there are occasions when property contiguous to one water supplier is more easily served by that supplier instead of the supplier actually having the jurisdiction. Otherwise the water supplier having jurisdiction may have to run an extensive distribution system to serve a relatively restricted number of customers. The Policies and Procedures of the agency accommodating this extended service should have a statement covering the services in this limited area and requiring the customers in this area to abide by all the rules and regulations of the water supplier providing the water service.

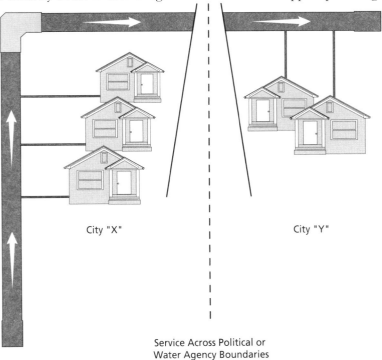

Figure 4.8
Service Across Political or Water Agency Boundaries

Service Across Political or Water Agency Boundaries

Typical High Hazard Services

The Policies and Procedures indicate typical services that constitute a hazard to the water supplier's distribution system. And, reference can be made to the more extensive list, which is given in Chapters 6 and 7 of this Manual. It should be explained in the Policies and Procedures at that any list should not be considered to be all-inclusive.

Summary

As the cross-connection control specialist continues to gain experience and knowledge in the field more should be added to the policies and procedures document of the administrative authority. This information will be useful to others involved in the cross-connection control program as well as the water users.

Elements of a Program

Chapter 5
Cross-Connection Control Surveys

Cross-Connection Control Surveys

Inside this Chapter

Cross-Connection Control Programs	110
Preparing for a Cross-Connection Control Survey	111
General	111
Notification	111
Existing Systems vs. New Construction	111
Typical Methods of Backflow Prevention	112
Before the Survey	112
The Site Survey	114
Documentation	115
Compliance	115

Cross-Connection Control Programs

Different administrative authorities may be responsible for various aspects of cross-connection control. They water supplier is typically the administrative authority that is responsible for system protection, that is the requirement of backflow protection at the meter or service connection in order to protect the water suppliers water system. Some water suppliers may also be responsible for internal protection. In this case they may also make requirements for backflow protection at each cross-connection within a facility. In many cases another agency, such as a local health agency, plumbing department or building department is responsible for internal protection. Some agencies may be responsible for cross-connection control requirements only during initial construction. Once the building is occupied, responsibility for internal protection may be transferred to a different administrative authority.

An administrative authority administering a system protection cross-connection control program has the primary responsibility to determine if hazards exist in the consumer's water system, which could be introduced into the water supplier's distribution system. For the administrative authority that conducts an internal protection program it is necessary to determine if hazards exist on the property, which could be introduced into the consumer's water system, degrading the quality of the water on the property.

Although one administrative authority may administer a system protection cross-connection control program and another may be administering an internal cross-connection control program, the survey itself will be similar with the end result differing. In the internal protection program, the administrative authority will make backflow protection requirements for each unprotected cross-connection found on the premises. In a system protection program the administrative authority will conduct the survey and make a determination as to which type of backflow prevention assembly will be required at the meter or service connection. In this case, the administrative authority should still be able to justify the requirement for backflow protection at the service connection, so a survey report detailing the various cross-connections on the premises will be necessary, even though the administrative authority is not requiring backflow protection at each of these cross-connections.

In making a determination of the backflow potential, consideration should be given to the fact that backflow may occur under any pressure differential, varying from vacuums to very high pressures. This means that backflow can occur from the consumer's premises into the public potable water system when the street main pressure is lower than the consumer's pressure, is at atmospheric pressure or is under a vacuum condition. Therefore, no specific positive or negative pressure range in the public system may be established that could be used as a yardstick for predicting the differential pressures where backflow may occur. It should be assumed that backflow into the public system can occur when a higher pressure can be attained in the consumer's system.

Internal cross-connections may present a serious health hazard, particularly when they are located in multi-storied buildings such as hotels, apartment houses, hospitals, medical and dental buildings, and other buildings having laboratories or other critical water uses. In such buildings the water supplier may find that there is a special danger to health or to the public water system because of the failure to install suitable backflow protection, or because of the removal or malfunctioning of such internally located backflow prevention assemblies. These assemblies may be on cross-connections to sewers, such as flush valve toilets, bedpan washers, etc. It may also be found there is no adequate installation or testing and maintenance program, and failure of such internal assemblies can therefore be anticipated. When this is found to be the case, the administrative authority may find it necessary to require protection at the service connection even though the individual internal cross-connections are presently protected in accordance with plumbing and health codes. In fact, the administrative authority in charge of system protection may require backflow protection at the service connection even though all internal cross-connections are protected properly. The fact that there are hazards on the premises may, in itself, warrant backflow protection at the service connection. Internal plumbing systems are changed regularly and the potential for creating a cross-connection should be considered when making the determination for backflow protection at the service connection.

Preparing for a Cross-Connection Control Survey

General
When manpower is available for making initial inspections and for maintaining an adequate inspection and re-inspection program on the consumer's premises, the administrative authority should make an evaluation of the degree of hazard based on an inspection of the water system, water uses and existing in-plant backflow protection. When deemed necessary, the installation of an approved backflow prevention assembly at an internal point(s) and/or at the service connection(s) must be required.

Even when a system of inspections and re-inspections are not practical or economically feasible, it is necessary to conduct an initial survey. Once a survey is complete, the administrative authority can determine the need for backflow protection at one or more points internally or at the service connection. Additional surveys may be required periodically depending upon the findings of the first survey. If the administrative authority is conducting a system protection program, it may not be necessary to conduct additional surveys if the backflow protection at the service connection is a reduced pressure principle assembly. In this case, the service connection protection would not be changed, even if more cross-connections were discovered internally. When considering a new or proposed plant, an evaluation should be made of the facility's proposed piping and a physical survey of the facility should be made before water service is granted.

The administrative authority should conduct cross-connection control surveys to determine where backflow protection is necessary. It is important to get as much information as possible about the survey site before the survey takes place. How this is handled may vary depending if the site is a new site, or an existing site.

Notification
Once it is determined that a survey will be needed, the owner of the facility should be notified. It will also be necessary to make arrangements with on site personnel to survey the facility. This would normally be the manager of the facility. The owner is notified of the need for the survey, and the on site manager is contacted to arrange the survey itself. The cross-connection control specialist conducting the survey should request the assistance of the facilities manager or someone knowledgeable of the on site system to act as a guide. This will ensure any questions that arise concerning the water system, or any other industrial systems on site can be answered expeditiously. The initial notification may take place via mail, but it will be necessary to follow up with a phone call in order to set up the appointment and arrange for the appropriate personnel to be available.

Existing Systems vs. New Construction
A survey should be made of the consumer's existing water system in order to determine the degree of hazard to both the consumer's system and the public potable water system. The administrative authority should determine the location and disposition of the water lines, the existence of cross-connections, the availability of auxiliary or used water supplies, the use or availability of pollutants, contaminants and other liquid, solid or gaseous substances. Thus the administrative authority will determine the degree of hazard.

When, in the opinion of the administrative authority having jurisdiction, there is a substantial hazard to the public system, no water connection should be installed or maintained between the public potable water supply and a consumer's premises without a backflow prevention assembly at the service connection.

Even though the administrative authority may conduct a system protection program it may determine whether internal protection will be sufficient of itself; or, whether, due to the general conditions within the premises, there is need for system protection at the meter plus the segregation of system and internal protection.

Even though there may be adequate internal protection an administrative authority conducting a system protection program, may require backflow protection at the service connection.

Contamination or pollution of a water system is usually brought about by a cross-connection to some sort of water using equipment or system. Chapter 7 has several examples of water using equipment explained including how they work and where they are used. Chapter 6 discusses facilities in a general manner and lists several different pieces of equipment which may be found in such facilities. The surveyor should become familiar with both of these chapters in order to understand which types of water using equipment exist and which may be expected to be located at the various types of facilities. It should be noted that Chapter 7 is not exhaustive.

Typical Methods of Backflow Prevention

When a degree of hazard has been determined and properly classified, effective steps should be taken to require correction of the condition or the installation of backflow prevention assembly(s) at the service connection(s) and/or internally to prevent the possible pollution or contamination of the consumer's and/or public potable water system. (See Chapter 3.) The rules or regulations of the administrative authorities require this in order to isolate within the consumer's water system(s) any pollution or contamination that may occur which could adversely affect the quality of the public or consumer's water system. Proper backflow protection should be maintained on each service connection or separation of internal systems, at the point of delivery and ahead of any outlet. Where the consumer's water line is divided at the point of delivery, an assembly must be installed on each leg or branch (i.e., industrial and domestic). The type and kind of backflow prevention assembly needed on each leg or branch is to be in accordance with the highest degree of hazard found on the premises.

It may be expedient or necessary under certain circumstances for the water supplier to approve the installation of a lead-in line into the premises a considerable distance beyond the service connection before installing the backflow protection. In this circumstance the administrative authority having system protection jurisdiction continues to have jurisdiction over the consumer's lead-in line as far as the backflow prevention assembly(s). It is the administrative authority's responsibility to see the consumer does not connect any lines between the service connection from the public potable water supply system and the backflow prevention assembly(s).

All backflow prevention assemblies should be installed above grade, where they are not subject to flooding and provided with freeze protection, where necessary. If a reduced pressure principle assembly is installed within a consumer's building there must be adequate floor drainage (not a dry well) beneath or near the assembly. Backflow prevention assemblies should only be installed in the orientation in which they were approved.

Before the Survey

One of the viable points of a cross-connection control program is having the cognizant water or health agency conducting site surveys or inspections of all commercial, industrial and if possible, residential facilities. The cognizant water or health agency should not make the specific selection of backflow prevention assembly solely on the basis of a description for a particular site or facility, but rather on the assessment of the degree of hazard at the facility or water usage from the site survey or inspection.

The cross-connection control personnel making the site survey must specify the level of backflow protection that is needed for a facility or water use.

Before the site survey or inspection, a review of the plans must be conducted, if the plans are available. These plans may include the water service maps and/or the detailed mechanical drawings.

As part of the plan check/review, it is advisable to review all plans for new and existing construction. The water service maps may help determine the on-site water uses. These maps are helpful in determining the building height and if there are pumps being used or planned. They will provide information on the intended use of the building. They can also show where the building or facility will be located in relation to the overall water system. For example, is the building or facility in the industrial, commercial or residential areas; or is it in a low-pressure or high-pressure region of the water system. They can also provide the size of the water meter or service connection. This may indicate the potential water usage.

Plan check/review can also help in determining what type of water uses are in the building or facility. They can provide information on fire lines or fire-pump connections, if there are any auxiliary water supplies, any lawn irrigation, solar systems and/or heating and cooling systems. They can help in determining where vaults may be located at each service connection such that one can identify if separate vaults are used for the water meters and the backflow prevention assemblies.

The water service maps are helpful in determining the number of existing services to a property. For example, if it is determined that there are more than one service connection, backflow prevention assemblies at each service connection would be necessary to prevent a flow-through condition, where water flows from one service connection to the other. Also, the water service maps will help in reviewing the elevation of piping within the building. A building with a high elevation of piping may cause a high static pressure and potential backpressure concerns. Finding a sewage ejector could indicate the need for nearby potable water lines for clean up and maintenance purposes.

Where adequate plans and specifications are not available and no realistic evaluation of the proposed water uses can be determined, the consumer, architect, engineer or other authorized person should be advised that eventually circumstances may require the installation of maximum backflow protection at the water service connection. Therefore, the responsible person should anticipate the possible need for a backflow prevention assembly. Drawings may show backflow prevention assemblies within an industrial facility where dedicated industrial water systems are planned. This is a clear indication that water using equipment will be in use, which may contain hazardous materials or chemicals.

The review of the detail drawings are helpful as part of the plan review/check. The detail drawings will provide detailing of the piping, plumbing fixtures and water using equipment within a building or facility. The detail drawings being reviewed may not reflect the actual condition of the building. Drawings that have been updated as changes were made to the facility are often called "As-Builts." If available these are the best drawings to review as they reflect the building as actually constructed. The drawings to be reviewed may include mechanical drawings, plumbing drawings, architectural drawings and/or landscape drawings.

The mechanical and plumbing drawings use many of the standard piping symbols. These piping symbols are used to reflect the many different pipe sections, piping fittings and plumbing fixtures. An example of these piping symbols is shown in figure 5.1.

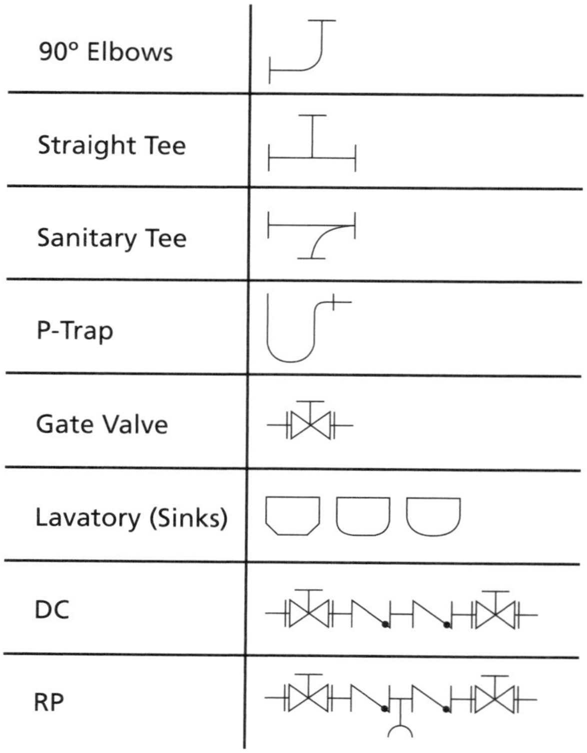

90° Elbows	
Straight Tee	
Sanitary Tee	
P-Trap	
Gate Valve	
Lavatory (Sinks)	
DC	
RP	

Figure 5.1
Typical piping symbols

Another helpful tool in the plan review/check process is the use of the North American Industry Classification System (NAICS) code (formerly the Standard Industry Code ([SIC]). This is a federal guideline that provides a five-digit code for each type of industry. The first two digits of the code refer to the major economic section. The third digit of the code refers to the economic sub-sector. The fourth digit of the code refers to the industry group. The fifth digit of the code refers to the NAICS industry group. If the companies NAICS code can be provided to the cross-connection control specialist, this will provide a good general idea of what to expect at the facility.

The cross-connection control program specialist must then be able to carry out the site surveys or inspections to assess the degree of hazard on a premise and decide what modifications should be made to control the cross-connections. The specialist must have a good knowledge of plumbing systems as well as the general uses of the potable and industrial water. The specialist must record and report the detailed findings. Often it is desirable to make detailed sketches to the exact location of the cross-connection and the type and size of backflow prevention assembly that is to be installed. Based on the findings from the site survey or inspection, the backflow protection will be specified commensurate with the degree of hazard.

The Site Survey

There are some essential tools that the surveyor should have. These include a note pad and clipboard to document the findings, any safety equipment that may be needed, such as a hard hat, safety glasses, ear protection, a flashlight, a laser pointer to trace the water piping. A camera or video camera if allowed by the on-site personnel may be helpful in documenting your findings.

Once on site with the appropriate personnel, the specialist must begin surveying the water uses on the premises. One method it to trace the water line beginning at the service connection and following the line through the facility to the various points of water use. With experience, the specialist may start going to specific points of water use, knowing what to expect at a specific type of facility. For example, if the specialist is surveying an office building, the only concern may be a centralized boiler, if one exists. The specialist can then ask if there is a boiler and then determine if there is a potable water make-up line.

If the specialist is surveying a facility with which he is not familiar, it will be necessary to learn as much as possible about what is done at the facility and it may be necessary to trace out the water lines to determine all of the water uses. Then, depending on whether the specialist is running a system protection program or an internal protection program a determination would be made for the backflow protection required at the service connection or at the various water uses, respectively. Of course the backflow protection must be commensurate with the degree of hazard and appropriate for the type of cross-connection.

Documentation

Once the survey is complete, a detailed report is to be prepared. The report should include the water uses, cross-connections, degree of hazard, any backflow protection, reference to any relevant codes and regulations. The report should conclude by indicating what backflow protection may be needed and where.

The title of the report should include the name of the facility, the date when the survey was done and any on-site personnel that were present during the survey. The water uses should be documented. This includes the listing of the water systems, plumbing fixtures and water using equipment. The water systems include the fire system, domestic system and irrigation system. Plumbing fixtures are those that are referenced in many of the plumbing codes and include fixtures such as flushometer valves, toilet tanks and faucets. Water using equipment would be any piece of equipment that uses water. See sample Field Survey Form , 8.17.

For each water use, the specialist should determine if there is a cross-connection. If there is a cross-connection the type should be determined: a direct cross-connection or an indirect cross-connection. For each water use the degree of hazard needs to be documented. Typically the two degrees of hazard are the health hazard (contaminant) and the non-health hazard (pollutant). A review of the substances that are being used is important especially if they are being mixed with water through a possible cross connection. Review of the labeling of the chemicals and the MSDS (Material Safety Data Sheets) will help in determining the degrees of hazards.

If during the survey or inspection backflow protection is found, then the documentation of that backflow protection and its purpose must be reported. The backflow protection documentation may include whether the protection is adequate, is it approved, is it installed properly and is it field tested and maintained on a periodic basis.

Information gained during the site survey should be detailed in the report or supporting notes. The documentation may include photos or drawings in order to detail all the information discovered. The report should be clear enough that another specialist could pick up the report, come back to the facility and find exactly what was discovered and indicated in the report. The ability to make quick, accurate sketches is a valuable tool for a specialist.

It is important in the report to include references to any codes, ordinances or regulations that would support the findings of the survey. For example, a cross-connection could be documented as a code violation by referencing to the specific plumbing code section.

Compliance

Finally a letter must be sent to the owner or manager detailing the findings and the requirements for correction. See sample compliance letters in Chapter 8. This may include a detailed report. Along with the report and the requirement for changes, by the installation of one or more backflow prevention assemblies, the letter should give a specific deadline for compliance. If compliance is not attained, it will be necessary to continue on in the process detailed in the administrative authority's non-compliance policy (see Chapter 4).

Cross-Connection Control Surveys

Chapter 6 — *Facilities*

Facilities

Inside this Chapter

Typical Facilities: Cross-Connections or Water Uses
    Which May Endanger the Public Water System **120**
    Services 120
    Manufacturing..121
    Food Processing/Service 123
    Medical Facilities...124
    Restricted 126
    Other Facilities..127

Typical Facilities: Cross-Connections or Water Uses Which May Endanger the Public Water System

The following facilities should be reviewed in all cases where the regulatory agency adopts a policy of requiring that the degree of hazard be based upon a complete survey of the consumer's water-using facilities. For convenience, these facilities have been divided into five groups. It should be noted that the following are intended to be representative; they are not all inclusive.

Backflow protection commensurate with the degree of hazard at each of these facilities should be required. The requirement should be based on the site survey. This chapter is intended to provide guidelines to the one conducting the site survey so the surveyor will know what to expect.

The specific backflow preventer may be determined by using the flowchart or table in figures 3.35 and 3.36 respectively. For system protection, an air gap, reduced pressure principle assembly or a double check valve assembly is normally required. For internal protection (i.e., protection at the point of water use), an air gap, reduced pressure principle assembly or a double check valve assembly, pressure vacuum breaker, spill-resistant vacuum breaker or an atmospheric vacuum breaker is normally required.

The categories listed below are intended to be general categories of all water using facilities. Some facilities may not fit perfectly into one of the categories listed. This list is intended to be a guideline, to guide the Specialist in preparing for a site survey. Other cross-connections may be discovered on a site survey, which are not listed under the categories listed here. The categories covered in this section are:

- Services
- Manufacturing
- Food Processing
- Medical
- Restricted
- Others

Services
Services may include such facilities as

- Car washes
- Film laboratories
- Laundry facilities

The hazards normally found in a facility of this type include cross-connections between the consumer's water system and the following:

Soap injection systems, recycling water systems, reservoirs, cooling towers and circulating systems which may be heavily contaminated with bird droppings, vermin, algae, bacterial slimes or toxic water treatment compounds such as copper sulfate, pentachlorophenol, chromates, metallic glucosides, compounds of mercury, quaternary ammonium compounds, etc.

Steam generating facilities and lines, which may be contaminated with boiler compounds such as those chemicals listed above (NOTE: a particular hazard is the possibility of steam getting back into the domestic system, causing either a system or a health hazard);

Water cooled equipment which may be sewer-connected such as compressors, heat exchangers, air conditioning equipment, etc. (NOTE: in multi-storied buildings of this type the supply line to the toilets, urinals, lavatories, laboratory sinks, tanks, etc., on the lower floors may be taken off the suction side of the house pump and, as a result, sewage or other contaminated substances may be drawn into the house supply line);

Sewer-connected plumbing fixtures such as flush valve toilets and urinals without atmospheric vacuum breakers or with improperly maintained atmospheric vacuum breakers (NOTE: this hazard is critical because little or no attention is given to the maintenance of an atmospheric vacuum breaker and frequently the working parts or the entire assembly may be removed from the line);

Hydraulically-operated equipment where the public potable water pressure is used directly and may be subject to backpressure;

Tanks, automatic film processing machines or other facilities used in processing films, which may be contaminated with chemicals such as acetic acid, potassium ferricyanide and/or one of the many different types of the aromatic series of organic chemicals;

Laundry machines having under-rim or bottom inlets;

Dye vats in which are used toxic chemicals and dyes;

Water storage tanks equipped with pumps and recirculating systems;

Shrinking, bluing and dyeing machines with direct connections to circulating systems;

Retention and mixing tanks (NOTE: some of these machines or pieces of equipment have pumps which can pump contaminated fluids through cross-connections into the public water supply);

Recycled water systems using recycled water for irrigation, toilets, urinals, trap primers, cooling towers, or other equipment.

Manufacturing
Manufacturing facilities may include the following:

- Aircraft manufacturing
- Automotive plants
- Chemical plants
- Metal works facilities
- Oil/gas refineries
- Plating plants
- Power plants
- Rubber manufacturing facilities
- Sand/gravel facilities

The hazards normally found in a facility of this type include cross-connections between the consumer's water system and the following:

Reservoirs, cooling towers and circulating systems which may be heavily contaminated with bird droppings, vermin, algae, bacterial slimes or toxic water treatment compounds such as copper sulfate, pentachlorophenol, chromates, metallic glucosides, compounds of mercury, quaternary ammonium compounds, etc.;

Steam generating facilities and lines which may be contaminated with boiler compounds such as those chemicals listed above (NOTE: a particular hazard is the possibility of steam getting back into the domestic system, causing either a system or a health hazard);

Plating facilities involving the use of highly toxic cyanides, heavy metals in solution (such as copper, cadmium, chrome, nickel, etc.), acids and caustic solutions;

Plating solution filtering equipment with pumps and circulating lines;

Tanks, vats, and other vessels used in painting, descaling, anodizing, cleaning, stripping, oxidizing, etching, passivating, pickling, dipping, rinsing operations, or other lines or facilities needed in the preparation or finishing of the product;

Water cooled equipment which may be sewer-connected such as compressors, heat exchangers, air conditioning equipment, etc. (NOTE: in multi-storied buildings of this type the supply line to the toilets, urinals, lavatories, laboratory sinks, tanks, etc., on the lower floors may be taken off the suction side of the house pump and, as a result, sewage or other contaminated substances may be drawn into the house supply line);

Industrial fluid systems and lines containing cutting and hydraulic fluids, coolants, hydrocarbon products, glycerine, paraffin, caustic and acid solutions, etc.;

Sewer-connected plumbing fixtures such as flush valve toilets and urinals without atmospheric vacuum breakers or with improperly maintained atmospheric vacuum breakers (NOTE: this hazard is critical because little or no attention is given to the maintenance of an atmospheric vacuum breaker and frequently the working parts or the entire assembly may be removed from the line);

Fire fighting systems, including storage reservoirs which may be treated for the prevention of scale formation, corrosion, algae, slime growths, etc., or fire systems which may be subject to contamination with antifreeze solutions, Foamite or other chemicals or chemical compounds used in fighting fire, or fire systems which are subject to contamination with auxiliary or used water supplies or industrial fluids;

Hydraulically-operated equipment where the public potable water pressure is used directly and may be subject to backpressure;

Sewer lines for the purpose of disposing of filter or softener backwash water or water from cooling systems or of providing for a quick drain for the building lines or of flushing or blowing out obstructions, etc. (NOTE: administrative authorities may require backflow protection at the service connection to any premise on which there is located a sewage ejector or pumping station, even though there are no cross-connections);

Retention and mixing tanks (NOTE: some of these machines or pieces of equipment have pumps which can pump contaminated fluids through cross-connections into the public water supply);

Steam boilers and lines; mud pumps and mud tanks; hydraulically-operated Tetrolite tanks; oil well casings (for dampening pressures);

Dehydration tanks and outlet lines from storage and dehydration tanks (for purging purposes);

Oil and gas tanks (to create hydraulic pressures and to hydraulically raise the oil and gas levels);

Oil and gas lines (for testing, evacuating and slugging purposes);

Pulp, bleaching, dyeing and processing facilities which may be contaminated with toxic chemicals;

Sand and gravel washing equipment supplied with a private with a private well or water pumped from sumps or retention basins on a recirculating basis. These cross-connections may be under pump pressure, in many cases considerably higher than the pressures in the public main.

Recycled water systems using recycled water for irrigation, toilets, urinals, trap primers, cooling towers, or other equipment.

Food Processing/Service

Services may include such facilities as

- Bakery
- Beverage manufacturing/bottling
- Brewery
- Cannery
- Dairy
- Frozen foods processing plant
- Packing house
- Restaurants
- Slaughterhouse

The hazards normally found in a facility of this type include cross-connections between the consumer's water system and the following:

Truck wash, forklift battery fill station, automatic hood wash systems, food mixing tanks, pressure washers, dish washing equipment, garbage disposals, pasteurization units, chlorine/sanitizer injector, lubrication lines to conveyor equipment

Reservoirs, cooling towers and circulating systems which may be heavily contaminated with bird droppings, vermin, algae, bacterial slimes or toxic water treatment compounds such as copper sulfate, pentachlorophenol, chromates, metallic glucosides, compounds of mercury, quaternary ammonium compounds, etc.;

Steam generating facilities and lines which may be contaminated with boiler compounds such as those chemicals listed above (NOTE: a particular hazard is the possibility of steam getting back into the domestic system, causing either a system or a health hazard);

Water cooled equipment which may be sewer-connected such as compressors, heat exchangers, air conditioning equipment, etc. (NOTE: in multi-storied buildings of this type the supply line to the toilets, urinals, lavatories, laboratory sinks, tanks, etc., on the lower floors may be taken off the suction side of the house pump and, as a result, sewage or other contaminated substances may be drawn into the house supply line);

Industrial fluid systems and lines containing cutting and hydraulic fluids, coolants, hydrocarbon products, glycerine, paraffin, caustic and acid solutions, etc.;

Fire fighting systems, including storage reservoirs which may be treated for the prevention of scale formation, corrosion, algae, slime growths, etc., or fire systems which may be subject to contamination with antifreeze solutions, Foamite or other chemicals or chemical compounds used in fighting fire, or fire systems which are subject to contamination with auxiliary or used water supplies or industrial fluids;

Auxiliary water supply. (**NOTE**: The rules or regulations of the administrative authority or the water supplier may require a backflow prevention assembly at the service connection to premises where there is an auxiliary water supply, even though there are no existing cross-connections.);

Steam connected facilities such as **pressure cookers, autoclaves, retorts, etc.; washers, cookers, tanks, lines, flumes and other equipment** used for storing, washing, cleaning, blanching,

Tanks, can and bottle washing machine lines where caustics, acids, detergents and other compounds are used in cleaning, sterilizing and flushing;

Laboratory equipment which may be chemically or bacteriologically contaminated;

Irrigation systems which may be equipped with pumps, injectors, pressurized tanks or vessels, or other facilities for injecting or aspirating into the irrigation system agricultural chemicals such as fungicides, pesticides, soil conditioners and other similar noxious, toxic or objectionable substances;

Irrigation systems subject to contamination from submerged inlets (sprinkler heads), auxiliary water supplies, ponds, reservoirs, swimming pools and other sources of stagnant, polluted or contaminated waters (agricultural chemicals which are broadcast on ground may contaminate water collecting around sprinkler heads);

Retention and mixing tanks (NOTE: some of these machines or pieces of equipment have pumps which can pump contaminated fluids through cross-connections into the public water supply);

Dehydration tanks and outlet lines from storage and dehydration tanks (for purging purposes);

Recycled water systems using recycled water for irrigation, toilets, urinals, trap primers, cooling towers, or other equipment.

Medical Facilities
Services may include such facilities as

- Convalescent Hospitals
- Dental Offices
- Hospitals
- Kidney Dialysis Centers
- Medical Laboratories
- Medical Offices
- Veterinary Offices

The hazards normally found in a facility of this type include cross-connections between the consumer's water system and the following:

Laboratory equipment which may be chemically or bacteriologically contaminated;
steam sterilizers, autoclaves, specimen tanks, physical therapy or whirlpool tubs, scope washers, bed pan washing stations, pipette washers, kidney dialysis equipment, showers/bath tubs with hoses attached, janitorial sinks, waste management equiment, lab faucets, autopsy and mortuary equipment (NOTE: this hazard is critical because little or no attention is given to the maintenance of atmospheric vacuum breakers and frequently they are removed from the line);

Reservoirs, cooling towers and circulating systems which may be heavily contaminated with bird droppings, vermin, algae, bacterial slimes or toxic water treatment compounds such as copper sulfate, pentachlorophenol, chromates, metallic glucosides, compounds of mercury, quaternary ammonium compounds, etc.;

Steam generating facilities and lines which may be contaminated with boiler compounds such as those chemicals listed above (NOTE: a particular hazard is the possibility of steam getting back into the domestic system, causing either a system or a health hazard);

Water cooled equipment which may be sewer-connected such as compressors, heat exchangers, air conditioning equipment, etc. (NOTE: in multi-storied buildings of this type the supply line to the toilets, urinals, lavatories, laboratory sinks, tanks, etc., on the lower floors may be taken off the suction side of the house pump and, as a result, sewage or other contaminated substances may be drawn into the house supply line);

Industrial fluid systems and lines containing cutting and hydraulic fluids, coolants, hydrocarbon products, glycerine, paraffin, caustic and acid solutions, etc.;

Sewer-connected plumbing fixtures such as flush valve toilets and urinals without atmospheric vacuum breakers or with improperly maintained atmospheric vacuum breakers (NOTE: this hazard is critical because little or no attention is given to the maintenance of an atmospheric vacuum breaker and frequently the working parts or the entire assembly may be removed from the line);

Fire fighting systems, including storage reservoirs which may be treated for the prevention of scale formation, corrosion, algae, slime growths, etc., or fire systems which may be subject to contamination with antifreeze solutions, Foamite or other chemicals or chemical compounds used in fighting fire, or fire **hydraulically-operated equipment** where the public potable water pressure is used directly and may be subject to backpressure;

Auxiliary water supply. (**NOTE**: The rules or regulations of the administrative authority or the water supplier may require a backflow prevention assembly at the service connection to premises where there is an auxiliary water supply, even though there are no existing cross-connections.);

Sewer lines for the purpose of disposing of filter or softener backwash water or water from cooling systems or of providing for a quick drain for the building lines or of flushing or blowing out obstructions, etc. (NOTE: administrative authorities may require backflow protection at the service connection to any premise on which there is located a sewage ejector or pumping station, even though there are no cross-connections);

Steam connected facilities such as **pressure cookers, autoclaves, retorts, etc.; washers, cookers, tanks, lines, flumes and other equipment** used for storing, washing, cleaning, blanching,

Tanks, can and bottle washing machine lines where caustics, acids, detergents and other compounds are used in cleaning, sterilizing and flushing;

Deionized water systems which may be used for supplying rinse water in surgery, in washing equipment or to blood analyzers in laboratories;

Tanks, automatic film processing machines or other facilities used in processing films, which may be contaminated with chemicals such as acetic acid, potassium ferricyanide and/or one of the many different types of the aromatic series of organic chemicals;

Contaminated or sewer connected equipment such as bedpan washers, flush valve toilets and urinals, autoclaves, specimen tanks, sterilizers, pipet tube washers, cuspidors, aspirators, autopsy and mortuary equipment, etc. (NOTE: It has been found that in this type of facility little or no attention is given to the maintenance of air gaps or atmospheric vacuum breakers. Also, in multi-storied buildings the supply line to the toilets, urinals, lavatories, laboratory sinks, etc., on the lower floors may be taken off the suction side of the house pump and, as a result, sewage or other contaminated substances may be drawn into the house supply line.);

Laundry machines having under-rim or bottom inlets; **sterilizers, autoclaves, specimen tanks, autopsy and morgue equipment; sewer connected plumbing fixtures** such as flush valve toilets and urinals without atmospheric vacuum breakers or with improperly maintained atmospheric vacuum breakers (NOTE: this hazard is critical because little or no attention is given to the maintenance of atmospheric vacuum breakers and frequently they are removed from the line);

Recycled water systems using recycled water for irrigation, toilets, urinals, trap primers, cooling towers, or other equipment.

Restricted
Services may include such facilities as

- Civil Works
- Classified
- Research Facilities

The hazards normally found in a facility of this type include cross-connections between the consumer's water system and the following:

Reservoirs, cooling towers and circulating systems which may be heavily contaminated with bird droppings, vermin, algae, bacterial slimes or toxic water treatment compounds such as copper sulfate, pentachlorophenol, chromates, metallic glucosides, compounds of mercury, quaternary ammonium compounds, etc.;

Steam generating facilities and lines which may be contaminated with boiler compounds such as those chemicals listed above (NOTE: a particular hazard is the possibility of steam getting back into the domestic system, causing either a system or a health hazard);

Reservoirs, cooling towers and circulating systems which may be heavily contaminated with bird droppings, vermin, algae, bacterial slimes or toxic water treatment compounds such as copper sulfate, pentachlorophenol, chromates, metallic glucosides, compounds of mercury, quaternary ammonium compounds, etc.;

Steam generating facilities and lines which may be contaminated with boiler compounds such as those chemicals listed above (NOTE: a particular hazard is the possibility of steam getting back into the domestic system, causing either a system or a health hazard);

Irrigation systems which may be equipped with pumps, injectors, pressurized tanks or vessels, or other facilities for injecting or aspirating into the irrigation system agricultural chemicals such as fungicides, pesticides, soil conditioners and other similar noxious, toxic or objectionable substances;

Irrigation systems subject to contamination from submerged inlets (sprinkler heads), auxiliary water supplies, ponds, reservoirs, swimming pools and other sources of stagnant, polluted or contaminated waters (agricultural chemicals which are broadcast on ground may contaminate water collecting around sprinkler heads);

Recycled water systems using recycled water for irrigation, toilets, urinals, trap primers, cooling towers, or other equipment.

Other Facilities

Services may include such facilities as

- Amusement Parks
- Aquaria
- Hotels/Motels
- Motion Picture Studios
- Schools and Colleges
- Waterfront Facilities
- Zoos

The hazards normally found in a facility of this type include cross-connections between the consumer's water system and the following:

Boilers for saunas, steambaths, foot baths, beauty salon sink, pools, fountains, laundry, pressure washers, automatic hood wash system, carbonated beverage dispensers, janitorial sinks, garbage disposals, dishwashing machines, recreational vehicle sewage dump station, film processing, swamp coolers;

Reservoirs, cooling towers and circulating systems which may be heavily contaminated with bird droppings, vermin, algae, bacterial slimes or toxic water treatment compounds such as copper sulfate, pentachlorophenol, chromates, metallic glucosides, compounds of mercury, quaternary ammonium compounds, etc.;

Steam generating facilities and lines which may be contaminated with boiler compounds such as those chemicals listed above (NOTE: a particular hazard is the possibility of steam getting back into the domestic system, causing either a system or a health hazard);

Water cooled equipment which may be sewer-connected such as compressors, heat exchangers, air conditioning equipment, etc. (NOTE: in multi-storied buildings of this type the supply line to the toilets, urinals, lavatories, laboratory sinks, tanks, etc., on the lower floors may be taken off the suction side of the house pump and, as a result, sewage or other contaminated substances may be drawn into the house supply line);

Industrial fluid systems and lines containing cutting and hydraulic fluids, coolants, hydrocarbon products, glycerine, paraffin, caustic and acid solutions, etc.;

Sewer-connected plumbing fixtures such as flush valve toilets and urinals without atmospheric vacuum breakers or with improperly maintained atmospheric vacuum breakers (NOTE: this hazard is critical because little or no attention is given to the maintenance of an atmospheric vacuum breaker and frequently the working parts or the entire assembly may be removed from the line);

Fire fighting systems, including storage reservoirs which may be treated for the prevention of scale formation, corrosion, algae, slime growths, etc., or fire systems which may be subject to contamination with antifreeze solutions, Foamite or other chemicals or chemical compounds used in fighting fire, or fire **hydraulically-operated equipment** where the public potable water pressure is used directly and may be subject to backpressure;

Auxiliary water supply. (NOTE: The rules or regulations of the administrative authority or the water supplier require a backflow prevention assembly at the service connection to premises where there is an auxiliary water supply, even though there are no existing cross-connections.

Sewer lines for the purpose of disposing of filter or softener backwash water or water from cooling systems or of providing for a quick drain for the building lines or of flushing or blowing out obstructions, etc. (NOTE: administrative authorities may require backflow protection at the service connection to any premise on which there is located a sewage ejector or pumping station, even though there are no cross-connections);

Steam connected facilities such as **pressure cookers, autoclaves, retorts, etc.; washers, cookers, tanks, lines, flumes and other equipment** used for storing, washing, cleaning, blanching,

Tanks, can and bottle washing machine lines where caustics, acids, detergents and other compounds are used in cleaning, sterilizing and flushing;

Laboratory equipment which may be chemically or bacteriologically contaminated; **steam sterilizers, autoclaves, specimen tanks, autopsy and mortuary equipment** (NOTE: this hazard is critical because little or no attention is given to the maintenance of atmospheric vacuum breakers and frequently they are removed from the line);

Tanks, automatic film processing machines or other facilities used in processing films, which may be contaminated with chemicals such as acetic acid, potassium ferricyanide and/or one of the many different types of the aromatic series of organic chemicals;

Contaminated or sewer connected equipment such as bedpan washers, flush valve toilets and urinals, autoclaves, specimen tanks, sterilizers, pipet tube washers, cuspidors, aspirators, autopsy and mortuary equipment, etc. (NOTE: It has been found that in this type of facility little or no attention is given to the maintenance of air gaps or atmospheric vacuum breakers. It is customary to bridge an air gap by means of a hose section. Also, in multi-storied buildings the supply line to the toilets, urinals, lavatories, laboratory sinks, etc., on the lower floors may be taken off the suction side of the house pump and, as a result, sewage or other contaminated substances may be drawn into the house supply line.);

Laundry machines having under-rim or bottom inlets;

Sterilizers, autoclaves, specimen tanks, autopsy and morgue equipment; sewer connected plumbing fixtures such as flush valve toilets and urinals without atmospheric vacuum breakers or with improperly maintained atmospheric vacuum breakers (NOTE: this hazard is critical because little or no attention is given to the maintenance of atmospheric vacuum breakers and frequently they are removed from the line);

Recycled water systems using recycled water for irrigation, toilets, urinals, trap primers, cooling towers, or other equipment.

Chapter 7 — *Equipment and Systems*

Equipment and Systems

Inside this Chapter

Equipment and Systems	132
Air Scrubber (Wet Scrubbing)	132
Aspirators	133
Autoclaves	134
Autopsy/Mortuary Tables	135
Boiler	136
Can and Bottle Washing Machines	137
Car Washing Machine (automatic type)	138
Carbonators	139
Chemical Dispenser	140
Cookers	141
Cooling Tower	142
Dental Vacuum Pump	143
Fire Sprinkler Systems	144
Heat Exchangers	145
Irrigation Systems	146
Kidney Dialysis Machine	147
Laboratory Equipment	148
Laundry Machines	149
Photographic Film Processing Machines	150
Portable Cleaning Equipment	151
Pumps	152
Recycled Water Systems	153
Sewage Ejector	154
Tanks/Vats	155
Toilet	156
Urinal	157
Water Softener	158

Equipment and Systems

This chapter is meant to give the reader a general understanding of various pieces of equipment, which may be found at various facilities. For each piece of equipment the equipment is described, a description of how it works is given then an explanation of where it may be used. It is important to realize that this is not an all-inclusive list. The listings here are typical pieces of equipment and systems that one may come across while conducting a site survey. The intent is to familiarize the reader with these pieces of equipment.

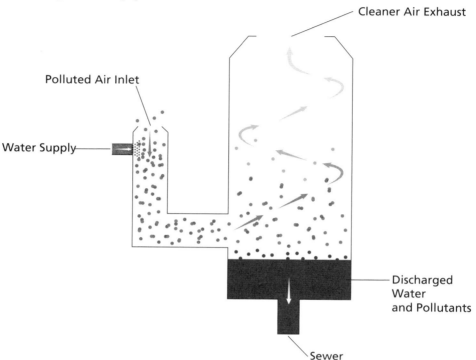

Air Scrubber (Wet Scrubbing)

A device or system that is used for air pollution control. These devices are used to remove particulates, pollutants and/dust particles from the air. One method of doing this is through a wet scrubbing system which utilizes water to clean the air.

How does it work?

An air scrubber works by the contact of the pollutants or dust particles with the scrubbing solution or water. As the air flows into the air scrubber system, water is then utilized to scrub the unwanted pollutants or dust particles from the air and this air flowing out of the air scrubber is cleaner air. As part of the scrubbing process, the water that has scrubbed the air is now discharged. The discharge line may be directly connected to the sewer. This discharged water is mixed in with the pollutants or dust particles.

Where is it used?

Air scrubbers may be used where industrial exhaust or air needs to be cleaned.

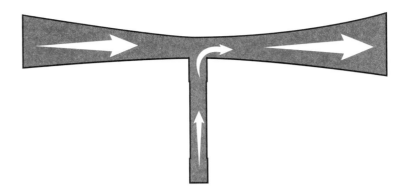

Aspirators
An aspirator is a piece of piping or tubing which is more narrow in the center than on either end.

How does it work?
As a fluid flows through an aspirator the velocity increases at the narrow section, thus developing a region of lower pressure. If a tube or pipe is connected to this lower region, suction occurs through the attached tube or pipe.

Where is it used?
Used commonly in laboratories or medical facilities to create a siphon. Is also used in irrigation systems to introduce fertilizer or chemicals into the system.

Equipment and Systems

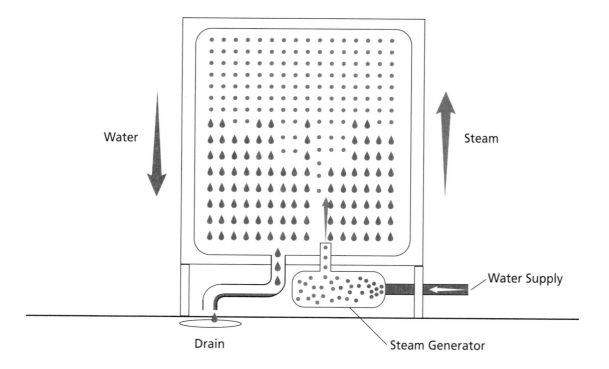

Autoclaves
A pressurized vessel for sterilization.

How does it work?
Uses steam under pressure to sterilize medical instruments. Some autoclaves generate steam for the sterilization process. In these cases a water line may be directly connected to the autoclave through a solenoid valve which opens and closes automatically as water is needed. However, some autoclaves have a reservoir, which must be manually filled, and do not have a water connection.

Where is it used?
Usually found in medical facilities.

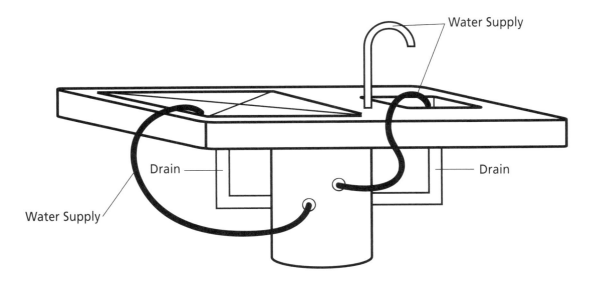

Autopsy/Mortuary Tables
Equipment used for conducting autopsies or funeral preparation and embalming.

How does it work?
Water feeds equipment for rinsing and aspirating purposes. Some aspirators may be equipped with reverse flow capabilities to clear any clogging, which may occur. There may also be water connected to a waste disposal unit. Many of these tables have vacuum breakers built-in as an integral part of the table.

Where is it used?
Morgues, hospitals, mortuaries, medical training facilities or medical laboratories.

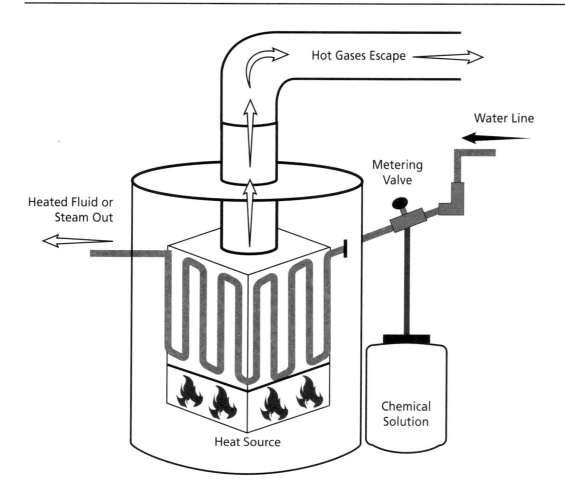

Boiler

A boiler is a closed vessel in which water and/or other fluids are heated under pressure. The heated fluid or steam is then circulated out of the boiler for use in various processes and heating applications.

How does it work?

As the water flows into the boiler, the water flows through various bends of tubes and fins. While the water is flowing through the tubes and fins, a source of heat is utilized to heat the water. This source of heat can vary from wood, coal, oil, natural gas, nuclear fission or electricity. Once the water is heated, it is then circulated and utilized for various processes and heating applications. Because the water is being heated the water may corrode or the minerals in the water may build up in these tubes or fins. Corrosion and/or these mineral deposits may act as thermal insulator that may restrict the flow of heat or these mineral deposits may act as a flow restrictor that may restrict the flow of water through the boiler. In order to control the corrosion and this build up of deposits from the water; the water may need to be treated. Treatment of the water may include softening the water with a water softener; addition of chemicals for anti-corrosion and chemicals for anti-scaling. Chemical pumps or chemical feed pots are utilized to introduce these chemicals into the boiler.

Additionally, as the water is heated it evaporates. A make-up water line is needed to re-supply the water that has been evaporated as a result of the boiling process.

Where is it used?

A boiler may be used anywhere heating processes or applications are utilized.

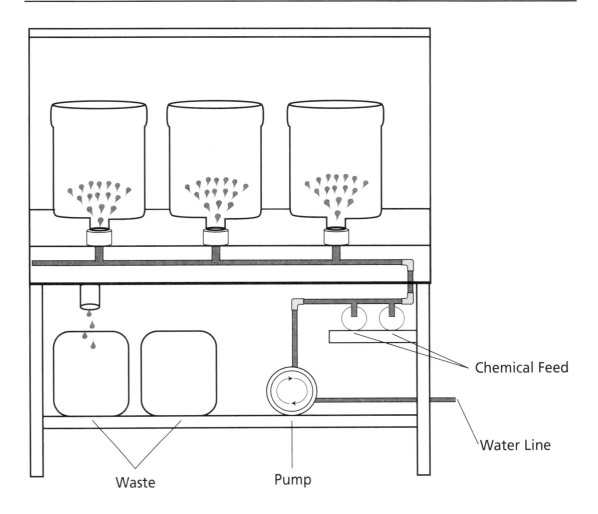

Can and Bottle Washing Machines
An automated machine for washing cans and bottles.

How does it work?
Water is used in the washing process using chemicals or detergents as well as the rinsing process. A pump in the machine may be used for increasing water pressure.

Where is it used?
Found in canneries, and bottling facilities

7 Equipment and Systems

Equipment and Systems

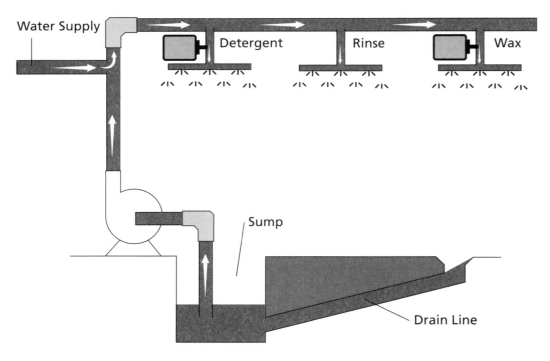

Car Washing Machine (automatic type)
A car wash is for washing motor vehicles.

How does it work?
Most car washes mix water with a detergent to spray on a vehicle and clean it with brushes or cloths. Then clean water is sprayed on the vehicle as a rinse. The detergent mixture is often recirculated for use on other vehicles. The recirculating system usually has a water make-up line. Since the recirculating system is under pressure, the mixture could be pressured through the water make-up line contaminating the potable water supply.

Where is it used?
Automatic Car Wash Machines are found at gas stations, car rental facilities, fleet service facilities, and car wash facilities

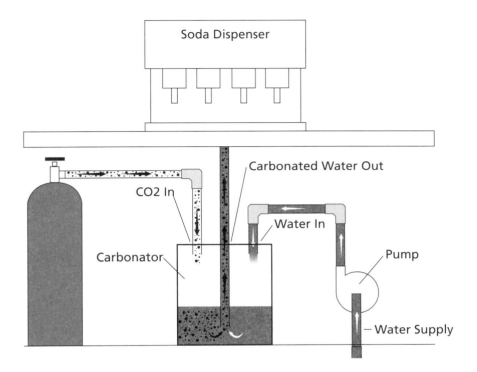

Carbonators
A carbonator is a piece of equipment used to add carbonation to drinks.

How does it work?
Carbon dioxide gas is added to a mixture of water and syrup to create sodas. Should the carbon dioxide backflow into the water line a problem could occur if the water line is made of copper. The carbon dioxide or carbonated water could leach the copper out of the piping presenting a health hazard.

Where is it used?
Used in restaurants or wherever non-pre-packaged carbonated beverages are sold.

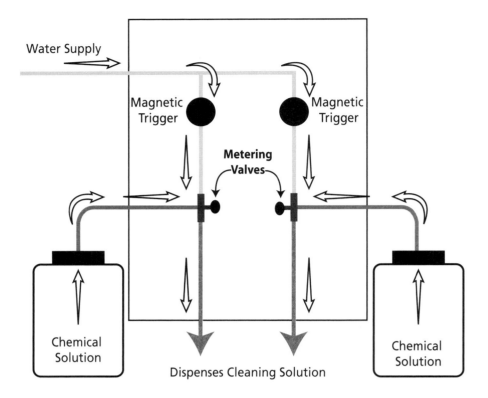

Chemical Dispenser
Chemical dispensers are devices used for mixing cleaning solutions. The chemical dispensers automatically dilute and dispense solutions into spray bottles, mop buckets or other containers.

How does it work?
The chemical dispensers use a mixing station that will dispense cleaning solutions by diluting a concentrate of chemical or soap with water. The chemical dispensers connect to existing water lines or faucets and draw or aspirate chemical concentrates to mix with the water to dispense the cleaning solution. The mixing stations may have several dispensers that dispense different solutions at different concentrations.

Where is it used?
Chemical dispensers are normally used in cleaning areas like a janitor room or kitchen.

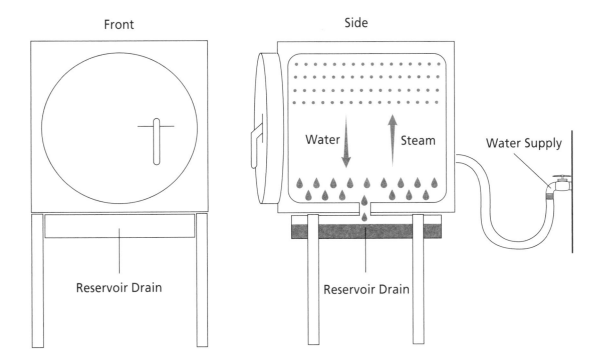

Cookers

Various pieces of equipment used in kitchens and food preparation areas. These may include steam kettles, steam cookers, pressure cookers, etc.

How does it work?

These either use steam or create steam to heat up or cook food. Many of these have no direct connection to the water lines and may be filled manually. Some may have water line connections controlled by solenoid valves to admit water into the heating chamber or reservoir.

Where is it used?

Restaurants, hotels, banquet halls or any facility that has food preparation areas.

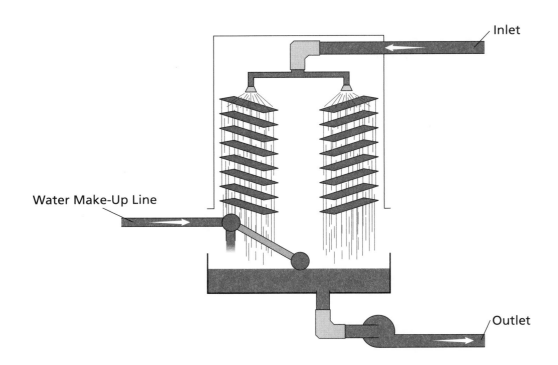

Cooling Tower

A cooling tower is an evaporative cooler used to cool water or another medium to ambient temperature.

How does it work?

Cooling towers use evaporation to remove the heat from water or fluids typically used in recirculating systems. Water typically drops through the "tower" to cool down. It then collects in a reservoir at the bottom of the cooling tower. A water make-up line normally feeds the reservoir through an air gap located inside of the tower itself.

Where is it used?

Cooling towers are used at power plants, refineries and any facility having a building cooling system.

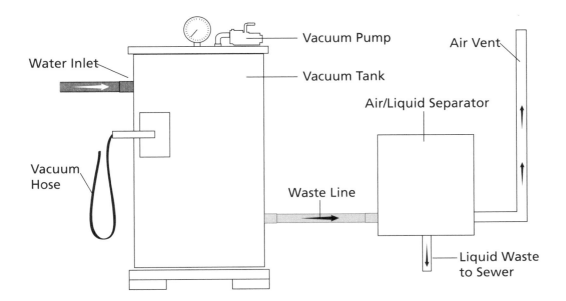

Dental Vacuum Pump

Dental vacuum pumps for conducting dental examinations and dental work. These pumps are used to siphon a patient's saliva and particles as a result of the dental work away from the mouth. There are different designs of the dental vacuum pump. Some are stand alone designs and some use an aspirator.

How does it work?

Dental systems that are of an aspirator design utilize water flow to create the vacuum condition. The water aspirator dental vacuum pump is essentially a pipe with a narrowing area. As water flows through that narrowing area, it speeds up and its pressure drops to a subatmospheric pressure. (See 1.9 or figure 3.16.) Some of these aspirators may be equipped with reverse flow capabilities to clear any clogging which may occur. The dental vacuum systems which use a vacuum pump may use water as a seal or to remove particulates from the vacuum flow.

Where is it used?

Dental vacuum pumps are found in dental offices.

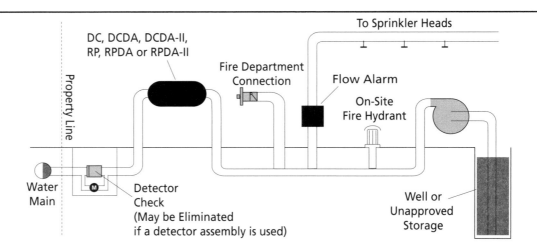

Fire Sprinkler Systems

NOTE: Although it is unlikely to find all of the components shown in the illustration on one fire sprinkler system, the components are shown so the surveyer is aware of the various components that may exist in a fire sprinkler system.

Fire sprinkler systems may include any number of several different components. As with all cross-connections the same principles apply. Backflow protection should be commensurate with the degree of hazard and the hydraulic condition. Since sprinkler systems are typically elevated from ground level, they should always be considered a backpressure hydraulic condition. So, the backflow preventer used depends on the degree of hazard present. In some sprinkler systems that use potable piping and have a flow through condition, where the water is continually flowing through the sprinkler system and then onto other uses, there may be no need for backflow protection.

Things to look for

Black Iron Piping: Many fire sprinkler systems use black iron piping. This material is not acceptable for potable water systems. Water that sits stagnant in these pipes will degrade to at least a pollutional hazard, and should be protected with the minimum of a double check valve assembly.

Detector check valve: Typically, this is a single check valve with a water meter on the bypass around the check valve. A detector check valve is designed to indicate if there is any unauthorized use of water, which may occur in a sprinkler system, since fire sprinkler systems are normally not metered. A detector check valve is not considered to be a backflow prevention assembly. In this configuration, water can backflow through the unprotected water meter.

Fire department connection: These are included in a fire protection system to allow a fire pumper truck to attach to the system and pump supplemental water into the sprinkler system. The connection should be downstream of any backflow protection.

Auxiliary water supply: An auxiliary water supply may be used to get large quantities of water to a fire very quickly. The quality of the water may or may not be acceptable as potable water. If it is not potable, appropriate backflow protection commensurate with the degree of hazard should be used.

Additives: Some fire sprinkler systems may have additives (e. g., anti-freeze or corrosion inhibitors) in the water or may be set up to allow the addition of additives (e .g., fire extinguishing foam) in order to fight specific fires, such as chemical fires. When additives are used, appropriate backflow protection will be required.

Detector assemblies: The double check detector assembly, reduced pressure principle detector assembly, double check detector assembly-type II and reduced pressure principle detector assembly-type II are backflow prevention assemblies designed to offer the appropriate level of backflow protection with the built in capability to detect unauthorized use of water.

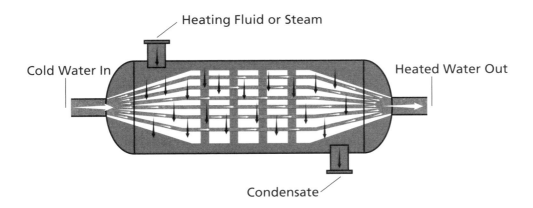

Heat Exchanger

A heat exchanger is a piece of equipment that is for heating or cooling a medium by transferring energy. When used to heat domestic water, typically there are two types of heat exchangers: a single-walled heat exchanger and a double-walled heat exchanger.

How does it work?

A single-walled heat exchanger uses two areas. Typically, one area would be filled with the heating medium or steam. Running through the steam are tubes with cold domestic water. As the water runs through these tubes, the water heats up and the steam cools down. The cooled steam, or condensate, is returned to the heating source (i.e., boiler) where steam is created and re-circulates through the system.

A double-walled heat exchanger is similar to a single-walled heat exchanger, except that the tubes containing the water have two walls, with a path to atmosphere (air-space) in between them. This still allows for the heat to transfer from the heating fluid/steam to the water, but should one of the tube walls rupture, the treated fluid/steam could not get into the drinking water system, but would go into the air-filled region and discharge through a port to the outside.

Where is it used?

A heat exchanger may be used anywhere heating and cooling systems are found.

Equipment and Systems

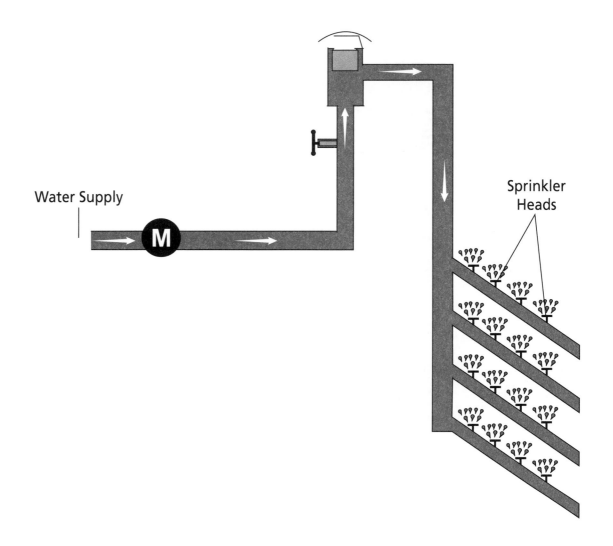

Irrigation Systems
Supplying of water to support vegetation.

How does it work?
Irrigation systems may take on many forms such as landscape sprinklers, drip irrigation, flood irrigation, etc. Irrigation systems may inject or aspirate herbicides, fertilizers or insecticides into the system. Agricultural irrigation which may be using untreated surface water or well water may not fall under the jurisdiction of the cross-connection control administrative authority.

Where is it used?
Universally used in most climates.

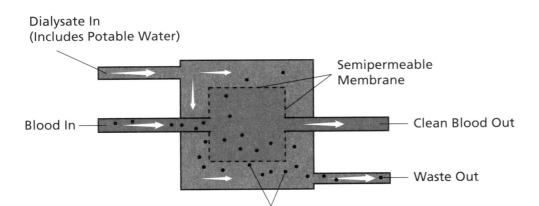

Kidney Dialysis Machine
A machine that is used to provide an artificial replacement for lost kidney function due to renal failure.

How does it work?
The kidney dialysis machine works by pumping the patient's blood through the blood compartment of the dialysis machine, exposing it to a semipermeable membrane. This membrane will not allow the passage of blood cells but will allow passage of small molecular weight toxins. The dialysis machine proportions the concentrates with water to make up dialysate. A water line may be installed to the kidney dialysis machine. The water is then sterilized and filtered so that it can be used to make up the dialysate.

Where is it used?
Kidney dialysis machines may be found in hospitals or medical facilities. Mobile units may be found in patients homes.

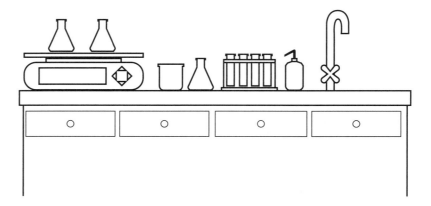

Laboratory Equipment

There is a wide variety of equipment used in various types of laboratories. Some laboratories may incorporate the use of chemicals, or biological contaminants; while others may not have any contaminants in the facility. One should consider the type of laboratory and the specific pieces of equipment in use when making a cross-connection control survey.

How does it work?

Some laboratory equipment will use water for mixing, washing/rinsing. One must determine if water is being used by the specific piece of laboratory equipment.

Where is it used?

Laboratory equipment is used almost anywhere including: medical facilities, chemical facilities, manufacturing facilities, educational facilities, etc.

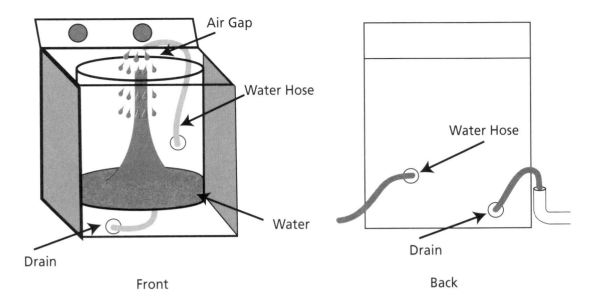

Front   Back

Laundry Machines
A laundry machine is a machine for washing and rinsing textiles.

How does it work?
Water along with detergent is mixed in the agitating tub, to clean the materials. The water is also used to rinse the materials of the detergent. Most machines have water lines, which are directly plumbed to the machine. Typically air gaps are built into these machines, however, some may have low inlets, which are not protected by air gaps

Where is it used?
Laundry machines may be found in commercial laundry facilities, or other facilities, which launder their own materials such as hotels, and hospitals. Laundry machines are also found in residential settings and at coin operated laundry facilities.

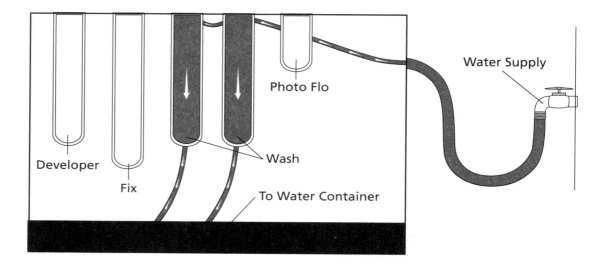

Photographic Film Processing Machines
A machine used to process photographic film, including x-rays.

How does it work?
Chemicals used in film processing may be mixed with water. Water may also be used as a rinse in the process. The water may be directly connected to the processing machine with automatic solenoid valves or there may be an indirect connection to a reservoir where the mixing takes place.

Where is it used?
Found in photo processing laboratories and medical facilities.

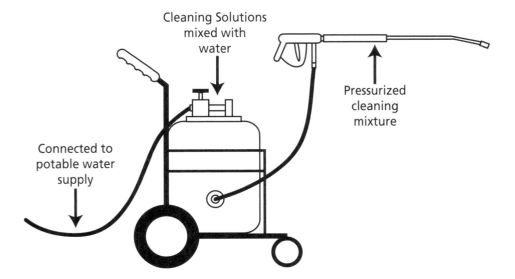

Portable Cleaning Equipment
A piece of equipment with a water connection used to clean various items with pressurized water.

How does it work?
Water is pressurized and mixed with cleaning solutions to produce a high pressure cleaning mixture dispensed from the spray wand.   Some of these may have a built in reservoir for water and not have a water connection; others may have a direct or indirect connection to the reservoir.

Where is it used?
This is used in a multitude of locations. For cleaning cars, carpets, concrete, bricks, buildings, etc.

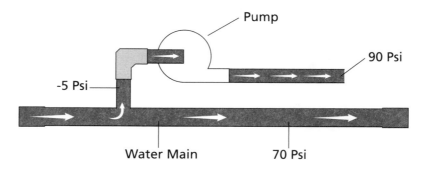

Pumps
A piece of equipment to increase pressure or circulate fluids.

How does it work?
A pump increases the pressure of the fluid in the system of which it is a part. Pumps may be used to increase water pressure in areas where the pressure is low, or they may be used to circulate fluids through a system to maintain a continuous flow or pressure. Pressure upstream of the pump may drop below atmospheric as the fluid is pulled into the pump. This may create the potential for backsiphonage upstream of the pump. Pumps may also create a pressure higher than the potable supply water pressure creating the potential for backpressure.

Where is it used?
Pumps may be used anywhere fluids are being used.

Equipment and Systems

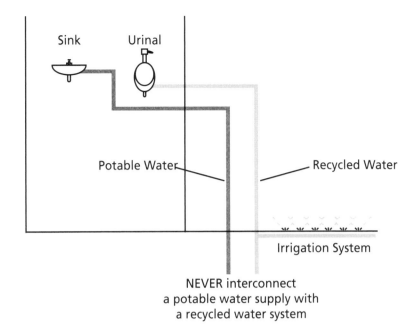

Recycled Water Systems

Recycled water systems use treated sewage water for specific purposes such as irrigation and flushing toilets. Recycled water systems may not be used for potable water purposes. Such systems must not have interconnection to the potable water system, even if backflow protection is provided. In facilities that are dual plumbed (where there is a domestic potable water supply and a recycled water line) all effort must be made to ensure that the lines are never interconnected. In these cases it is normal to require a shut down test at least annually to ensure that the lines are not interconnected. This may involve shutting down the recycled system, then opening the recycled fixtures (a flushometer valve for example) to see if any water comes through in the system. If water is continually provided, even though the recycled system is shut down, an interconnection to the potable system is indicated. Other similar tests may involve dying the recycled water system to see if the potable water system's water becomes dyed, again indicating an interconnection.

How does it work?
Wastewater that is treated for beneficial uses that save potable water, such as irrigation, cooling tower make up water, toilet and urinal flushing and trap priming

Where is it used?
Essentially all states have rules and regulations governing the production and use of recycled water.

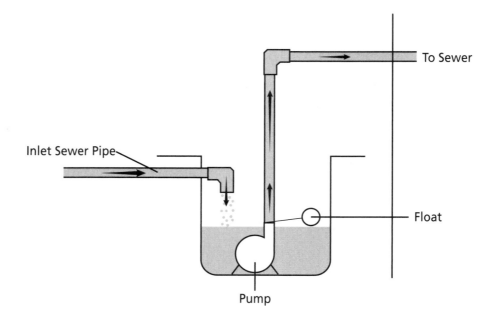

Sewage Ejector
A pump used to lift sewage to a sanitary sewer at a higher elevation.

How does it work?
Wastewater from a building will typically flow by gravity to a sealed tank or basin in the basement of a building. When the wastewater in the tank/basis reaches a predetermined level, a float switch will be activated, and a pump will transfer the wastewater from a lower elevation to the waste system (i.e., sewer) at a higher elevation.

Normally there is no potable water interconnected to the sewage ejector. Older style pumps that would need to be primed (filled with water), may have a potable water line directly connected to the discharge piping of the pump.

Hose bibbs may be in the vicinity of the ejector for wash down purposes, but the hose should never be interconnected to the ejector system

Where is it used?
Used in basements and other locations that are located at an elevation below the waste system (i.e., sewer).

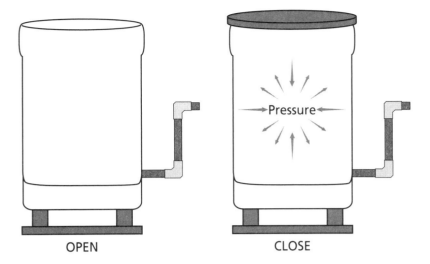

OPEN   CLOSE

Tanks/Vats
Vessels used in various processes. May be used as dye vats, pickling tanks, plating tanks, shrinking tanks, soaking tanks, storage tanks, etc.

How does it work?
Tanks or vats may be open (not able to be pressurized), or they may be closed meaning the tank may be pressurized above atmospheric pressure. Tanks may be used for mixing, storing, heating, cooling. These may contain anything including, chemicals, dyes, acids, caustics, etc. Water may be added to tanks for mixing or other purposes. Water may be introduced into the tank via an air gap, a submerged inlet, or a low inlet.

Where is it used?
Found in any industrial facility.

7 Equipment and Systems

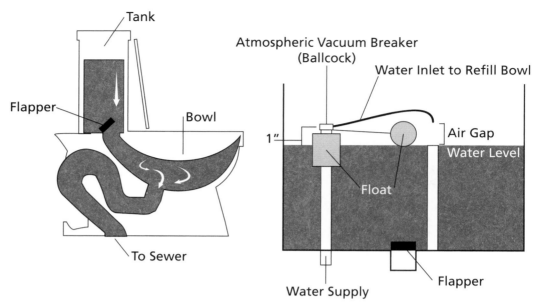

Toilet

A toilet is a plumbing fixture that is primarily used for the disposal of the bodily wastes - urine and fecal matter.

How does it work?

When the toilet is activated or flushed, water will flow into the bowl and carry the liquid and solid waste into the waste system (i.e., septic or sewer). Water is typically indirectly connected to the toilet through a ballcock in an elevated tank/cistern installed above the bowl, or through a flush valve (i.e., flushometer).

Dry 'pit' toilets generally do not use water to dispose of the waste material

Where is it used?

In all locations where sanitary handling of bodily waste is necessary.

Equipment and Systems

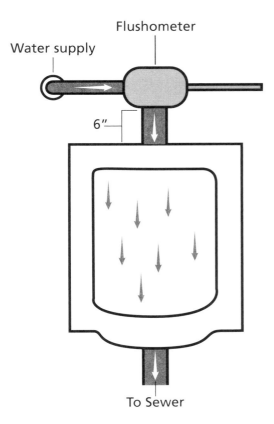

Urinal
A urinal is a plumbing fixture that is primarily used for the disposal of the liquid bodily waste (urine).

How does it work?
When the urinal is activated or flushed, water will flow into the fixture and carry the liquid waste into the waste system (i.e., septic or sewer). Water is typically indirectly connected to the urinal through a ballcock in an elevated tank/cistern installed above the urinal, or through a flush valve (i.e., flushometer).

Waterless urinals do not have a water connection, since there is no water needed to dispose of the liquid waste material.

Where is it used?
In all locations where sanitary handling of liquid bodily waste is necessary.

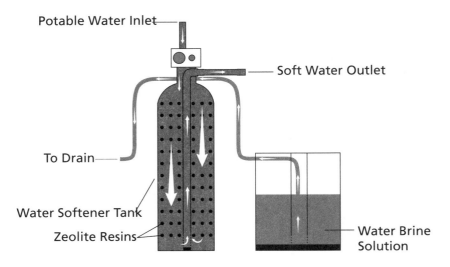

Water Softener
A device that reduces the calcium and magnesium ion (harder ions) concentration in hard water and replaces it with sodium ions (softer ions) to produce soft water.

How does it work?
The hard water passes through a bed of resins. These resins are known as zeolites. These zeolites initially contain sodium ions and will exchange the sodium ions with the calcium and magnesium ions in the water. The resulting water is then softer. As the zeolite resins exchange the ions; they then need to be regenerated. A concentrated water-brine solution is then passed through the resins to regenerate. The water that is used for regeneration is then discharged into the drain. This discharge line must drain to the sewer through an air gap.

Where is it used?
Water softeners may be used where there are problems with scale build up. For example, these softeners may be used as make up water lines to boilers, chillers and cooling towers. Additionally, areas that have private wells with hard water may use a water softener to soften the water from the well.

Chapter 8 — *Samples and Forms*

Samples and Forms

Inside Chapter 8

Approved Backflow Prevention Assemblies (service protection) 162

Installation and Maintenance Requirements 164

Guidelines for Parallel Installation of
　　Backflow Prevention Assemblies 173

Periodic Field Test and Maintenance Report with Report Form 175

Non-compliance Notice to
　　Install Backflow Prevention Assembly 177

Non-compliance of Periodic Field Test and Maintenance 178

Notice of Discontinuance of Water Supply 179

Air Gap Periodic Test and Maintenance Report 180

Notice of Appointment of Water Supervisor 182

Resolution Relative to Backflow Prevention Assembly Testers 183

Application for a Backflow Prevention Assembly
　　Tester Certificate 185

List of Certified Backflow Prevention Assembly Testers 187

Notice of Shutdown 188

Model Ordinance 189

Minimum Requirements for Backflow Prevention
　　Assembly Tester Certification Program 193

Minimum Requirements for Cross-Connection Control
　　Specialist Certification Programs 195

Field Survey Form 196

Backflow Incident Report 197

Approved Backflow Prevention Assemblies (internal protection) 199

8 Samples and Forms

NOTE: The following documents may be found on the accompaning compact disc found on the inside of the back cover. All forms are meant to be samples to be used by the end-user. It is not permissable to include the following forms in any type of media for resale.

8.1 Approved Backflow Prevention Assemblies (service protection)

The letter may be issued to notify a consumer that an air gap or approved backflow prevention assembly is required at water service connection. The letter cites the Federal Safe Drinking Water Act Amendments of 1996 and can be modified to local and water and health agency rules and regulations. In addition, it addresses some installation requirements for approved backlow prevention assemblies.

Administrative Authority
Street Address
City, State, ZIP

Phone
Fax

Email
URL

Date

John Doe
Title
Company Name
4321 First Street
Anytown, State ZIP

Dear Mr. **Doe**,

You are herewith informed that you must install on (**certain designated**) water service(s) to your premises either an air gap or an approved backflow prevention assembly (reduced pressure principle assembly, double check valve assembly). This action is taken in accordance with the Federal Safe Drinking Water Act Amendments of 1996 and with the State of (**state**) and (**local**) Cross-Connection Control (**rules or regulations**). Under these (**rules or regulations**) the (**water supplier or administrative authority**) has the primary responsibility of protecting the public potable water from backflow of dangerous substances which would endanger the public health or physically damage the public water system.

On (**date**), as part of our program to see that the (**rules or regulations**) are complied with, (**name or program specialist**) conducted a survey of your plumbing system. This survey revealed potential/actual cross-connections of the following conditions:

    (**List Conditions**)

The above conditions present backflow hazards to the (**on premise system**) and (**to the public supply**). To correct these conditions the (**water supplier or administrative authority**) requires the following:

Install an approved backflow prevention assembly downstream of the water meter a minimum of 12 inches above grade and accessible for field testing and maintenance.

This letter does not address internal protection requirements. We suggest you contact the agency having jurisdiction to ensure your water system complies with plumbing codes.

(Use the following paragraph only when appropriate)

A water supervisor shall be appointed who shall be responsible for the installation and use of pipelines and equipment in a manner that avoids and eliminates cross-connections in accordance with applicable laws and regulations. The (**water supplier or administrative authority**) shall be kept informed of the identity of the person appointed as water supervisor.

It is necessary to shut off the flow of water through a backflow prevention assembly during the time it is being field tested and/or repaired. If the complete interruption of water through a given service is critical to your operation, we recommend you install backflow prevention assemblies in parallel. This will allow one assembly to continue serving water while the other is being field tested or repaired. A check should be made with your engineer or plumber to be sure that assemblies are properly sized for desired flows.

Note that installation of a backflow prevention assembly will prevent release of on-site pressure to the utility water mains. Therefore, it is important that a temperature/pressure relief valve and/or thermal expansion tank be properly installed to relieve any excessive increase in on-site pressure due to hot water heating systems or other activities.

Attached is a list of backflow prevention assemblies that have been evaluated and approved by the Foundation for Cross-Connection Control and Hydraulic Research of the University of Southern California. The assemblies listed thereon have been adopted by this (**water supplier or administrative authority**) as the only assemblies approved for use on the water lines under our jurisdiction.

You will be allowed (**typically 30 to 60 days**) days from the date of this letter to provide the corrective measures previously outlined.

For additional information regarding this matter you may either write to (**name of person**) at (**address**) or telephone (**phone number**) between the hours of (**specify times**). Please contact (**name of person**) as soon as the work is done or if for any reason you cannot comply with the (**number**) day installation period or for clarification of any cross-connection control requirements discussed in this letter.

Sincerely yours,

CCC Program Specialist/Administrative Authority

8.2 Installation and Maintenance Requirements

The letter may be issued to inform a consumer about installation and maintenance of an air gap or approved backflow prevention assembly. Illustrations depicting installation requirements for air gaps, reduced pressure principle assemblies, double check valve assemblies and pressure vacuum breaker assemblies are included. Furthermore, illustrations depicting service connection installation requirements and installation of backflow prevention assemblies on irrigation systems are also included.

You have been informed that you must install and maintain a(an):

- ❏ air gap
- ❏ reduced pressure principle assembly
- ❏ double check valve assembly
- ❏ pressure vacuum breaker

on the water service(s) (**to or within**) your premises. This action was taken in accordance with the Federal Safe Drinking Water Act Amendments of 1996 and with the State of (**state**) and (**local**) Cross-Connection Control (**rules or regulations**). Under these (**rules or regulations**) the (**water supplier or administrative authority**) has the primary responsibility for protecting the public potable water from backflow of any pollution or contamination.

AIR GAP RECEIVING TANK WITH PUMP OR GRAVITY DISTRIBUTION

A receiving tank shall be installed on the property side of and adjacent to the water meter. The supply line between the meter and the tank must be permanently exposed for inspection purposes. There must be no outlet, tee, tap or connection of any kind to or from the supply pipe between the water meter and the opening from which the water is discharged into the receiving tank. The discharge inlet into the tank must be located at a distance of not less than two times the cross-sectional diameter of the inlet pipe above the top or overflow rim of the tank. Where required, a wind guard should be installed to prevent water loss or damage by wind-driven spray. The tank should be elevated (50 feet or more is recommended) to give an adequate gravity head; or, as an alternate, a pump may be installed to provide adequate head.

MECHANICAL BACKFLOW PREVENTION ASSEMBLIES

An approved backflow prevention assembly shall be installed on the property side of and adjacent to the water meter in accordance with (refer to Figs. 8.1, 8.2, 8.3 and 8.4) or an approved pressure vacuum breaker shall be installed adjacent to the specified point of use and in accordance with (refer to Fig. 8.6). Where construction or equipment location present citing problems for the above noted assembly a deviation may be granted by (**administrative authority**) providing such request is made in writing prior to the installation of the assembly. On the service line there must be no outlet, tee, tap or connection of any sort to or from the supply pipe line between the water meter and the protective assembly.

GENERAL

In the event you elect to install an air gap, it will be necessary for you to report at the end of each year to us that this method of protection has remained operative and effective throughout the year without being by-passed. In the event you desire to install a mechanical backflow prevention assembly, it will be necessary for you to notify us immediately after installation in order that a field test may be made of the assembly to determine its operational characteristics. We will make the initial field test; however, any servicing required to make the assembly operate satisfactorily will be your responsibility. The (**agency**) personnel making the inspection and initial field test of the assembly will give you specific instructions as to our requirements for future field test and servicing of the assembly, and will answer any questions you may have regarding rendition of service where backflow protection is required. Under the regulations of the (**administrative authority**) you are required to maintain this assembly in a continuous state of good repair and to field test the assembly at intervals of one year, unless the condition of the assembly indicates the necessity for more frequent field tests and servicing. We will provide you with test report forms 30 days in advance of your next periodic test date. It will be your responsibility to have it tested by a certified backflow prevention assembly tester. A report form showing the condition of the assembly and repairs made, if any, shall be prepared and forwarded to us within the 30-day period.

The enclosed drawings (enclose copies of Figs. 8.1, 8.2, 8.3, 8.4, and 8.6) provide installation requirements. The type of assembly or method of protection as well as any deviation from the forgoing requirements must be approved by us before installation. We will supply a list of approved backflow prevention assemblies upon request.

For additional information regarding this matter, you may either write to (**name of person**) or call at (**telephone number**).

Sincerely yours,

CCC Program Specialist/Administrative Authority

Samples and Forms

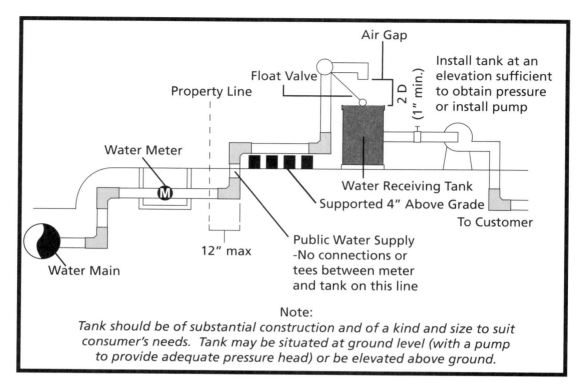

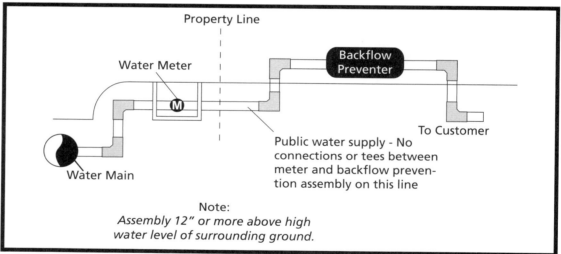

Figure 8.1
Installation Guidelines for Air Gap and Backflow Prevention Assemblies at Water Service Connection

Above Ground

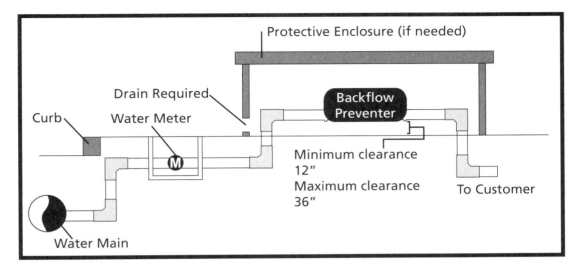

In Building

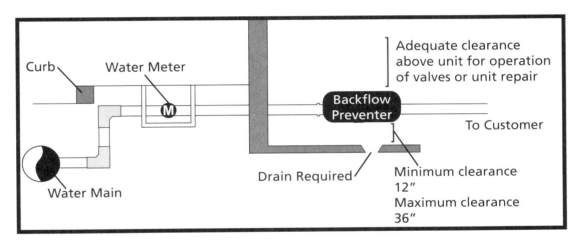

In Basement

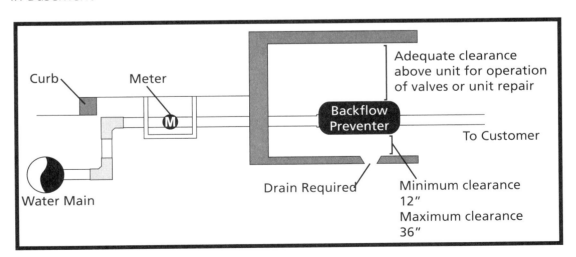

Figure 8.2
Typical Installations with Minimum Clearances

Plan View

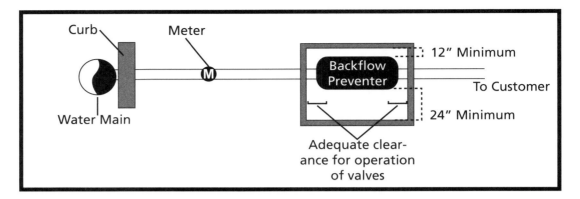

Figure 8.3
Typical Installations with Minimum Clearances, Plan View

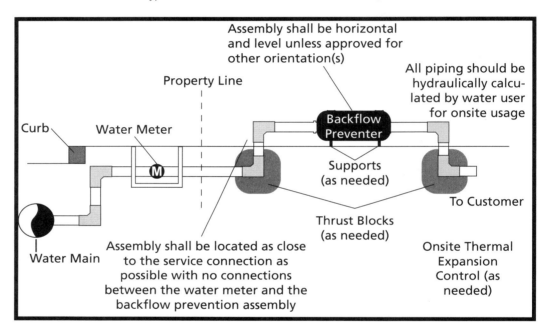

Figure 8.4
Service Connection Installation Guidelines

Backflow Prevention Assemblies are to be used within their rated operating conditions:

Pressure: Backflow prevention assemblies typically have maximum working water pressure (MWWP) of 150 psi (1034 KPa) or 175 psi (1206KPa). Assemblies are designed to operate continuously at this pressure, which is identified on the assembly.

Temperature: Backflow prevention assemblies are designed to operate continuously at their maximum working water temperature (MWWT), which is identified on the assembly.

Rate of Flow: Backflow prevention assemblies are designed to operate continuously up to their rated flow (i.e., gallons per minute- GPM; or liters per second- L/s) per Table 10-1.

NOTE: All installations of backflow prevention assemblies must be in compliance with state and local plumbing and building codes. Contact local administrative authority for detailed requirements.

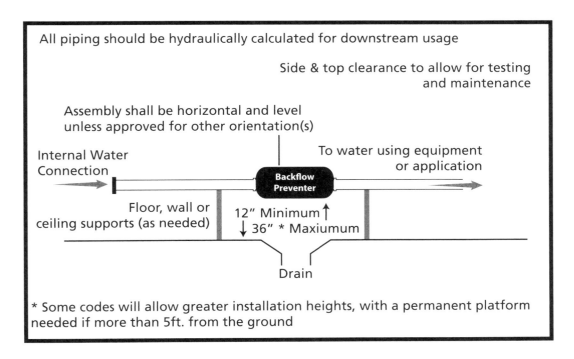

Figure 8.5
Internal Connection Installation Guidelines

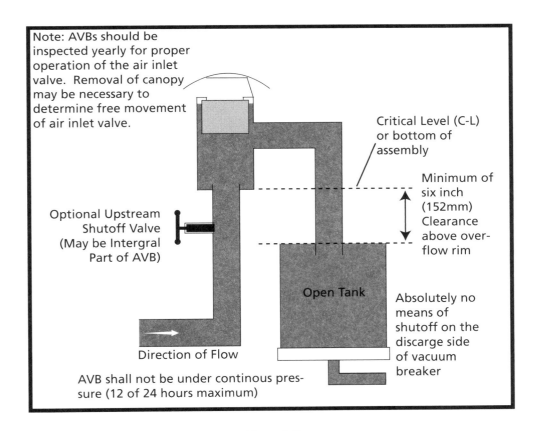

Figure 8.6
Typical AVB Installation

Samples and Forms

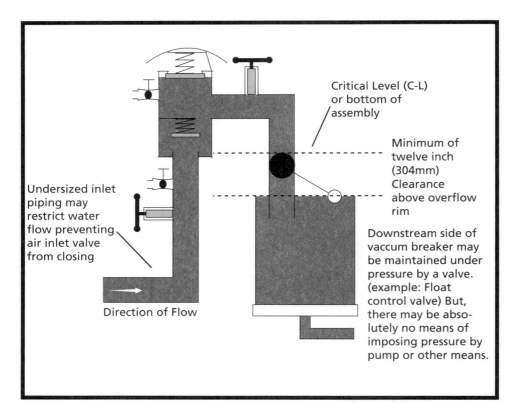

*Figure 8.7
Typical PVB Installation*

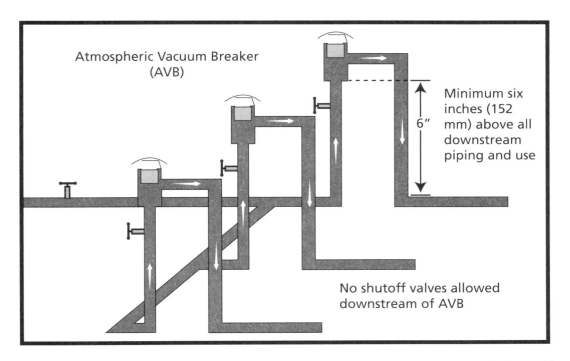

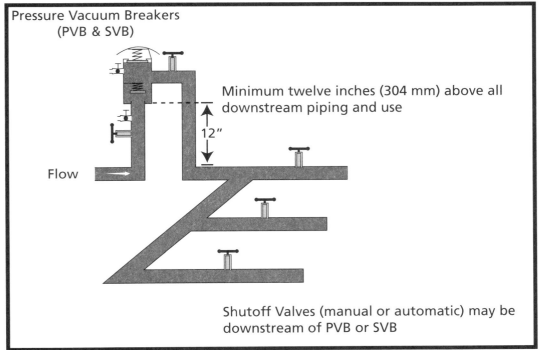

*Figure 8.8
Multi-Zone Irrigation Systems*

Samples and Forms

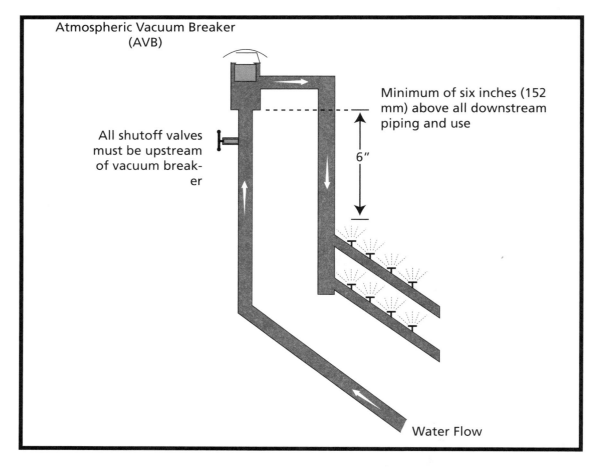

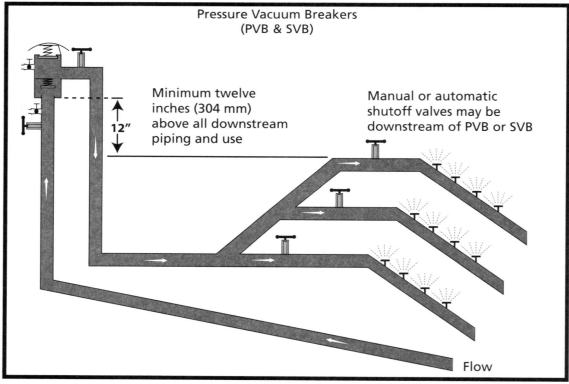

Figure 8.9
Irrigation Systems Feeding Uphill

8.3 Guidelines for Parallel Installation of Backflow Prevention Assemblies

Two or more backflow prevention assemblies of the same type may be installed in parallel to provide uninterrupted water service to the water user, typically known as a critical service. When it is necessary to install backflow prevention assemblies in parallel, there are several hydraulic conditions which may be considered.

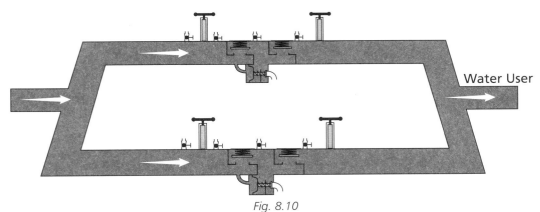

Fig. 8.10
Example of RP's Installed in Parallel

8.3.1 Full Line Size
If it is necessary to maintain maximum rated flow to the water user, then it may be necessary that each of the backflow prevention assemblies be individually sized to handle the rated flow of the inlet piping. If one of the assemblies is shut off for field testing or maintenance, the other assembly(s) can maintain the full flow capacity of the water user's system.

8.3.2 Hydraulic Sizing
If the hydraulic design of the installation does not require full flow capacity through each of the assemblies, then the assemblies may be hydraulically sized so that the combined flow capacity of the assemblies in parallel will equal or exceed the flow capacity of the inlet piping.

As an example:
If two 4-inch assemblies are installed in parallel in the place of a single 6-inch unit, then the combined flow capacity would be adequate.

- Two 4-inch assemblies ----- 2 x 500 gpm = 1000 gpm
(per Table 10-1; rated flow of 4-inch DC or RP is 500 gpm)

- Rated flow of single 6-inch = 1000 gpm

With both assemblies in operation in this installation, the water user has full flow capacity. However, during routine field testing or maintenance of one of the assemblies the water user should reduce his downstream flow demand, so the flow of the assembly remaining on-line does not exceed its maximum rated flow.

8.3.3
If it is only necessary to keep the water user's system pressurized during the field test or maintenance of the primary assembly then the smaller secondary assembly only needs to be sized to keep pressure in their system. This may include incidental use of water such as flushing toilets or make up water to a cooling tower, etc.

8.3.4 Manifold Assembly

Two or more backflow prevention assemblies may be incorporated into a manifold assembly (see 10.1.1.2.18) to provide uninterrupted water service to the water user. The manifold assembly is comprised of backflow prevention assemblies (DC or RP) of the same manufacturer, model, and size. Manifold adapter fittings on both the inlet and outlet of the manifold assembly are considered integral components. The size of the manifold assembly is determined by the inlet and outlet connections of the manifold adapter fittings.

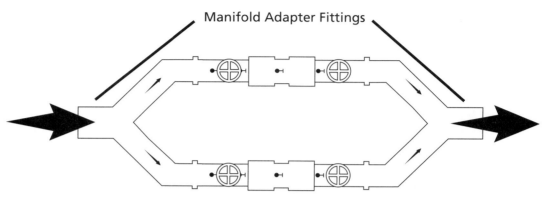

Figure 8.11
Example of Manifold Assembly

It shall be noted that when two or more assemblies are installed in parallel, the multiple assemblies do not act in unison during normal flow conditions. One of the assemblies will act as a primary assembly, while the other(s) operates as a secondary. Only the primary assembly will open when the flow of water begins, and the secondary assembly(s) will begin to open only when the flow rate increases to a higher level. The rate of flow when the transition takes place, from one to multiple assemblies, will depend upon the loading of the check valves in the assemblies. The assembly with the lowest check valve opening points (or cracking pressures) will generally become the primary assembly in the manifold assembly.

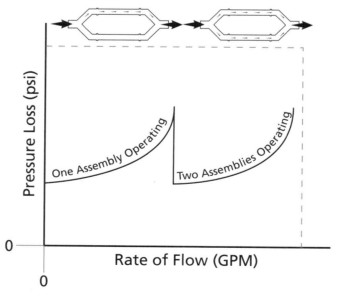

Figure 8.12
Sample Flow Rate versus Pressure Loss Curve for Manifold Assembly

8.4 Periodic Field Test and Maintenance Report with Report Form

The letter may be issued to notify consumer of periodic field testing and maintenance required per rules or regulations. The letter may make reference to a list of certified backflow prevention assembly testers. (See 8.12 Sample List: Backflow Prevention Assembly Testers.) It also addresses the steps that must be taken if the backflow prevention assembly fails its field test. Included with the letter is a sample field test and maintenance form.

Administrative Authority
Street Address
City, State, ZIP

Phone
Fax

Email
URL

Date

John Doe
Title
Company Name
4321 First Street
Anytown, State ZIP

Dear Mr. **Doe**,

      PERIODIC FIELD TEST AND MAINTENANCE REPORT
      Backflow Prevention Assembly - Water Service
      Connection No. (**number**)

The backflow prevention assembly described on the reverse side hereof is due for its periodic field test, as required under our (**rule or regulation**). Will you please have this field test performed by a certified backflow prevention assembly tester possessing a valid Certification issued by the (**administrative authority**).

If the field test discloses that the assembly is not operating satisfactorily, please have the necessary repairs made and the assembly retested by a Certified tester. On completion of a field test showing that the assembly is operating satisfactorily, the Certified tester shall complete the Field Test and Maintenance Report form on the reverse side hereof and forward it to this office no later than (**due date**).

Additional information relative to this matter may be obtained by writing to our (**specify the office**) or by calling (**phone number**).

Sincerely yours,

CCC Specialist/Administrative Authority

8 Samples and Forms

Samples and Forms

Manufacturer	Model	Size	Serial Number	RP ❑
Service Number:		Location:		DC ❑
Service Name/Address		Owner Name/Address:		PVB ❑

Detector Assembly: Water Meter Reading: Before Test _____ After Test _____

SVB ❑
DCDA ❑
RPDA ❑
DCDA-II ❑
RPDA-II ❑

Reduced Pressure Principle Assembly

Double Check Valve Assembly

	Check Valve #1	Check Valve #2	Relief Valve	PVB/SVB
Initial Test	Closed Tight ❑ _____PSID Leaked ❑	Closed Tight ❑ _____PSID Leaked ❑	Opened at_____PSID Did Not Open ❑	**Air Inlet** Opened at____PSID Did Not Open ❑ Opened Fully? Yes ❑ No ❑ **Check Valve** Held at_____PSID Leaked ❑
Repairs: Details	❑ Cleaned ❑ Replaced _____ _____ _____	❑ Cleaned ❑ Replaced _____ _____ _____	❑ Cleaned ❑ Replaced _____ _____ _____	❑ Cleaned ❑ Replaced _____ _____ _____
Final Test	_____PSID	_____PSID Closed Tight ❑	Opened at _____PSID	Air Inlet _____PSID Opened Fully? Yes ❑ No ❑ Check Valve Held at _____PSID

Comments: _____

Initial Test	Date_____ Time_____	Certified Tester No._____	❑ Pass ❑ Fail
	Test by (Signature) _____	Print Name _____	
Repair	Date_____ Time_____	Certified Tester No._____	
	Test by (Signature) _____	Print Name _____	
Final Test	Date_____ Time_____	Certified Tester No._____	❑ Pass ❑ Fail
	Test by (Signature) _____	Print Name _____	

Onsite Contact Acknowledged _____
Signature

Figure 8.13
Field-Test Form

8.5 Non-compliance Notice to Install Backflow Prevention Assembly

The letter may be issued to notify a consumer who has not complied with the requirement to install an approved backflow prevention assembly. The letter makes it clear that an approved backflow prevention assembly is required according to administrative authority's rules and regulations. If no action is taken the consumer is notified that further action may be taken to correct the matter.

Administrative Authority
Street Address
City, State, ZIP

Phone
Fax

Email
URL

Date

John Doe
Title
Company Name
4321 First Street
Anytown, State ZIP

Dear Mr. **Doe**,

On (**date**) we mailed to (**name**) a letter stating the requirement that a approved backflow prevention assembly be installed on your premises. This action is taken in accordance with the Federal Safe Drinking Water Act Amendments of 1996 and with the State of (**state**) and (**local**) Cross-Connection Control (**rules or regulations**). Under these (**rules or regulations**) the (**water supplier or administrative authority**) has the primary responsibility of protecting the public potable water from backflow of dangerous substances which would endanger the public health or physically damage the public water system.

Our request is in compliance with our (**rule or regulation**) number (**number**) and your failure to carry out your responsibility in this matter by (**date**) will result in discontinuance of water service to your property.

Should you desire instructions or additional information in this matter, please contact (**name**) at (**phone**).

Sincerely yours,

CCC Specialist/Administrative Authority

8.6 Non-compliance of Periodic Field Test and Maintenance

The letter may be issued to notify a consumer who has not complied with the requirement for the periodic field test and maintenance of a backflow prevention assembly. The letter makes it clear that a completed field test and maintenance form must be submitted to comply with the rules and regulations set forth by the administrative authority.

Samples and Forms

Administrative Authority
Street Address
City, State, ZIP

Phone
Fax

Email
URL

Date

John Doe
Title
Company Name
4321 First Street
Anytown, State ZIP

Dear Mr. **Doe**,

On (**date**) we mailed to (**name**) a combined letter and Field Test and Maintenance Report form requesting that you have performed the necessary periodic field test for the backflow prevention assembly identified on the form. This was to have been returned to us no later than (**date**). As of today we have neither heard from you nor received your report.

Our request is in compliance with our (**rule or regulation**) number (**number**) and your failure to carry out your responsibility in this matter by (**date**) will result in discontinuance of water service to your property.

Should you desire instructions or additional information in this matter, please contact (**name**) at (**phone**).

Sincerely yours,

CCC Specialist/Administrative Authority

8.7 Notice of Discontinuance of Water Supply

The letter below may be issued to a consumer who has not complied with the requirement to install a backflow prevention assembly or field test a backflow prevention assembly. This letter will notify the consumer that the water supply to the premises will be discontinued. This letter is the final culmination to letters 8.4, 8.5 and 8.6. This letter should be sent in such a way that proof of delivery may be confirmed.

Administrative Authority
Street Address
City, State, ZIP

Phone
Fax

Email
URL

Date

John Doe
Title
Company Name
4321 First Street
Anytown, State ZIP

Dear Mr. **Doe**,

You are hereby notified that in accordance with (**rule or regulation**), the water supply to your premises, located at

(**address**)

will be discontinued on (**date - normally five to ten days from current date**) and shall remain discontinued until you have complied with the requirements of this (**name of agency and office**). The discontinuance of your water service until you have so complied is required by (**rule or regulation**).

Sincerely yours,

CCC Specialist/Administrative Authority

8.8 Air Gap Periodic Test and Maintenance Report

The letter may be issued to notify a consumer of a periodic field inspection report required per rules and regulations set forth by the administrative authority concerning an air gap. In addition to the letter a sample field inspection report form is included.

Administrative Authority
Street Address
City, State, ZIP

Phone
Fax

Email
URL

Date

John Doe
Title
Company Name
4321 First Street
Anytown, State ZIP

Dear Mr. **Doe**,

This will advise that the air gap described on this form is due for its periodic inspection and report as required under the (**rules or regulations**).

Will you please have this field test performed by a certified backflow prevention assembly tester possessing a valid Certification issued by the (**administrative authority**).

Additional information relative to this matter may be obtained by writing to (**office**) or by calling (**phone**).

Sincerely yours,

CCC Specialist/Administrative Authority

(**Water Supplier**)
(**Address**)

Attn: CCC Specialist

Subject: Air Gap Field Inspection
Report Form

Gentlemen:

I hereby certify that the air gap between potable water system and the service connection No. (**number**) complies with the requirements of an approved air gap and has not been bypassed, or otherwise made ineffective.

Date _____ Time _____

Cert. Tester No. _____

Test by (Signature) _____ Print Name _____

Figure 8.14
Air Gap Field Inspection Report Form

8.9 Notice of Appointment of Water Supervisor

The letter may be used to notify the water supplier of the appointment of a water supervisor at the consumer's facility. A water supervisor is responsible for maintaining the consumer's water system(s) on the property free from unprotected cross-connections and other sanitary defects, as required by regulations and laws. (See 1.71)

(Water Supplier)
(Address)

Attn: CCC Specialist

In accordance with the provisions of the (**rules or regulations**), (**name**) has been appointed Water Supervisor. He/She will be responsible for maintaining the water system on the property located at

    (Address)

free from unprotected cross-connections and other sanitary defects, as required by the plumbing code, regulations and laws.

_____
(Firm or company)

By:

_____ _____
        (Signature)   **(Title)**

_____
(Address)

Phone: (   ) _____

Date:_____

8.10 Resolution Relative to Backflow Prevention Assembly Testers

The resolution addresses the steps required by the administrative authority for an individual to become a backflow prevention assembly tester. It clearly states that no person shall be considered capable of performing field tests and making reports on backflow preventions assemblies, unless the person has proven his qualifications to the satisfaction of the administrative authority.

RESOLUTION RELATIVE TO BACKFLOW PREVENTION ASSEMBLY TESTERS ADOPTED BY
(Administrative Authority Name)

In order to comply with requirements of the (**rules or regulations**), under which the health agency has the overall responsibility for preventing water from unapproved sources from entering the potable water systems within water consumers' premises or the public water supply directly; and to ensure that such assemblies will be adequately field tested and reliable test reports made to the {**administrative authority(s)**}, the following Resolution is adopted:

No person shall be deemed to be capable of performing field tests and making reports on backflow prevention assemblies, as required in the (**rules or regulations**), unless the person has proven his qualifications to the satisfaction of the (**administrative authority**) as required. To determine such qualifications the (**administrative authority**) shall conduct examinations to determine the ability of any person desiring to field test and make reports on backflow prevention assemblies for the purpose of complying with (**administrative authority**) requirements. The (**administrative authority**) may accept third-party certification programs which comply with the (**administrative authority's**) "minimum requirements." The provisions of the (**administrative authority**) regulations shall not be deemed to have been complied with unless the field tests have been performed and the report completed by a person who has satisfactorily passed such examinations given by the (**administrative authority**) and who has received from the (**administrative authority**) a certificate or license. A third party certificate may be acceptable in lieu of the (**administrative authority**) certificate or license.

The (**administrative authority**) shall make available to owners of properties on which assemblies requiring tests are maintained, a list of certified backflow prevention assembly testers qualified to make the required field tests.

APPLICATION AND QUALIFICATION OF TESTERS

APPLICATIONS
Applications shall be submitted to the (**administrative authority**) on forms prepared by the (**administrative authority**).

Each application for certification shall be in the name of the individual desiring to take the examinations for certification.

EXAMINATIONS
The examinations administered by the (**administrative authority**) shall be sufficiently comprehensive to determine that the applicant is fully capable to make accurate field tests and reports on all types of assemblies which are approved for use in the (**administrative authority's**) jurisdiction.

CLASSIFICATION AND QUALIFICATION OF TESTERS:

A. A **General Backflow Prevention Assembly Tester** shall mean any person duly certified under the (**rules and regulations**) of the (**administrative authority**) to commercially engage in the business of field testing any type of backflow prevention assembly.

QUALIFICATIONS: Shall have a valid certificate of qualifications as a journey man plumber and shall be, or shall work under a person who has a certificate of registration as a master plumber, or who possesses a valid state contractor's license qualifying him to do plumbing contracting, or have adequate qualifications in the opinion of the (**administrative authority**).

B. A **Manufacturer's Agent** shall mean any person duly certified under the (**rules and regulations**) of the (**administrative authority**) to field test any type of backflow prevention assembly made or manufactured by one company.

QUALIFICATIONS: Shall be a full-time employee or representative of the company which makes or distributes the one make of assembly for which the applicant is certified and shall be fully qualified by experience and training, as determined by the (**administrative authority**), to field test the assemblies made or distributed by the company of which the person is an employee or representative.

C. A **Limited Backflow Prevention Assembly Tester** shall mean any person duly certified under the (**rules and regulations**) of the (**administrative authority**) to field test any or all types of backflow prevention assemblies at the premises owned or controlled by the individual or company, the location of which shall be specified on this certificate.

QUALIFICATIONS: Shall be a journeyman plumber, or have other qualifications which, in the opinion of the (**administrative authority**), fully qualify the applicant for the functions performed under this category.

D. **Single Assembly Tester** shall mean any person duly certified under the (**rules and regulations**) of the (**administrative authority**) to field test a single backflow prevention assembly at a specific location.

QUALIFICATIONS: Shall have demonstrated ability to perform the field test and to perform the duties of a Water Supervisor, as required by (**rules or regulations**).

RESPONSIBILITIES AND ADDITIONAL REQUIREMENTS OF TESTERS

Each applicant for certification as a tester of backflow prevention assemblies shall furnish evidence to show that he has available the necessary tools and equipment to properly field test such assemblies. The tester shall be responsible for the accuracy of all tests and reports prepared by the tester.

REVOCATION OF CERTIFICATION OF TESTER

The certificate issued to any tester may be revoked or suspended by the (**administrative authority**) for improper field testing or reporting. Proceedings for suspension or revocation shall be in accordance with (**rules or regulations**).

I hereby certify that the above Resolution was adopted by the (**administrative authority**) at its meeting held (**date**).

8.11 Application for a Backflow Prevention Assembly Tester Certificate

The letter may be used as an application for those individuals applying to take a certification exam. The application covers the basics including contact information, present and previous employment and experience.

Name:_____ Date _____

Address:_____

_____

Telephone:_____

PRESENT EMPLOYMENT:

List here enough information to show fulfillment of the experience requirement. Other qualifications which you feel would fulfill the eligibility requirement should be listed on Page two under "Other Experience and Qualifications."

From:_____ To:_____

Firm Name_____ Telephone_____
Address_____
_____

Type of work_____
_____

PREVIOUS EMPLOYMENT:

From:_____ To:_____

Firm Name_____ Telephone_____
Address_____
_____

Type of work_____

REGISTRATION:

Master Plumber Registration Number   _____
Journeyman Certificate of
        Qualification Number   _____
Plumbing Contractor's License Number   _____
Other   _____

OTHER EXPERIENCE OR QUALIFICATIONS:
_____
_____
_____
_____

Check in the appropriate square the certification desired:

 ❏ General Certificate   Any type or make of backflow prevention assembly may be field tested.

 ❏ Manufacture's Agent   Any type of backflow prevention assembly manufactured by one company may be field tested.

 ❏ Limited Certificate   Restricts the tester to field test a single-type backflow prevention assembly.

 ❏ Single Assembly Tester   Restricts the tester to field test ONE backflow prevention assembly.

Indicate your choice of date and location for examination(s):

	Date	Location
1.	_____	_____
2.	_____	_____
3.	_____	_____

           _____
               Signature

Samples and Forms

8.12 List of Certified Backflow Prevention Assembly Testers

This template may be used for producing a list of certified backflow prevention assembly testers which may be given to consumers as needed. Those on this list may be able to perform field test and maintenance in accordance with the rules and regulations set forth by the administrative authority.

BACKFLOW PREVENTION ASSEMBLY TESTERS CERTIFIED AS OF (**date**)

(**List of Certified Testers with addresses**)

Name	Certification No.	Address

Figure 8.15 List of Certified Testers

NOTE: The above Certified Backflow Prevention Assembly Testers have demonstrated their ability to field test backflow prevention assemblies to the satisfaction of the (**administrative authority**). The certifying agency assumes no liability for the abilities, performance or quality of work performed by individuals listed. The company or individual utilizing the backflow prevention assembly tester should determine that the Tester maintains liability insurance.

8.13 Notice of Shutdown

The letter may be used to advise the consumer of a pending water system shutdown due to code compliance, emergency repairs or preventative maintenance.

Administrative Authority
Street Address
City, State, ZIP

Phone
Fax

Email
URL

Date

John Doe
Title
Company Name
4321 First Street
Anytown, State ZIP

Dear Mr. **Doe**,

Please be advised of the necessity for shut down of (**specify water service**) due to:

      ❑ Code compliance
      ❑ Emergency repairs
      ❑ Preventative maintenance

This shut down will affect (**specify location within plant**) on (**date**) from (**hour to hour**).

This work is essential and must be implemented. However, if this schedule is too disruptive to your work schedule a postponement may be possible. If you have need for rescheduling please telephone (**name**) at (**extension**) before (**date**).

Thank you for your cooperation.

Sincerely yours,

CCC Specialist/Administrative Authority

8.14 Model Ordinance

This document is a model ordinance regarding a cross-connection control program. The administrative authority may require certain modifications for local purposes.

AN ORDINANCE FOR THE CONTROL OF BACKFLOW AND CROSS-CONNECTIONS

Amendments to the (**local or state authority**) Code of (**city or state**)

Section 1. CROSS-CONNECTION CONTROL — GENERAL POLICY

1.1 Purpose. The purpose of this Ordinance is:

    1.1.1 To protect the public potable water supply of (**political jurisdiction**) from the possibility of contamination or pollution by isolating within the consumer's internal distribution system(s) or the consumer's private water system(s) such contaminants or pollutants which could backflow into the public water systems; and,

    1.1.2 To promote the elimination or control of existing cross-connections, actual or potential, between the consumer's in-plant potable water system(s) and non-potable water system(s), plumbing fixtures and industrial piping systems; and,

    1.1.3 To provide for the maintenance of a continuing Program of Cross-Connection Control which will systematically and effectively prevent the contamination or pollution of all potable water systems.

1.2 Responsibility. The (**Water Commissioner or State Health Official**) shall be responsible for the protection of the public potable water distribution system from contamination or pollution due to the backflow of contaminants or pollutants through the water service connection. If, in the judgment of said (**Water Commissioner or Health Official**) an approved backflow prevention assembly is required (at the consumer's water service connection; or, within the consumer's private water system) for the safety of the water system, the (**Water Commissioner or Health Official**) or his designated agent shall give notice in writing to said consumer to install such an approved backflow prevention assembly(s) at a specific location(s) on his premises. The consumer shall immediately install such an aproved backflow prevention assembly(s) at the consumer's own expense; and, failure, refusal or inability on the part of the consumer to install, have tested and maintained said assembly(s), shall constitute grounds for discontinuing water service to the premises until such requirements have been satisfactorily met.

Section 2. DEFINITIONS
(See Chapter 1, Definitions)

Section 3. REQUIREMENTS

3.1 Water System

    3.1.1 The water system shall be considered as made up of two parts: The Water Supplier's System and the Consumer's System.

    3.1.2 Water Supplier's System shall consist of the source facilities and the distribution system; and shall include all those facilities of the water system under the complete control of the utility, up to the point where the consumer's system begins.

3.1.3 The source shall include all components of the facilities utilized in the production, treatment, storage, and delivery of water to the distribution system.

3.1.4 The distribution system shall include the network of conduits used for the delivery of water from the source to the consumer's system.

3.1.5 The consumer's system shall include those parts of the facilities beyond the termination of the water supplier distribution system which are utilized in conveying potable water to points of use.

3.2 Policy

3.2.1 No water service connection to any premise shall be installed or maintained by the Water Supplier unless the water supply is protected as required by (**political jurisdiction**) laws and regulations and this (**name of legal document**). Service of water to any premises shall be discontinued by the Water Supplier if a backflow prevention assembly required by this (**name of legal document**) is not installed, tested and maintained, or if it is found that a backflow prevention assembly has been removed, bypassed, or if an unprotected cross-connection exists on the premises. Service will not be restored until such conditions or defects are corrected.

3.2.2 The consumer's system should be open for inspection at all reasonable times to authorized representatives of the (**Water or Health agency name**) to determine whether unprotected cross-connections or other structural or sanitary hazards, including violations of these regulations, exist. When such a condition becomes known, the (**Water Commissioner or Health Officer**) shall deny or immediately discontinue service to the premises by providing for a physical break in the service line until the consumer has corrected the condition(s) in conformance with the (**political jurisdiction**) statutes relating to plumbing and water supplies and the regulations adopted pursuant thereto.

3.2.3 An approved backflow prevention assembly shall also be installed on each service line to a consumer's water system at or near the property line or immediately inside the building being served; but, in all cases, before the first branch line leading off the service line wherever the following conditions exist:

a. In the case of premises having an auxiliary water supply which is not or may not be of safe bacteriological or chemical quality and which is not acceptable as an additional source by the (**Water Commissioner or Health Officer**), the public water system shall be protected against backflow from the premises by installing an approved backflow prevention assembly in the service line commensurate with the degree of hazard.

b. In the case of premises on which any industrial fluids or any other objectionable substance is handled in such a fashion as to create an actual or potential hazard to the public water system, the public system shall be protected against backflow from the premises by installing an approved backflow prevention assembly in the service line commensurate with the degree of hazard. This shall include the handling of process waters and waters originating from the water supplier's system which have been subject to deterioration in quality.

c. In the case of premises having (1) internal cross-connections that can not be permanently corrected or protected against, or (2) intricate plumbing and piping arrangements or where entry to all portions of the premises is not readily accessible for inspection purposes, making it impracticable or impossible to ascertain whether or not dangerous cross-connections exist, the public water system shall be protected against backflow from the premises by installing an approved backflow prevention assembly in the service line.

3.2.4 The type of protective assembly required under subsections 3.2.3a, b, and c shall depend upon the degree of hazard which exists as follows:

a. In the case of any premise where there is an auxiliary water supply as stated in subsection 3.2.3.a of this section and it is not subject to any of the following rules, the public water system shall be protected by an approved air gap or an approved reduced pressure principle backflow prevention assembly.

b. In the case of any premise where there is water or substance that would be objectionable but not hazardous to health, if introduced into the public water system, the public water system shall be protected by an approved double check valve backflow prevention assembly.

c. In the case of any premise where there is any material dangerous to health, which is handled in such a fashion as to create an actual or potential hazard to the public water system, the public water system shall be protected by an approved air gap or an approved reduced pressure principle backflow prevention assembly. Examples of premises where these conditions will exist include sewage treatment plants, sewage pumping stations, chemical manufacturing plants, hospitals, and mortuaries and plating plants.

d. In the case of any premise where there are unprotected cross-connections, either actual or potential, the public water system shall be protected by an approved air gap or an approved reduced pressure principle backflow prevention assembly at the service connection.

e. In the case of any premise where, because of security requirements or other prohibitions or restrictions, it is impossible or impractical to make a complete in-plant cross-connection survey, the public water system shall be protected against backflow from the premises by either an approved air gap or an approve reduced pressure principle backflow prevention assembly on each service to the premise.

3.2.5 Any backflow prevention assembly required herein shall be a make, model and size approved by the (**Water Commissioner or Health Official**). The term "Approved Backflow Prevention Assembly" shall mean an assembly that has been manufactured in full conformance with the standards established by the American Water Works Association entitled: AWWA/ANSI C510-2007 Standard for Double Check Valve Backflow Prevention Assemblies; AWWA/ANSI C511-2007 Standard for Reduced Pressure Principle Backflow Prevention Assemblies; and, have met completely the laboratory and field performance standard of the Foundation for Cross-Connection Control and Hydraulic Research of the University of Southern California (USC FCCCHR) established in: Standards of Backflow Prevention Assemblies Chapter 10 of the most current edition of the Manual of Cross-Connection Control. Said AWWA and USC FCCCHR standards have been adopted by the (**Water Commissioner or Health Official**). Final approval shall be evidenced by a "Certificate of Compliance" for the said AWWA standards; or the appearance of the specific model and size on the List of Approved Backflow Prevention Assemblies published by the USC FCCCHR along with a "Certificate of Approval" for the said USC FCCCHR Standards; issued by an approved testing laboratory. The following testing laboratory has been qualified by the (Water Commissioner or Health Officer) to test and approve backflow prevention assemblies:

> Foundation for Cross-Connection Control and Hydraulic Research
> University of Southern California
> Los Angeles, California 90089-2531

Testing laboratories other than the laboratory listed above will be added to an approved list as they are qualified by the (**Water Commissioner or Health Officer**).

Backflow preventers which may be subjected to backpressure or backsiphonage that have been fully tested and have been granted a Certificate of Approval by said qualified laboratory and are listed on the laboratory's current list of approved backflow prevention assemblies may be used without further test or qualification.

3.2.6 It shall be the duty of the consumer at any premise where backflow prevention assemblies are installed to have a field test performed by a certified backflow prevention assembly tester upon installation and at least once per year. In those instances where the (**Water Commissioner or Health Officer**) deems the hazard to be great enough he may require field tests at more frequent intervals. These tests shall be at the expense of the water user and shall be performed by (**Water Supplier**) personnel or by a certified tester approved by the (**Water Commissioner or Health Officer**). It shall be the duty of the (**Water Commissioner or Health Officer**) to see that these tests are made in a timely manner. The consumer shall notify the (**Water Commissioner or Health Officer**) in advance when the tests are to be undertaken so that an official representative may witness the field tests if so desired. These assemblies shall be repaired, overhauled or replaced at the expense of the consumer whenever said assemblies are found to be defective. Records of such tests, repairs and overhaul shall be kept and made available to the (**Water Commissioner or Health Officer**).

3.2.7 All presently installed backflow prevention assemblies which do not meet the requirements of this section but were approved devices for the purposes described herein at the time of installation and which have been properly maintained, shall, except for the field testing and maintenance requirements under subsection 3.2.6, be excluded from the requirements of these rules so long as the (**Water Commissioner or Health Officer**) is assured that they will satisfactorily protect the water purveyor's system. Whenever the existing device is moved from the present location or requires more than minimum maintenance or when the (**Water Commissioner or Health Officer**) finds that the maintenance constitutes a hazard to health, the unit shall be replaced by an approved backflow prevention assembly meeting the requirements of this section.

3.2.8 The (Water Commissioner or Health Officer) is authorized to make all necessary and reasonable rules and policies with respect to the enforcement of this ordinance. All such rules and policies shall be consistent with the provisions of this ordinance and shall be effective (**number**) days after being filed with the (**clerk or secretary**) of the (**political jurisdiction**).

The foregoing ordinance was first read at the meeting of the (**name of governing body**) of the (**political jurisdiction**) of _____ on the _____ day of _____, 20____ and adopted by the following called vote on motion of (Official).

    Ayes:
    Noes:
    Abstaining: Approver_____
    Absent: **Official Title**

    Attest:

    _____
    Clerk or Secretary

8.15 Minimum Requirements for Backflow Prevention Assembly Tester Certification Program

This document is an outline of what a backflow prevention assembly tester certification program should include. Training course information include a sylabus of topics that must be covered, and examination requirements are summarized.

Course of Study

A. Course typically requires a minimum of 40 to 60 hours
   1. Approximately: 50% lecture/50% lab

B. Need to know criteria/Lecture topics
   1. History of backflow & cross-connection control
   2. Hydraulics and theory of backflow
   3. Cross connection control programs
   4. Products and component performance standards
   5. Field testing and diagnostics of backflow prevention assemblies
   6. Backflow prevention assembly maintenance
   7. Applicable codes and regulations

C. Lab work
   1. Field testing actual reduced pressure principle backflow prevention assemblies (RP) or reduced pressure principle detector assemblies (RPDA) with appropriate backflow field test equipment demonstrating failure modes as well as properly operating assemblies.
   2. Field testing actual double check valve assemblies (DC) or double check detector assemblies (DCDA) with appropriate backflow field test equipment demonstrating failure modes as well as properly operating assemblies.
   3. Field testing actual pressure vacuum breaker assemblies (PVB) with appropriate backflow field test equipment demonstrating failure modes as well as properly operating assemblies.
   4. Field testing actual spill-resistant pressure vacuum breakers assemblies (SVB) with appropriate backflow field test equipment demonstrating failure modes as well as properly operating assemblies.

Examination for Tester Certification

A. Minimum 100 multiple choice written questions related to lecture portion of class.
   1. Applicant shall demonstrate basic understanding of lecture topics by achieving a passing grade of 70% or better on this portion of the exam.
   2. A score of less than 70% shall result in a failure of this portion of the exam.

B. Applicant shall satisfactorily complete a practical (hands-on) examination on the DC, RP, PVB & SVB.
   1. Applicant shall demonstrate proficiency with basic field test procedures via practical exam and demonstrate understanding of basic diagnostics and assembly failure procedures. The practical exam shall consist of field-testing one of each type of backflow prevention assembly. Applicants must successfully complete the field test procedure on each individual assembly before moving on to the next assembly. If the applicant makes an error on an individual assembly, applicant will be given a second opportunity on the same assembly. An error during the second attempt on the same assembly will terminate the exam.

2. An error shall be any action or omission by the applicant that could affect the outcome of the field test procedure, whether or not it affected the outcome of the applicant's field test procedure. Any action or omission by the applicant that could not affect the outcome of the field test procedure shall not be considered an error.
3. The diagnostic portion of the practical exam if written will require a passing grade of 70%. A grade less than 70% will terminate the applicant's exam.
4. Practical exams shall be conducted and witnessed by a qualified proctor.

Rules and Regulations Governing Tester Certification Program

A. Rules and regulations shall contain the following:
1. Applicable Definitions
2. General Qualifications
3. General Certifications Requirements
4. Certification Fees
5. Examination Protocol
6. Re-certification Requirements
   a. Maximum duration between re-certification
7. Revocation of Certification
8. Appeal Process
9. Certification without Exam
10. Revision of Rules Process

B. Rules and regulations governing tester certification program shall be made available prior to exam upon request of applicant.

8.16 Minimum Requirements for Cross-Connection Control Specialist Certification Programs

This document is an outline of what a specialist certification program should include. Training course information includes a syllabus, of topics that must be covered and examination requirements are summarized.

Course of Study
- A. This course typically requires a minimum of 40 to 60 hours.
    1. Approximately 75% Lecture and 25% Lab/Field

- B. Lecture topics/need to know criteria
    1. Backflow & Cross-Connection Control review.
    2. Regulations & Codes. (State, federal and local)
    3. Cross-Connection Control surveys
        a. Plan review
    4. Identification of water using equipment
        a. Commercial, medical, industrial, etc.
    5. Recycled water use sites
    6. Gray water use sites
    7. Fire sprinkler systems
    8. Cross-Connection Control Specialist's responsibilities

- C. Lab/Field work
    1. Plan reading to identify hazards
        a. Review mechanical plans and identify potentially hazardous water using equipment or plumbing systems.
        b. Conduct site survey
    2. Survey actual site, record actual/potential hazards
    3. Prepare report on recommendations to correct or abate hazards

Examination for Cross-Connection Control Specialists Certification
- A. Minimum 100 multiple-choice questions related to lecture and lab/field work
    1. Applicant shall demonstrate basic understanding of lecture, and lab/field work topics by achieving a passing grade of 70% or better on the exam.
    2. Optional field survey report.

Rules and Regulations Governing Cross-Connection Control Specialist Certication Program
- A. Rules and regulations shall contain the following
    1. Applicable Definitions.
    2. General Qualifications
    3. General Certifications Requirements
    4. Certification Fees
    5. Examination Protocol
    6. Re-certification Requirements
        a. Maximum duration between re-certification
    7. Revocation of Certification
    8. Appeal Process
    9. Certification without Exam
    10. Revision of Rules Process

- B. Rules and regulations governing specialist certifications shall be made available prior to exam upon request of applicant.

8.17 Field Survey Form

This template may be used for field surveys being conducted by a cross-connection control specialist or administrative authority. The form lays out the necessary information that needs to be recorded to identity the overall degree of hazard at a location. Water usage, degree of hazard and type of existing protection are some of the fields inluded on the survey form.

FIELD SURVEY FORM

LOCATION _____ DATE _____

	WATER USE	DEG OF HAZARD **	CROSS-CONN		TYPE OF EXISTING PROTECTION*	PROTECTION				INSTALLATION COMMENTS
			Direct	Indirect		Approved		Adequate		
						Yes	No	Yes	No	
1										
2										
3										
4										
5										
6										
7										
8										
9										
10										

*Detailing of Protection

	MAKE	MODEL	SIZE	SERIAL NO.	COMMENTS
1					
2					
3					
4					
5					
6					
7					
8					
9					
10					

** Detailing of Hazards (i.e., chemicals, etc.)

1		6	
2		7	
3		8	
4		9	
5		10	

Code References

1		6	
2		7	
3		8	
4		9	
5		10	

Conclusion:

Include any sketches, which may be necessary

Sample.Forms.8.17.doc

Figure 8.16
Field Survey Form

8.18 Backflow Incident Report

This document may be used to collect vital information about a backflow incident. Location, nature of the incident and actions taken are some of the fields included on the incident report.

Received by _____ Report # _____ Date __/__/__ Time _____

Person Reporting Incident _____ PH #_____
Address _____

Nature of Incident as reported:
_____
_____
_____
_____

Location of Incident:
_____
_____
_____
_____

Immediate Action Taken:
_____
_____
_____
_____

Name and Phone number of Agency & individuals contacted (i.e., Health Dept., Fire Dept., etc.):
_____
_____
_____
_____

Field Report by_____ Date ___/___/___ Time_____

Action taken in field to determine and isolate affected area:
_____
_____
_____
_____

Source of Contamination (if found):
_____
_____
_____
_____

Continued on the next page

Degree of Hazard - Health _____ Non-Health _____

Type of Backflow Protection (if any) at source of contamination:
_____
_____

Date of latest Backflow Prevention Assembly Field Test ___/___/___

Please attach copy of field test report.

Corrective action taken to restore water quality:
_____
_____
_____
_____
_____
_____
_____

Recommendations to prevent similar incidents:
_____
_____
_____
_____

Comments
List of attachments (i.e., compliance letters, notifications, reports, etc.):
_____
_____
_____
_____

Signature _____ Date ___/___/___
Signature _____ Date ___/___/___
Signature _____ Date ___/___/___
Signature _____ Date ___/___/___

Samples and Forms

8.19 Approved Backflow Prevention Assemblies (internal protection)

The letter may be issued to notify a consumer that an air gap or approved backflow prevention assembly is required for internal protection. In addition, it addresses some installation requirements for approved backlow prevention assemblies.

Administrative Authority
Street Address
City, State, ZIP

Phone
Fax

Email
URL

Date

John Doe
Title
Company Name
4321 First Street
Anytown, State ZIP

Dear Mr. **Doe**,

You are herewith informed that you must install on (**certain designated**) water lines within your premises either an air gap or an approved backflow prevention assembly (**reduced pressure principle assembly, double check valve assembly, pressure vacuum breaker, spill-resistant pressure vacuum breaker, atmospheric vacuum breaker**). This action is taken in accordance with the State of (**state**) and (**local**) Cross-Connection Control (**rules or regulations**) and/or applicable (**plumbing code**) Under these (**rules or regulations**) the (**administrative authority**) has the primary responsibility of protecting the public potable water from backflow of dangerous substances which would endanger the public health or physically damage the public water system.

On (**date**), as part of our program to see that the (**rules or regulations**) are complied with, (**name or program specialist**) conducted a survey of your plumbing system. This survey revealed potential/actual cross-connections of the following conditions:

    (**List Conditions**)

The above conditions present backflow hazards to the (**on premise system**) and (**to the public supply**). To correct these conditions the (**administrative authority**) requires the following:

    (**List Corrections**)

This letter addresses protection of certain cross-connections detected in our survey. We do not however, accept responsibility to guarantee that all cross-connections will be protected or for cross-connections that may be created in the future, due to repair or alterations made in your water system.

It is necessary to shut off the flow of water through a backflow prevention assembly during the time it is being field tested and/or repaired. If the complete interruption of water is critical to your operation, we recommend you install backflow prevention assemblies in parallel. This will allow one assembly to continue serving water while the other is being field tested or repaired. A check should be made with your engineer or plumber to be sure that assemblies are properly sized for desired flows.

Note that installation of a backflow prevention assembly will prevent release of downstream pressure to the on-site plumbing system. Therefore, it is important that a temperature/pressure relief valve and/or thermal expansion tank be properly installed to relieve any excessive increase in on-site pressure due to hot water heating systems or other activities.

Attached is a list of backflow prevention assemblies that have been evaluated and approved by the Foundation for Cross-Connection Control and Hydraulic Research of the University of Southern California. The assemblies listed thereon have been adopted by this (**administrative authority**) as the only assemblies approved for use on the water lines under our jurisdiction.

You will be allowed (**typically 30 to 60 days**) days from the date of this letter to provide the corrective measures previously outlined.

For additional information regarding this matter you may either write to (**name of person**) at (**address**) or telephone (**phone number**) between the hours of (**specify times**). Please contact (**name of person**) as soon as the work is done or if for any reason you cannot comply with the (**number**) day installation period or for clarification of any cross-connection control requirements discussed in this letter.

Sincerely yours,

Administrative Authority

Chapter 9: Field Test Procedures

Field Test Procedures

Inside this Chapter

Preliminary Steps 204

Field Test Reporting 205

Maintenance and Repairs 206

Manual of Cross-Connection Control Tenth Edition Modifications 206

Reduced Pressure Principle Backflow Prevention Assembly (RP)
Using Five Needle Valve Test Kit 210
    Field Test Procedure 210
    Diagnostics 224

Reduced Pressure Principle Backflow Prevention Assembly (RP)
Using Two Needle Valve Field Test Kit 230
    Field Test Procedure 231
    Diagnostics 246

Reduced Pressure Principle Backflow Prevention Assembly (RP)
Using Three Needle Valve Field Test Kit 253
    Field Test Procedure 254
    Diagnostics 270

Double Check Valve Backflow Prevention Assembly (DC) 277
    Field Test Procedure 278
    Diagnostics 284

Pressure Vacuum Breaker Assembly (PVB) 290
    Field Test Procedure 291
    Diagnostics 301

Spill-Resistant Pressure Vacuum Breaker Assembly (SVB) 307
    Field Test Procedure 308
    Diagnostics 315

Reduced Pressure Principle-Detector
Backflow Prevention Assembly (RPDA) 318

Double Check-Detector Backflow Prevention Assembly (DCDA) 320

Reduced Pressure Principle-Detector Backflow
Prevention Assembly-Type II (RPDA-II) 322

Double Check-Detector Backflow
Prevention Assembly-Type II (DCDA-II) 324

Backflow Prevention Assembly Field Test Procedures

As part of a complete cross-connection control program, proper field test procedures for the required initial and subsequent annual tests must be used. In this the field test procedures are detailed as follows:

9.1 Preliminary Steps and Recommendations
9.2 Reduced Pressure Principle Backflow Prevention Assembly (RP)
    9.2.1 Five needle valve field test procedure
    9.2.2 Two needle valve field test procedure
    9.2.3 Three needle valve field test procedure
9.3 Double Check Valve Backflow Prevention Assembly (DC)
9.4 Pressure Vacuum Breaker Backsiphonage Prevention Assembly (PVB)
9.5 Spill-Resistant Pressure Vacuum Breaker Backsiphonage Prevention Assembly (SVB)
9.6 Reduced Pressure Principle: Detector Backflow Prevention Assembly (RPDA)
9.7 Double Check: Detector Backflow Prevention Assembly (DCDA)
9.8 Reduced Pressure Principle – Detector Backflow Prevention Assembly-Type II (RPDA-II)
9.9 Double Check – Detector Backflow Prevention Assembly-Type II (DCDA-II)

A properly calibrated field test kit is essential to ensure accurate data acquisition. Methods to inspect the accuracy of the field test kits used in the above field test procedures are supplied in Appendix A (Field Test Kit Accuracy Verification). Field test kits should be checked for accuracy at least once a year, and re-calibrated when inaccuracy is greater than ±0.2 psid. Local administrative authorities should be consulted regarding possible local accuracy verification requirements.

9.1 Preliminary Steps

9.1.1
The certified backflow prevention assembly tester (i.e., tester) (see 1.18) must observe the condition of the field test kit during all steps of the following field test procedures. Visually inspect the field test kit for obvious leaks or damage. The field test kit should zero out when not pressurized; needle valves and fittings must be drip tight; field test kit should be drained after field testing to protect against freezing.

9.1.2
The tester must observe general safety procedures during all aspects of the field test and maintenance procedures. Including, but not limited to:

- Personal Safety – Issues such as personal protective equipment: hard hat, gloves, ear protection, safety glasses, etc.
- Confined Spaces – Backflow prevention assemblies may be located in areas which could be defined as a confined space. Tester should understand and follow all confined space requirements [see Occupational Safety and Health Standards: 29 Code of Federal Regulation (CFR) 1910.146.]
- Elevated platforms – Backflow prevention assemblies may be installed in elevated locations requiring access by ladders or stairways.
- Utility hazards – Backflow prevention assemblies may be located in areas containing other utility services (i.e., steam, natural gas, electrical, telecommunications, etc.)
- Tools – Commercially available and specialized.

9.1.3
As a prelude to each of the field test procedures it is essential to follow some basic steps.

9.1.3.1 Notification
Owner of the assembly and/or the on-site personnel must be notified that water service will be shut off during the field test procedure. Special arrangements may have to be made so that interruption of service will not create a hardship on the user.

NOTE: Fire Sprinkler Systems
The tester should request that the owner or occupant notify the authority having jurisdiction, the fire department, if required, and the alarm receiving facility before shutting down a fire sprinkler system or its water supply. The notification prior to testing should include the purpose for the shutdown of the system, the component(s) involved, and the estimated time required. Per NFPA Standards (NFPA 25-2008, Chapter 13.6.2), a forward flow test should be performed at the time of the annual field test.

9.1.3.2 Identify
Make sure that the proper assembly is being field tested by checking identification on the assembly for make (manufacturer), model, size and serial number. Record all of this information, as well as the field test data, before leaving the location.

9.1.3.3 Inspect
Inspect the assembly for the required components for the field test procedure (i.e., upstream and downstream shutoff valves, open and closed shutoff valves and properly located test cocks).

9.1.3.4 Observe
Carefully observe area around the assembly for telltale signs of leakage (i.e., moss or algae growth, plant life, or soil erosion). This should supply the tester with additional information regarding the condition of the assembly before the field test is performed. Example: Wet spot under relief valve port of reduced pressure principle backflow prevention assembly is an indication of relief valve activity, possibly from pressure fluctuations or fouling of the assembly. Proper field testing will define the problem. Also, inspect the installation for unsafe conditions such as an assembly installed where water discharge or leakage may create a hazard. This is especially important for assemblies installed inside buildings.

9.1.4 Field Test Reporting

9.1.4.1 Physical Identification
Physical identification of the backflow prevention assembly must be recorded. This information should include, but not be limited to:

- Assembly: Manufacturer, model, serial number, size
- Location: Address, physical location

9.1.4.2 Field Test Results
The field test results from the field test procedure must be accurately recorded on the appropriate field test form (See Fig. 8.4 for example of field test form.) The field test form will typically detail the following:

- Initial Field Test Results
- Repair/Maintenance/replacement performed
- Retest Field Test Results

As a prelude to each of the field test procedures it is essential to follow some basic steps: Notify, Identify, Inspect and Observe.

9.1.4.3 Tester Information

Tester information must be accurately recorded on the appropriate field test form. The field test form will typically require the following:

- Tester identification (i.e., certification No. or license No.)
- Tester's signature and printed name, phone number
- Date and time of field test

9.1.4.4 Distribution of Field Test Forms

Following the completion of the field test, copies (i.e., multi-copy form, photocopies, electronic, etc.) must be distributed to the following within the designated time frame (typically 10-30 days):

- Administrative Authority(s)
- Owner
- Tester

9.1.5 Maintenance and Repairs

Consult the manufacturer's repair/maintenance manuals before attempting any disassembly; large spring loads may be present in some designs.

To maintain the backflow prevention assembly in proper operating condition, the tester/repair person must follow some basic steps.

- Use properly operating and calibrated field test kit.
- Use proper field test procedures (see 9.2, 9.3, 9.4, 9.5, 9.6, 9.7, 9.8, and 9.9)
- Consult manufacturer's repair/maintenance manuals when disassembly is required. Specialized tools (i.e., not commercially available) may be required for some repair/maintenance procedures.
- Use only original manufacturer's authorized replacement parts.
- Immediately following repair, maintenance, or replacement procedures, the backflow prevention assembly must be retested and reports properly distributed.
- When returning an assembly to normal operating condition, slowly open shutoff valve No. 1 to repressurize the assembly (with the shutoff valve No. 2 in the closed position). Once the assembly is pressurized, slowly open shutoff valve No. 2 to pressurize the downstream piping and equipment to avoid water hammer.

9.1.6 Manual of Cross-Connection Control Tenth Edition Modifications

To give the tester more detailed information about the field test procedures, a number of improvements have been incorporated in this Tenth Edition of the Manual of Cross-Connection Control. Please note the following:

Illustrated field test procedures for the 2, 3, and 5 needle valve configuration field test kits have been included for all assemblies.

- For the RP there are three separate sections:
  - 9.2.1 uses the five needle valve configuration
  - 9.2.2 uses the two needle valve configuration
  - 9.2.3 uses the three needle valve configuration
- For the DC, PVB, and SVB:
  - Only those specific steps of each field test procedure where the field test kit needle valves are opened or closed will be highlighted with a separate illustrated insert.

The backflow prevention assembly tester must observe the condition of the field test kit during all steps of the field test procedures

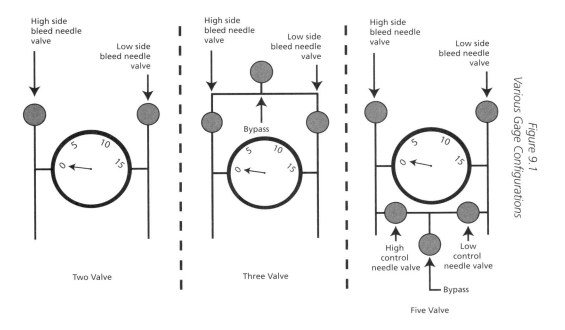

Figure 9.1
Various Gage Configurations

NOTE: Gages may have the high and low bleed needle valves located on either the right or left hand side.

RP (9.2)

- Additional detailing for each of the tests has been added.
- Test cock flushing procedure has been modified to reduce risk of activating relief valve.
- Elimination of recommended 3 psi buffer between the No. 1 check valve and relief valve readings.
- No. 1 check valve reading changed to a minimum value of 5 psid.

DC (9.3)

- Detailing has been added to clarify when it is critical to locate field test kit at the proper elevation.

PVB (9.4)

- The field test procedure has been modified to attach the bleed-off valve arrangement at the beginning of the test.
- Detailing has been added to clarify when it is critical to locate field test kit at the proper elevation.

SVB (9.5)

- The field test procedure has been modified to evaluate the condition of the check valve first, then the air inlet valve.
- The field test procedure has been modified to attach the bleed-off valve arrangement at the beginning of the test.
- Detailing has been added to clarify when it is critical to locate field test kit at the proper elevation.

RPDA and DCDA (9.6 and 9.7)

- New language has been added to help verify detection of flow through the bypass.

RPDA-II and DCDA-II (9.8 and 9.9)

- New sections.

Illustrated Field Test Procedures

The field test procedures have been illustrated to enhance the understanding of the procedures. To reflect more accurately the physical position of the check valve(s), differential pressure relief valve, test cocks, vent valve or air inlet valve during the field test procedures, the following illustrations will be utilized. The Legend below will be helpful in understanding the procedures on the following pages.

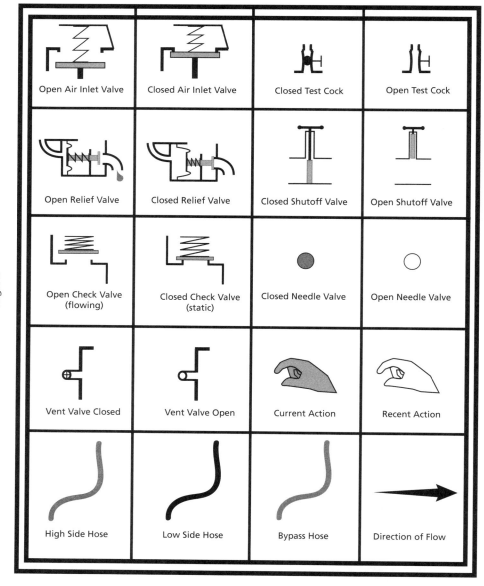

Figure 9.2
Legend

Flushing test cocks/vent valve

One of the initial steps for each of the field test procedures is to open and close the test cocks/vent valve to flush out any foreign material, and to verify that pressure is available at the test cock(s)/vent valve. Care must be taken when flushing a test cock/vent valve so to direct the flowing water in a safe manner. Opening the test cock/vent valve to its fully open position may not be necessary. Care must be taken when the vent valve of a SVB is a machine screw, such that the vent valve is not unscrewed too far (i.e., machine screw disengages from body.)

Appendix

Additional field-testing information for each of the field test procedures has been included in Appendix A.

RP

Guidance is provided for obtaining direction of flow values for the check valve No. 2.

PVB and SVB

A means of determining backpressure from the piping downstream of the PVB and SVB is detailed.

Bleed-Off Valve Arrangement

Guidance is provided for the use of multiple bleed valve arrangements for larger test cocks.

RPDA, DCDA, RPDA-II, DCDA-II

Operation of bypass. Guidance has been provided for the detection of plugged piping, reading water meter before/after field test, periodic activation of water meter.

RP, DC, PVB, SVB

- Abbreviated Field Test Procedures – Training Aid

Field Test Kit Periodic Test Recommendations

- Accuracy verification
- Leak test

One of the initial steps for each of the field test procedures is to open and close the test cocks/vent valve to flush out any foreign material, and to verify that pressure is available at the test cock(s)/vent valve.

9.2.1 Reduced Pressure Principle Backflow Prevention Assembly (RP) Using Five Needle Valve Field Test Kit

9.2.1.1 Equipment Required:

a. Differential pressure field test kit with minimum range 0-15 psid (0.1 or 0.2 psid graduations)
b. Temporary bypass hose (comparable to the size of the test cocks)
c. Four brass 1/4" IPS x 45° SAE flare connectors
d. Brass Adapter fittings for each test cock size: 1/8" x 1/4", 1/4" x 1/2", 1/4" x 3/4"

9.2.1.2 Field Test Procedure

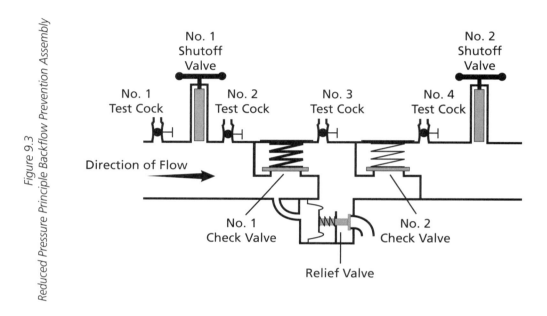

Figure 9.3
Reduced Pressure Principle Backflow Prevention Assembly

Test No. 1   Relief Valve Opening Point

Purpose: To test the operation of the differential pressure relief valve.

Requirement: The differential pressure relief valve must operate to maintain the zone between the two check valves at least 2 psi less than the supply pressure.

NOTE: It is important that the tester does not cause the relief valve to discharge before step j below. (See diagnostics 9.2.1.3.1: Exercising the Differential Pressure Relief Valve.)

NOTE: If the differential pressure relief valve is discharging continuously, See diagnostics 9.2.1.3.2: Continuously Discharging Differential Pressure Relief Valve.

Some assemblies are equipped with the adapter fittings which fit the field test kit hoses as an integral part of the test cocks. Some types of test cocks are opened when the hose from the field test kit are attached.

Steps:
Follow all preliminary steps detailed in 9.1

a. Bleed water through test cocks to eliminate foreign material. Open No. 4 test cock to establish flow through the unit, then open test cocks No. 3, No. 2 (open No. 2 test cock slowly), and No. 1. Then close test cocks No. 1, No. 2, No. 3., and No. 4. Be careful not to activate the differential pressure relief valve while bleeding the test cocks.

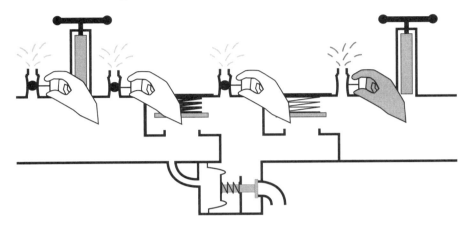

b. Install appropriate fittings to test cocks.

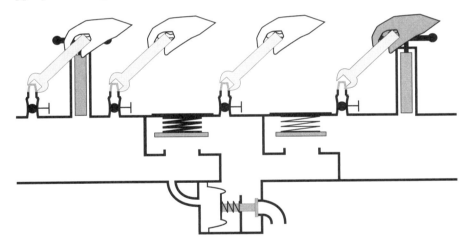

NOTE: Some assemblies are equipped with the adapter fittings, which fit the field test kit hoses as an integral part of the test cocks.

Some types of test cocks are opened when the hose from the field test kit is attached. If assembly contains this type of test cock, see Appendix A.2.1 for hose attachment procedure.

c. Attach hose from the high side of the field test kit to the No. 2 test cock.

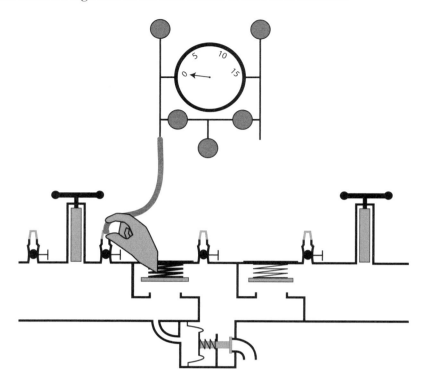

NOTE: Some types of test cocks are opened when the hose from the field test kit is attached. If assembly contains this type of test cock, see Appendix A.2.1 for hose attachment procedure.

d. Attach hose from the low side of the field test kit to the No.3 test cock.

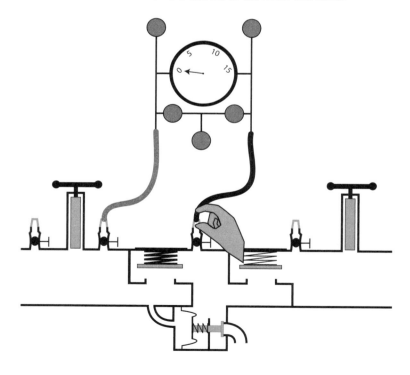

e. Slowly open test cock No. 3 fully and then bleed all air from the low side of the field test kit by opening the low side bleed needle valve.

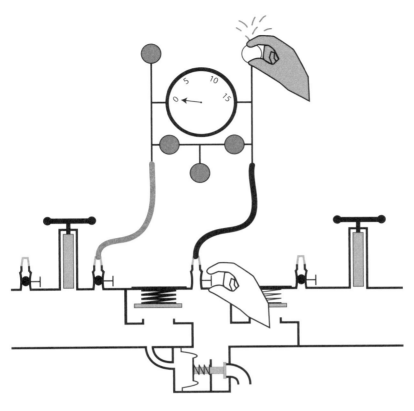

f. Maintain the low side bleed needle valve in the open position and slowly open test cock No. 2 fully to pressurize the field test kit.

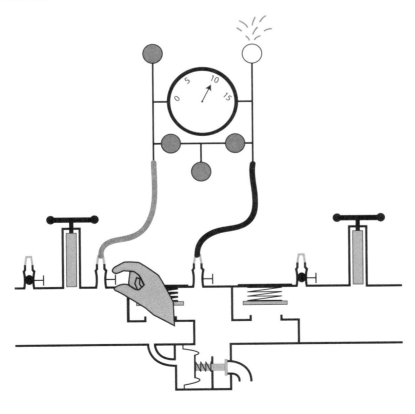

g. Open the high side bleed needle valve to bleed all air from the high side of the field test kit.

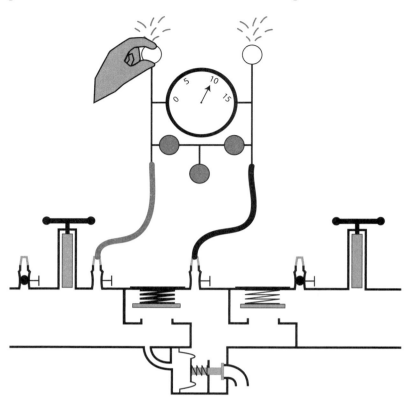

h. Close No. 2 shutoff valve.

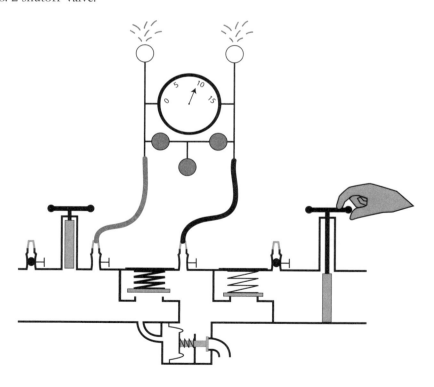

i. Close the high side bleed needle valve. After the field test kit reading has reached the upper end of the scale, slowly close the low side bleed needle valve.

- If the reading remains above the differential pressure relief valve opening point, (i.e., relief valve does not discharge) then observe the reading. This is the apparent differential pressure across the No. 1 check valve. During Tests No. 1, No. 2, and No. 3 of this procedure the field test kit is on-line showing the differential pressure across the No. 1 check valve. Proceed to step j.
- If the reading drops to the low end of the scale and the differential pressure relief valve discharges continuously, then the No. 1 check valve is leaking, and should be recorded as such. Tests No. 1, No. 2 and No. 3 may not be completed. (See diagnostics 9.2.1.3.3: Leaking No. 1 Check Valve.)

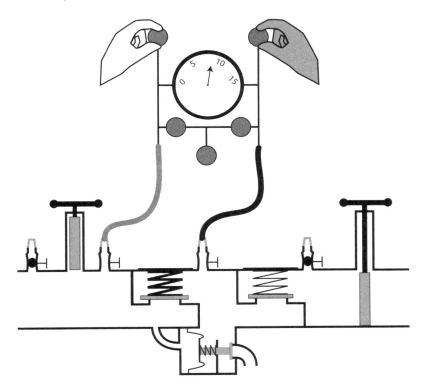

It is important that the tester does not cause the relief valve to discharge before step j.

j. Open the high side control needle valve approximately one turn, and then slowly open the low side control needle valve no more than one-quarter turn to bypass water from the No. 2 test cock to the No. 3 test cock. Observe the differential pressure reading as it slowly drops to the relief valve opening point. Record this opening point value when the first discharge of water is detected.

- If the low side control needle valve is opened one-quarter turn and the reading does not decrease to the relief valve opening point, then See diagnostics 9.2.1.3.4: Instructions for Leaking No. 2 Shutoff Valve.
- If the differential pressure reading drops to zero, and the differential pressure relief valve does not discharge any water, record that the differential pressure relief valve did not open. Proceed to Step k.

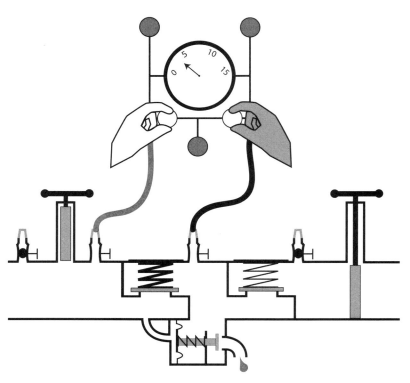

k. Close the low side control needle valve.

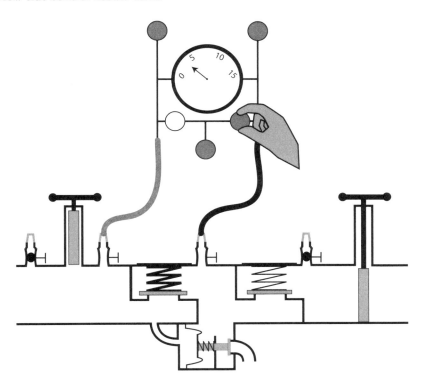

Field Test Procedures

Test No. 2   Tightness of No. 2 Check Valve

Purpose: To test the No. 2 check valve for tightness against backpressure.

Requirement: The No. 2 check valve shall be tight against backpressure.

Steps:

a. Maintain the No. 2 shutoff valve in the closed position, and the high side control needle valve in the open position (from Test No. 1).

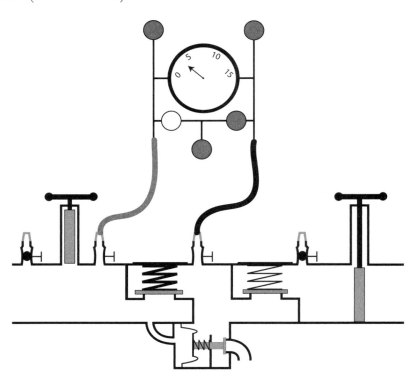

b. Bleed all of the air through the bypass hose by opening the bypass control needle valve. Close the bypass control needle valve only.

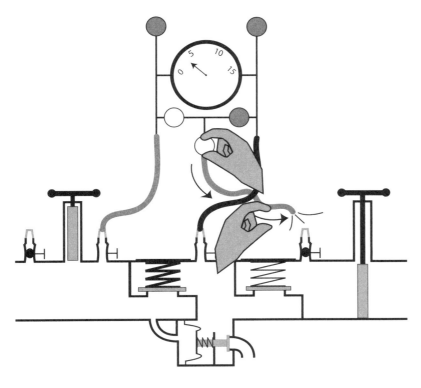

c. Attach the bypass hose from the field test kit to the No. 4 test cock, then fully open the No. 4 test cock.

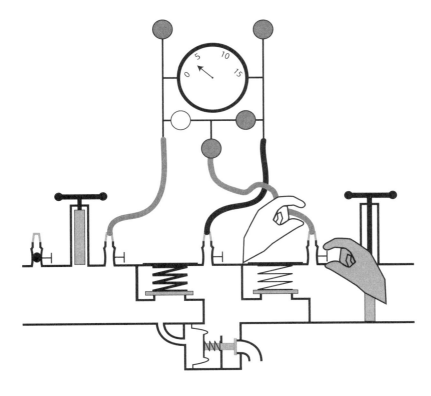

d. Open low side bleed needle valve. Once the reading reaches a value above the apparent No. 1 check valve differential pressure reading, slowly close the low side bleed needle valve.

Field Test Procedures

e. Open the bypass control needle valve.

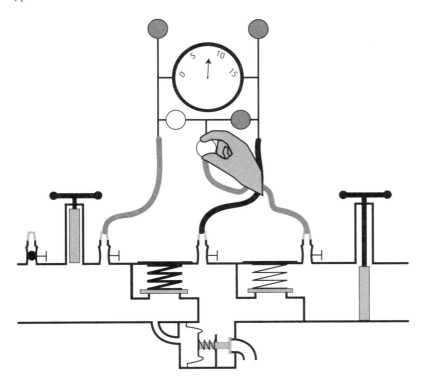

- If the differential pressure reading remains steady then the No. 2 check valve is recorded as "closed tight." Go to Test No. 3.
- If the differential pressure reading decreases, but stabilizes above the relief valve opening point, the No. 2 check valve can still be recorded as "closed tight." See the Diagnostic 9.2.1.3.5: Disc Compression: Second Check Valve. Go to Test No. 3.
- If the differential pressure reading falls to the relief valve opening point, open the low side bleed needle valve until the reading reaches a value above the apparent No. 1 check valve reading, then slowly close the low side bleed needle valve.
  - If the differential pressure reading settles above the relief valve opening point (relief valve does not open), record the No. 2 check valve as "closed tight," and proceed to Test No. 3. See diagnostics 9.2.1.3.5: Disc Compression, Second Check Valve
  - If the differential pressure reading falls to the relief valve opening point again, then the No. 2 check valve is recorded as "leaking," and Test No. 3 below cannot be completed. Go to Test No. 3, step b.
- If the differential pressure reading falls to zero and the relief valve did not open (as recorded in Test No. 1, step j), open the low side bleed needle valve until the reading reaches a value above the apparent No. 1 check valve reading, then slowly close the low side bleed needle valve.
  - If the differential pressure reading settles at a value above zero (0.0 psid) and relief valve does not open, record the No. 2 check valve as "closed tight," and proceed to Test No. 3. (See diagnostics 9.2.1.3.5: Disc Compression, Second Check Valve)
  - If the differential pressure reading falls to zero (0.0 psid) again and relief valve does not open, then the No. 2 check valve is recorded as "leaking," and Test No. 3 below cannot be completed. Go to Test No. 3, step b.
- If the differential pressure reading increases, see diagnostics 9.2.1.3.6: Backpressure Condition.

Test No. 3   Tightness of No. 1 Check Valve

Purpose: To determine the tightness of No. 1 check valve, and to record the static differential pressure across No. 1 check valve.

Requirement: The static differential pressure across check valve No. 1 shall be greater than the relief valve opening point (Test No. 1), and at least 5.0 psid.

Steps:
a. With the bypass hose connected to No. 4 test cock as in step e of Test No. 2 (high side control needle valve and bypass control needle valve remaining open), open the low side bleed needle valve on the field test kit until the reading reaches a value above the apparent No. 1 check valve differential pressure. Slowly close the low side bleed needle valve. After the reading settles, the differential pressure reading indicated (reading is not falling on the field test kit) is the actual static (i.e., no flow) differential pressure across check valve No. 1 and is to be recorded as such.

Field Test Procedures

b. Close all test cocks, slowly open shutoff valve No. 2, and remove all test equipment and fittings.

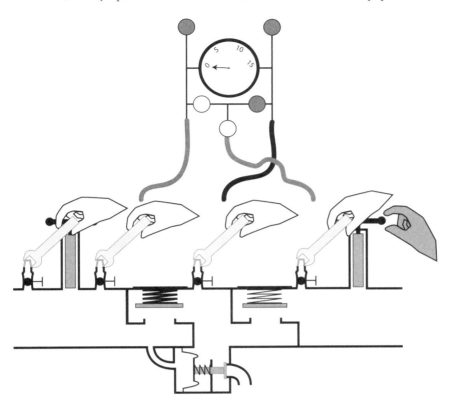

9.2.1.3 Diagnostics

9.2.1.3.1 Exercising the Differential Pressure Relief Valve

It is one of the objectives of the field test procedure to determine the opening point value of the differential pressure relief valve, the first time that it opens. If the relief valve is activated (caused to open) before step j of Test No. 1, then a misleading value may be recorded. By causing the relief valve to open prematurely, the exercising of the moving components in the relief valve will generally produce higher relief valve opening point values. If the initial opening point would have been below the minimum 2.0 psid, but the tester activates the relief valve and then records the opening point value as greater than 2.0 psid, the tester may inadvertently pass an assembly which is not functioning properly.

In normal field operation, the differential pressure relief valve may not get exercised prior to the occurrence of a backflow condition. Therefore, the corresponding field test should evaluate the unit under the same conditions.

9.2.1.3.2 Continuously Discharging Differential Pressure Relief Valve.

If the differential pressure relief valve is observed to be continuously discharging upon arrival, perform steps a through i of Test No. 1.

- If the relief valve stops discharging continuously then a backpressure condition is present with a leaking No. 2 check valve. Perform Test No. 1, step j. Record No. 2 check valve as leaking and the apparent differential pressure reading, Test No. 1 step i, as the actual static pressure across check valve No. 1. Go to Test No. 3, step b.
- If relief valve continues discharging after step i.
  - If No. 1 check valve reading is 5.0 psid or greater, this indicates a leaking relief valve. Record relief valve as leaking. Appropriate repair must be made before testing the check valves. Go to Test No. 3 step b.
  - If No. 1 check valve reading is less than 5.0 psid this indicates a failing No. 1 check valve. See diagnostics 9.2.1.3.3, Leaking No. 1 Check Valve.

9.2.1.3.3 Leaking No. 1 Check Valve

If during step i of Test No. 1, the No. 1 check valve is found to be leaking, this will not allow the tester to proceed with the field test procedure. Since the relief valve is already discharging the relief valve opening point cannot be accurately determined. However, the indicated differential pressure reading with the relief valve discharging is an approximate value of the relief valve opening point.

The No. 2 check valve can not be evaluated since the field test kit reading is already at the relief valve opening point. Go to Test No. 3, step b.

The following should be recorded:

- No. 1 Check Valve: Leak
- No. 2 Check Valve: (No test result)
- Differential Pressure Relief Valve: (No test result)
- Overall Assembly Status: Fail

If the same condition occurs after repairing the No. 1 check valve, then there may be a leak in the differential pressure relief valve diaphragm.

9.2.1.3.4 Leaking No. 2 Shutoff Valve (See Fig. 9.4)

If the differential pressure reading on the field test kit during Test No. 1, step j, does not decrease to the relief valve opening point, and the low side control needle valve is opened one-quarter turn, it is likely that the No. 2 shutoff valve is leaking. Close high side and low side control needle valves. The

No. 2 shutoff valve should be re-opened and closed in an effort to get a better seal. This may particularly occur when testing units, which do not have resilient seated shutoff valves. Make sure that hose fittings are drip-tight. Return to Test No. 1, step j. If this does not correct the leak in the No. 2 shutoff valve, then continue below.

> **NOTE**: In tests No. 1, No. 2, and No. 3 above, a leaking No. 2 shutoff valve may affect the accuracy of the recorded values. A small leak in the No. 2 shutoff valve can be tolerated as long as the hose capacity is enough to satisfy the leak of the No. 2 shutoff valve. In Test No. 1, step j, if the low side control needle valve must be opened more than one-quarter turn then the above procedure may provide invalid data. If the volume of the leak of the No. 2 shutoff requires more than one-quarter turn of the needle valve, then fittings should be placed in the No. 1 and No. 4 test cocks to accommodate an additional temporary bypass hose from the No. 1 test cock to the No. 4 test cock. A 1/2"∅ or 3/4"∅ hose may be needed to satisfy the leak in larger assemblies.

Close low side control needle valve (high side control needle valve remains open). Attach temporary bypass hose to No. 1 test cock and bleed all air from the temporary bypass hose by opening No. 1 test cock, then close. Attach the other end of the temporary bypass hose to No. 4 test cock. Fully open No. 1 test cock to pressurize the temporary bypass hose. Slowly open No. 4 test cock to the fully open position, observing the reading on the field test kit.

- If the differential pressure reading stabilizes above the relief valve opening point, re-open the low side control needle valve no more than one-quarter turn.

  - If the differential pressure reading decreases, observe the differential pressure reading as it slowly drops to the relief valve opening point. Record the opening point value when the first discharge of water is detected. Proceed to 9.2.1.3.4.a below.

  - If the differential pressure reading drops to zero, and the differential pressure relief valve does not discharge any water, record that the differential pressure relief valve did not open. Proceed to 9.2.1.3.4.a below.

    9.2.1.3.4.a: Close the low side control needle valve. Open the low-side bleed needle valve, once the reading reaches a value above the apparent No. 1 check valve reading, close the low side bleed needle valve. The No. 2 check valve is recorded as "closed tight", and the reading indicated is the actual static differential pressure across the No. 1 check valve and recorded as such. The No. 2 shutoff valve should be recorded as leaking. Proceed to Test No. 3, step b.

  - If the reading on the field test kit still will not drop to the relief valve opening point, then the leak through the No. 2 shutoff valve is greater than the capacity of the temporary bypass hose, and should be recorded as such. Repair or replacement of the No. 2 shutoff valve is necessary before an accurate test can be completed. Close low side control needle valve. Proceed to Test No. 3, step b.

- If the differential pressure reading decreases to the low end of the scale and the differential pressure relief valve discharges continuously, record the opening point value when the first discharge of water is detected. Open low side bleed needle valve, once the reading reaches a value above the apparent No. 1 check valve reading, close the low side bleed needle valve.

  - If the differential pressure reading remains above the relief valve opening point, then the No. 2 check valve is recorded as "closed tight", and the reading indicated is the actual static pressure drop across the No. 1 check valve and recorded as such. The No. 2 shutoff valve should be recorded as leaking. Proceed to Test No. 3, step b.

- If the differential pressure reading decreases to the relief valve opening point again, then either the No. 1 check valve and/or No. 2 check valve are leaking. It is not possible to determine which check valve is leaking. This concludes the field test, proceed to Test No. 3, Step b; appropriate repairs must be made. The No. 2 shutoff valve should be recorded as leaking.

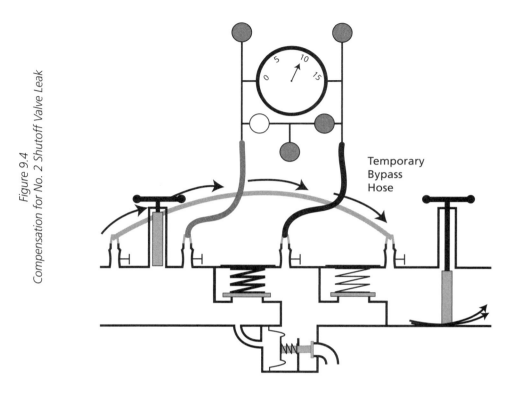

Figure 9.4
Compensation for No. 2 Shutoff Valve Leak

9.2.1.3.5 Disc Compression: Second Check Valve

As high side pressure from the No. 2 test cock is being transferred to the No. 4 test cock (i.e., backside of the No. 2 check valve) during step e of Test No. 2, the differential pressure reading on the field test kit may drop. This lowering of the No. 1 check valve reading can be caused by the backpressure created behind the No. 2 check valve. This backpressure will cause the No. 2 check valve seat to imbed more deeply into the elastomer disc.

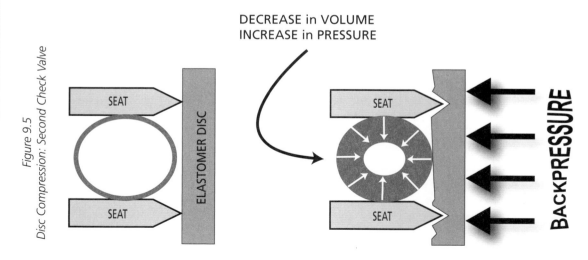

Figure 9.5
Disc Compression: Second Check Valve

This decreases the volume between the two check valves (i.e., the zone of reduced pressure) and, in turn, increases the pressure in the zone. An increase of the zone pressure will lower the differential pressure across the No. 1 check valve. To eliminate this false reading on the field test kit, the excess pressure built up in the zone must be bled off. This is done by opening the low side bleed needle valve (see step a, Test No. 3).

The amount of disc compression will be dependent upon the hardness of the elastomer disc, the profile of the seat or the design of the check valve. With some assemblies the amount of disc compression will cause the differential pressure reading to lower to the relief valve opening point (Test No. 2, step e), and give a false impression that the No. 2 check valve is leaking.

9.2.1.3.6 Backpressure Condition

As the bypass needle valve is opened during step e of Test No. 2, the differential pressure reading on the field test kit may increase. This may be caused by a pressure downstream of the No. 2 check valve, which is greater than the line pressure at the No. 2 test cock (i.e., backpressure condition). The backpressure condition would pass through the bypass hose from the No. 4 test cock and raise the pressure on the high side of the field test kit, and therefore increase the reading on the field test kit. See figure 9.6. This backpressure will cause a backflow condition through the field test kit if the field test kit is left online. For the backpressure condition to reach the No. 4 test cock, the No. 2 shutoff valve would have to be leaking.

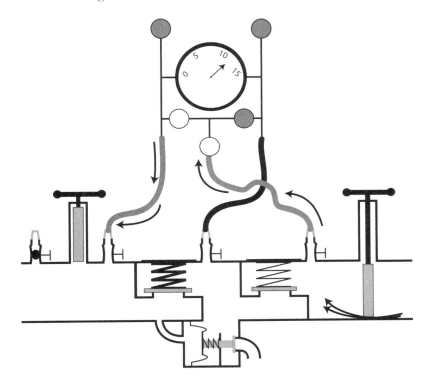

Figure 9.6
Backflow Condition due to Backpressure

Field Test Procedures

To determine if there is a backpressure condition, close the No. 2 test cock (see Fig. 9.7), then observe the reading:

- If the differential pressure reading remains steady or drops to the low end of the scale, backpressure is not present. This false indication of backpressure may have been caused by pressure fluctuations. Open No. 2 test cock. Return to step e, Test No. 2.
- If the differential pressure reading increases then backpressure is present, close the No. 4 test cock. The following information can be recorded:

  - No. 2 Check Valve is holding tight against a backpressure, evidenced by the relief valve not discharging upon arrival.
  - No. 1 Check Valve is holding tight, and the apparent differential pressure reading, Test No. 1, step i can be recorded as the actual static differential pressure across check valve No. 1. At the time the apparent reading was observed, there was no flow going through the No. 1 Check Valve. Record No. 2 shutoff valve as leaks with backpressure. Return to Test No. 3, step b.

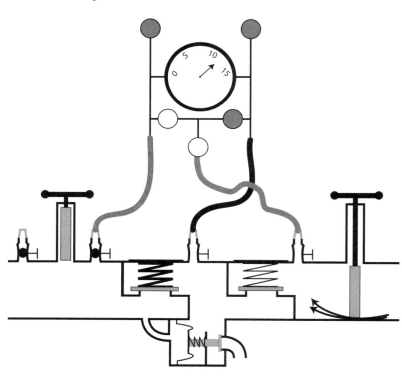

Figure 9.7
Backflow Condition due to Backpressure

9.2.1.3.7 Diagnostics Summary

NOTE: Many problems can be corrected by cleaning the internal components of the backflow prevention assemblies. Carefully observe condition of components.

Problem	May be caused by
Relief valve discharges continuously	1. Faulty No. 1 check valve 2. Faulty No. 2 check valve with backpressure condition 3. Faulty relief valve 4. Plugged relief valve sensing line(s) 5. Leak through RV diaphragm
Relief valve discharges intermittently	1. Properly working assembly with large pressure fluctuations 2. No. 1 check valve reading is close to RV opening point, with small line pressure fluctuation. 3. Water hammer 4. No. 1 check valve leak with intermittent flow through assembly

Relief Valve discharges after No. 2 shutoff valve is closed (Test No. 1)	1. Normally indicates faulty No. 1 check valve
	  • Dirty or damaged disc
	  • Dirty or damaged seat
	2. Leak through relief valve diaphragm

Relief valve would not open; differential on the field test kit would not drop (Test No. 1)	1. Leaky No. 2 shutoff valve with flow through the assembly
Relief valve would not open, differential drops to zero (Test No. 1)	1. Relief valve stuck closed due to corrosion or scale 2. Relief valve sensing line(s) plugged

Relief valve opens too high (with sufficiently high No. 1 check reading)	1. Faulty relief valve
	  • Dirty or damaged disc
	  • Dirty or damaged seat

No. 1 check reading too low; less than 5.0 (Tests No. 1 and No. 3)	1. Dirty or damaged disc 2. Dirty or damaged seat 3. Guide members hanging up 4. Damaged spring
Leaky No. 2 Check (Test No. 2)	1. Dirty or damaged disc 2. Dirty or damaged seat 3. Guide members hanging up 4. Damaged spring

NOTE: Lubricants should only be used to assist with the reassembly of components, and must be non-toxic.

Many problems can be corrected by cleaning the internal components of the backflow prevention assemblies. Carefully observe condition of components.

9.2.2 Reduced Pressure Principle Backflow Prevention Assembly (RP) Using Two Needle Valve Field Test Kit

9.2.2.1 Equipment Required:

a. Differential pressure field test kit with minimum range 0-15 psid (0.1 or 0.2 psid graduations)
b. Temporary bypass hose (comparable to the size of the test cocks)
c. Four brass 1/4" IPS x 45° SAE flare connectors
d. Brass adapter fittings for each test cock size: 1/8" x 1/4", 1/4" x 1/2", 1/4" x 3/4"

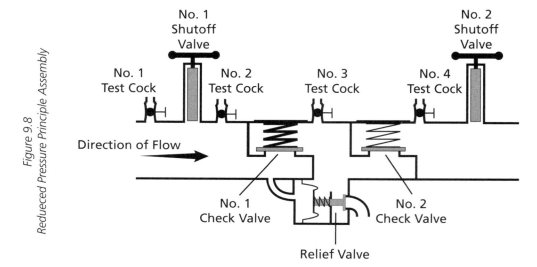

Figure 9.8
Redueced Pressure Principle Assembly

9.2.2.2 Field Test Procedure

Test No. 1   Relief Valve Opening Point

Purpose: To test the operation of the differential pressure relief valve.

Requirement: The differential pressure relief valve must operate to maintain the zone between the two check valves at least 2 psi less than the supply pressure.

> **NOTE**: It is important that during this test the tester does not cause the relief valve to discharge before step k below. (See diagnostics 9.2.2.3.1: Exercising the Differential Pressure Relief Valve.)
>
> **NOTE**: If the differential pressure relief valve is discharging continuously, see diagnostics 9.2.2.3.2: Continuously Discharging Differential Pressure Relief Valve.

Steps:
Follow all preliminary steps detailed in 9.1

a. Bleed water through test cocks to eliminate foreign material. Open No. 4 test cock to establish flow through the unit, then open test cocks No. 3, No. 2 (open No. 2 test cock slowly), and No. 1. Then close test cocks No. 1, No. 2, No. 3 and No. 4. Be careful not to activate the differential pressure relief valve while bleeding the test cocks.

b. Install appropriate fittings to test cocks.

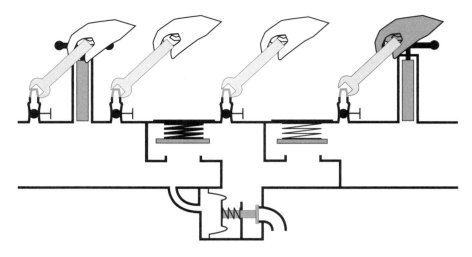

NOTE: Some assemblies are equipped with the adapter fittings which fit the field test kit hoses as an integral part of the test cocks.

c. Attach hose from the high side of the field test kit to the No. 2 test cock.

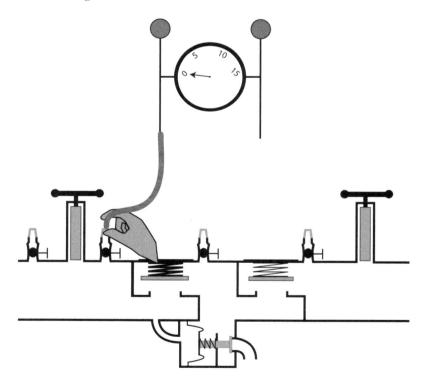

NOTE: Some types of test cocks are opened when the hose from the field test kit are attached. If assembly contains this type of test cock, see Appendix A.2.1 for hose attachment procedure.

d. Attach hose from the low side of the field test kit to the No. 3 test cock.

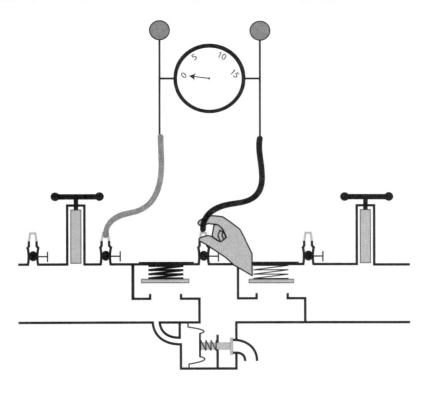

e. Attach bypass hose to low side bleed needle valve. Slowly open test cock No. 3 fully and then bleed all air from the low side of the field test kit by opening the low side bleed needle valve.

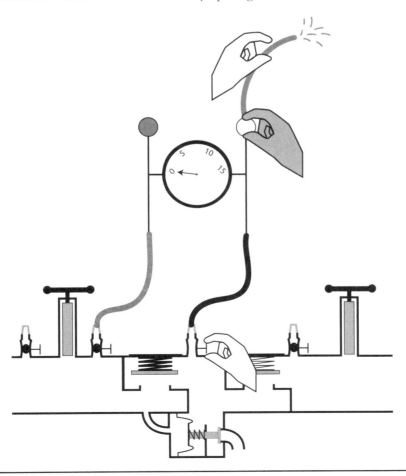

f. Maintain the low side bleed needle valve in the open position and slowly open test cock No. 2 fully to pressurize the field test kit.

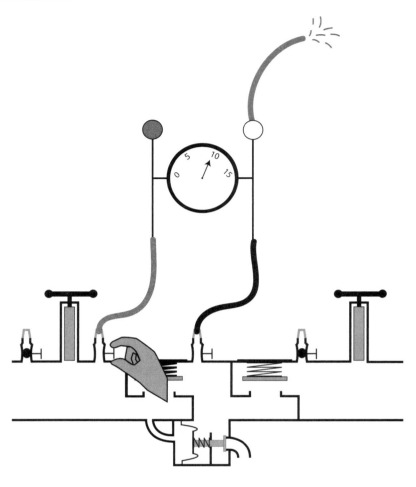

g. Open the high side bleed needle valve to bleed all air from the high side of the field test kit.

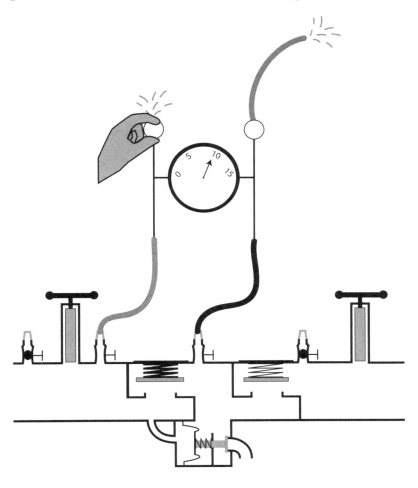

h. Close No. 2 shutoff valve.

i. Close the high side bleed needle valve. After the field test kit reading has reached the upper end of the scale, slowly close the low side bleed needle valve.

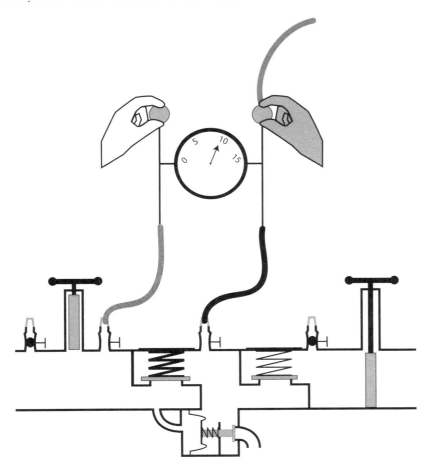

- If the reading remains above the differential pressure relief valve opening point (i.e., relief valve does not discharge), then observe the reading. This is the apparent differential pressure across the No. 1 check valve. During Tests No. 1, No. 2, and No. 3 of this procedure the field test kit is on-line showing the differential pressure across the No. 1 check valve. Proceed to step j.
- If the reading drops to the low end of the scale and the differential pressure relief valve discharges continuously, then the No. 1 check valve is leaking , and should be recorded as such. Tests No. 1, No. 2 and No. 3 may not be completed. (See diagnostics 9.2.2.3.3: Leaking No. 1 Check Valve.)

j. Attach bypass hose from low side bleed needle valve to the high side bleed needle valve

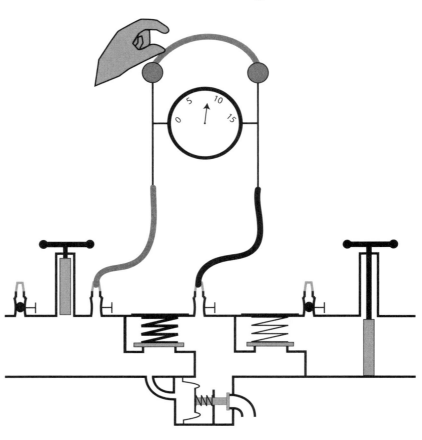

k. Open the high side bleed needle valve approximately one turn, and then slowly open the low side bleed needle valve no more than one-quarter turn to bypass water from the No. 2 test cock to the No. 3 test cock. Observe the differential pressure reading as it slowly drops to the relief valve opening point. Record this opening point value when the first discharge of water is detected.

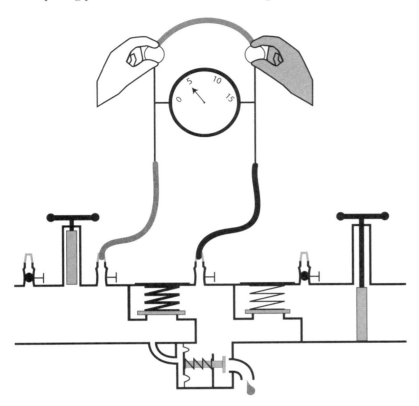

- If the low side bleed needle valve is opened one-quarter turn and the reading does not decrease to the relief valve opening point, then see diagnostics 9.2.2.3.4: Instructions for Leaking No. 2 Shutoff Valve.
- If the differential pressure reading drops to zero, and the differential pressure relief valve does not discharge any water, record that the differential pressure relief valve did not open. Proceed to step l.

1. Close both of the needle valves. Detach the bypass hose from the low side bleed needle valve.

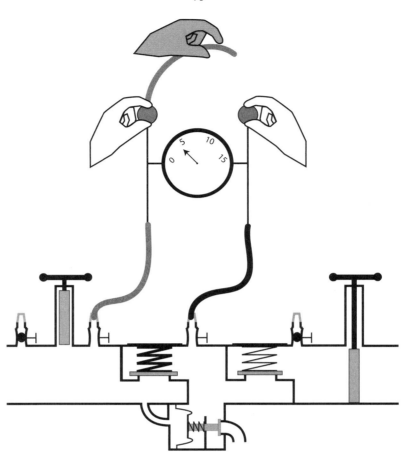

Field Test Procedures

Test No. 2   Tightness of No. 2 Check Valve

Purpose: To test the No. 2 check valve for tightness against backpressure.

Requirement: The No. 2 check valve shall be tight against backpressure.

Steps:
a. Maintain the No. 2 shutoff valve in the closed position (from Test No. 1).

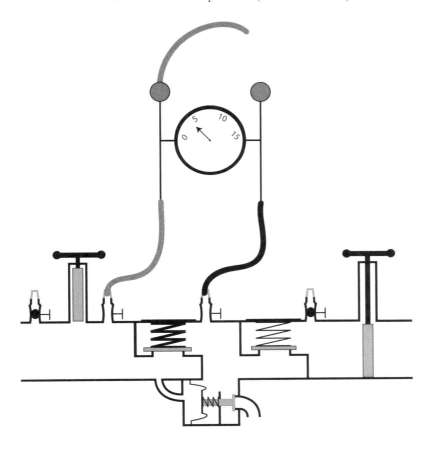

b. Attach the bypass hose from the high side bleed needle valve on the field test kit to the No. 4 test cock.

c. Open the No. 4 test cock

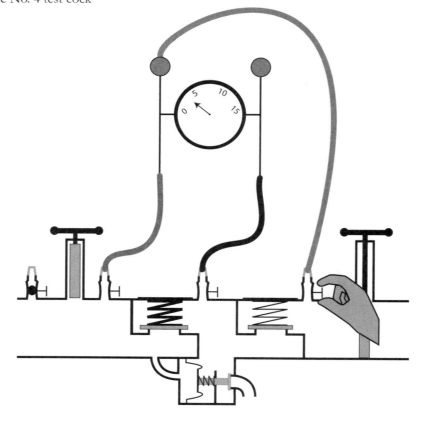

d. Open low side bleed needle valve. Once the reading reaches a value above the apparent No. 1 check valve differential pressure reading, slowly close the low side bleed needle valve.

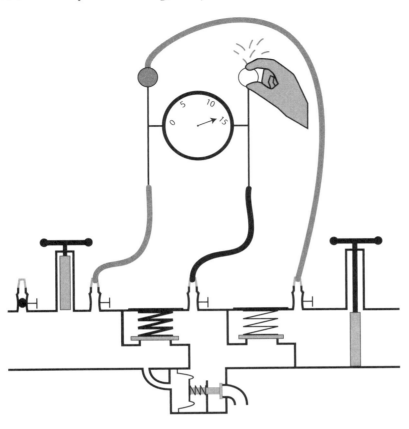

e. Open the high side bleed needle valve.

Field Test Procedures

- If the differential pressure reading remains steady then the No. 2 check valve is recorded as "closed tight." Go to Test No. 3.
- If the differential pressure reading decreases, but stabilizes above the relief valve opening point, the No. 2 check valve can be recorded as "closed tight." See the diagnostics 9.2.2.3.5: Disc Compression, Second Check Valve. Go to Test No. 3.
- If the differential pressure reading falls to the relief valve opening point, open the low side bleed needle valve until the reading reaches a value above the apparent No. 1 check valve reading, then slowly close the low side bleed needle valve.
  - If the differential pressure reading settles above the relief valve opening point (relief valve does not open), record the No. 2 check valve as "closed tight," and proceed to Test No. 3. (see diagnostics 9.2.2.3.5: Disc Compression, Second Check Valve)
  - If the differential pressure reading falls to the relief valve opening point again, then the No. 2 check valve is recorded as "leaking," and Test No. 3 below cannot be completed. Go to Test No. 3, step b.
- If the differential pressure reading falls to zero, and the relief valve did not open (as recorded in Test 1, step k), open the low side bleed needle valve until the reading reaches a value above the apparent No. 1 check valve reading, then slowly close the low side bleed needle valve.
  - If the differential pressure reading settles at a value above zero (0.0 psid) and relief valve does not open, record the No. 2 check valve as "closed tight," and proceed to Test No. 3. (see diagnostics 9.2.2.3.5: Disc Compression, Second Check Valve)
  - If the differential pressure reading falls to zero (0.0 psid) and relief valve does not open again, then the No. 2 check valve is recorded as "leaking," and Test No. 3 below cannot be completed. Go to Test No. 3, step b.
- If the differential pressure reading field test kit increases, see diagnostics 9.2.2.3.6: Backpressure Condition.

Test No. 3   Tightness of No. 1 Check Valve

Purpose: To determine the tightness of No. 1 check valve, and to record the static differential pressure across No. 1 check valve.

Requirement: The static differential pressure across check valve No. 1 must be greater than the relief valve opening point (Test No. 1), and at least 5.0 psid.

Steps:
a. With the bypass hose connected to No. 4 test cock as in step e of Test No. 2 (high side bleed needle valve remaining open), open the low side bleed needle valve on the field test kit until the reading exceeds the apparent No. 1 check valve differential pressure. Slowly close the low side bleed needle valve. After the reading settles, the differential pressure reading indicated (reading is not falling on the field test kit) is the actual static (i.e., no flow) differential pressure across check valve No. 1 and is to be recorded as such.

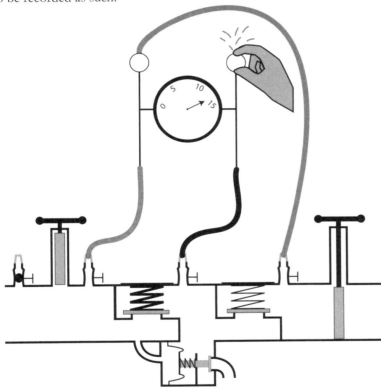

b. Close all test cocks, slowly open shutoff valve No. 2, and remove all test equipment and fittings.

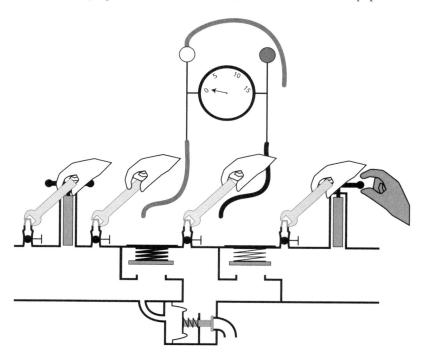

9.2.2.3 Diagnostics

9.2.2.3.1 Exercising the Differential Pressure Relief Valve

It is one of the objectives of the field test procedure to determine the opening point value of the differential pressure relief valve, the first time that it opens. If the relief valve is activated (caused to open) before step k of Test No. 1, then a misleading value may be recorded. By causing the relief valve to open prematurely, the exercising of the moving components in the relief valve will generally produce higher relief valve opening point values. If the initial opening point would have been below the minimum 2.0 psid, but the tester activates the relief valve and then records the opening point value as greater than 2.0 psid, the tester may inadvertently pass an assembly which is not functioning properly.

In normal field operation, the differential pressure relief valve may not get exercised prior to the occurrence of a backflow condition. Therefore, the corresponding field test should evaluate the unit under the same conditions.

9.2.2.3.2 Continuously Discharging Differential Pressure Relief Valve.

If the differential pressure relief valve is observed to be continuously discharging upon arrival, perform steps a through i of Test No. 1.

- If the relief valve stops discharging continuously then a backpressure condition is present with a leaking No. 2 check valve. Perform steps j and k of Test No. 1. Record No. 2 check valve as leaking, and the apparent differential pressure reading, Test No. 1, step i, as the actual static differential pressure across check valve No. 1. Go to Test No. 3, step b.
- If relief valve continues discharging after step i.
  - If No. 1 check valve reading is 5.0 psid or greater this indicates a leaking relief valve. Record relief valve as leaking. Appropriate repair must be made before testing the check valves. Go to Test No. 3, step b.
  - If No. 1 check valve reading is less than 5.0 psid this indicates a failing No. 1 check valve. See diagnostics 9.2.2.3.3: Leaking No. 1 Check Valve.

9.2.2.3.3 Leaking No. 1 Check Valve

If during step i of Test No. 1, the No. 1 check valve is found to be leaking, this will not allow the tester to proceed with the field test procedure. Since the relief valve is already discharging the relief valve opening point can not be accurately determined. However, the indicated differential pressure reading with the relief valve discharging is an approximate value of the relief valve opening point.

The No. 2 check valve can not be evaluated since the field test kit reading is already at the relief valve opening point. Go to Test No. 3, step b.

The following should be recorded:

- No. 1 Check Valve – Leak
- No. 2 Check Valve – (No test result)
- Differential Pressure Relief Valve – (No test result)
- Overall Assembly Status: Fail

If the same condition occurs after repairing the No. 1 check valve, then there may be a leak in the differential pressure relief valve diaphragm.

9.2.2.3.4 Leaking No. 2 Shutoff Valve (See Fig. 9.9)

If the differential pressure reading on the field test kit during Test No. 1, step k, does not decrease to the RV opening point, and the low side bleed needle valve is opened one quarter turn, it is likely that the No. 2 shutoff valve is leaking. Close high side and low side bleed needle valves. The No. 2 shutoff valve should be re-opened and closed in an effort to get a better seal. This may particularly occur when testing units which do not have resilient seated shutoff valves. Make sure that hose fittings are drip-tight. Return to Test No. 1, step k. If this does not correct the leak in the No. 2 shutoff valve, then continue below.

> **NOTE**: In tests No. 1, No. 2, and No. 3 above, a leaking No. 2 shutoff valve may affect the accuracy of the recorded values. A small leak in the No. 2 shutoff valve can be tolerated as long as the hose capacity is enough to satisfy the leak of the No. 2 shutoff valve. In Test No. 1, step k, if the low side bleed needle valve must be opened more than one-quarter turn then the above procedure may provide invalid data. If the volume of the leak of the No. 2 shutoff requires more than one-quarter turn of the needle valve, then fittings should be placed in the No. 1 and No. 4 test cocks to accommodate an additional temporary bypass hose from the No. 1 test cock to the No. 4 test cock. A $1/2"\varnothing$ or $3/4"\varnothing$ hose may be needed to satisfy the leak in larger assemblies.

Close low side bleed needle valve (high side control needle valve remains open). Attach temporary bypass hose to No. 1 test cock and bleed all air from the temporary bypass hose by opening No. 1 test cock, then close. Attach the other end of the temporary bypass hose to No. 4 test cock. Fully open No. 1 test cock to pressurize the temporary bypass hose. Slowly open No. 4 test cock to the fully open position, observing the reading on the field test kit.

- If the differential pressure reading stabilizes above the relief valve opening point, re-open the low side bleed needle valve no more than one-quarter turn.

    - If the differential pressure reading decreases, observe the differential pressure reading as it slowly drops to the relief valve opening point. Record the opening point value when the first discharge of water is detected. Proceed to 9.2.2.3.4.a below.

    - If the differential pressure reading drops to zero, and the differential pressure relief valve does not discharge any water, record that the differential pressure relief valve did not open. Proceed to 9.2.2.3.4.a below.

9.2.2.3.4.a: Close the low side bleed needle valve. Bleed water through test cock No. 3 by loosening the low side hose on the No. 3 test cock, once the reading reaches a value above the apparent No. 1 check valve reading, tighten the low side hose on the No. 3 test cock. The No. 2 check valve is recorded as "closed tight", and the reading indicated is the actual static differential pressure across the No. 1 check valve and recorded as such. The No. 2 shutoff valve should be recorded as leaking. Proceed to Test No. 3, step b.

- If the reading on the field test kit still will not drop to the relief valve opening point, then the leak through the No. 2 shutoff valve is greater than the capacity of the temporary bypass hose, and should be recorded as such. Repair or replacement of the No. 2 shutoff valve is necessary before an accurate test can be completed. Close low side bleed needle valve. Proceed to Test No. 3, step b.

- If the differential pressure reading decreases to the low end of the scale and the differential pressure relief valve discharges continuously, record the opening point value when the first discharge of water is detected. Bleed water through test cock No. 3 by loosening the hose on test cock No. 3, once the reading reaches a value above the apparent No. 1 check valve reading, tighten the hose on test cock No. 3.

    - If the differential pressure reading remains above the relief valve opening point, then the No. 2 check valve is recorded as "closed tight", and the reading indicated is the actual static pressure drop across the No. 1 check valve and recorded as such. The No. 2 shutoff valve should be recorded as leaking. Proceed to Test No. 3, step b.

    - If the differential pressure reading decreases to the relief valve opening point again, then either the No. 1 check valve and/or No. 2 check valve are leaking. It is not possible to determine which check valve is leaking. This concludes the field test, proceed to Test No. 3, Step b; appropriate repairs must be made. The No. 2 shutoff valve should be recorded as leaking.

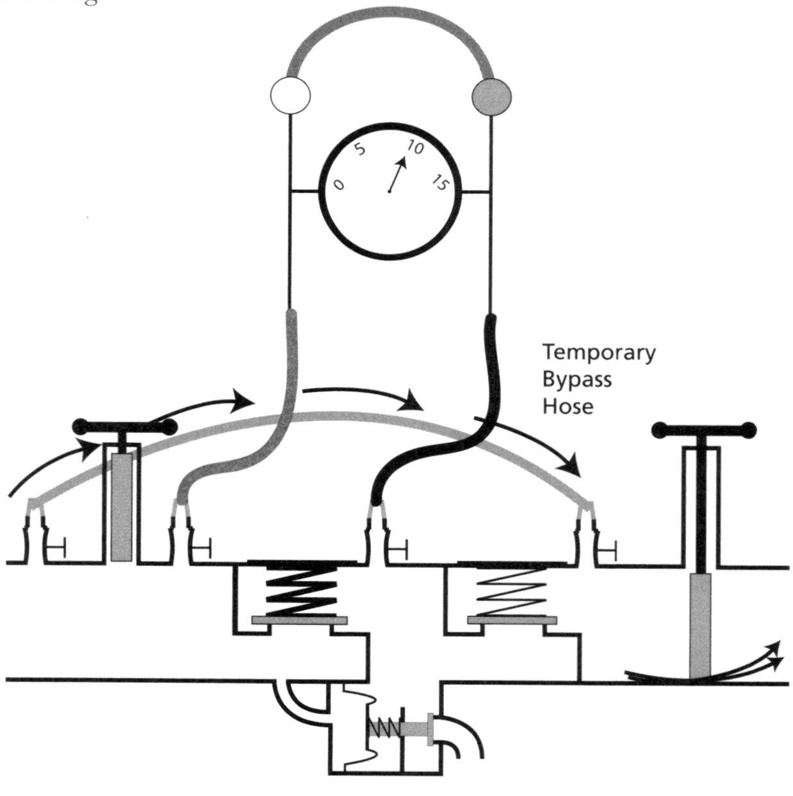

Figure 9.9
Compensation for No. 2 Shutoff Valve Leak

9.2.2.3.5 Disc Compression: Second Check Valve (see Fig. 9.10)

As high side pressure from the No. 2 test cock is being transferred to the No. 4 test cock (i.e., backside of the No. 2 check valve) during step e of Test No. 2, the differential pressure reading on the field test kit may drop. This lowering of the No. 1 check valve reading can be caused by the backpressure created behind the No. 2 check valve. This backpressure will cause the No. 2 check valve seat to imbed more deeply into the elastomer disc.

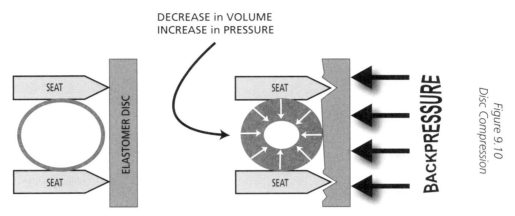

Figure 9.10
Disc Compression

This decreases the volume between the two check valves (i.e., the zone of reduced pressure) and, in turn, increases the pressure in the zone. An increase of the zone pressure will lower the differential pressure across the No. 1 check valve. To eliminate this false reading on the field test kit, the excess pressure built up in the zone must be bled off. This is done by opening the low side bleed needle valve (see step a, Test No. 3).

The amount of disc compression will be dependent upon the hardness of the elastomer disc, the profile of the seat, or the design of check valve. With some assemblies the amount of disc compression will cause the differential pressure reading to lower to the relief valve opening point (Test 2, step e), and give a false impression that the No. 2 check valve is leaking.

9.2.2.3.6 Backpressure Condition

As the bypass needle valve is opened during step e of Test No. 2, the differential pressure reading on the field test kit may increase. This may be caused by a pressure downstream of the No. 2 check valve which is greater than the line pressure at the No. 2 test cock (i.e., backpressure condition). The backpressure condition would pass through the bypass hose from the No. 4 test cock and raise the pressure on the high side of the field test kit, and therefore increase the reading on the field test kit (see Fig. 9.11). This backpressure will cause a backflow condition through the field test kit if the field test kit is left online. For the backpressure condition to reach the No. 4 test cock, the No. 2 shutoff valve would have to be leaking.

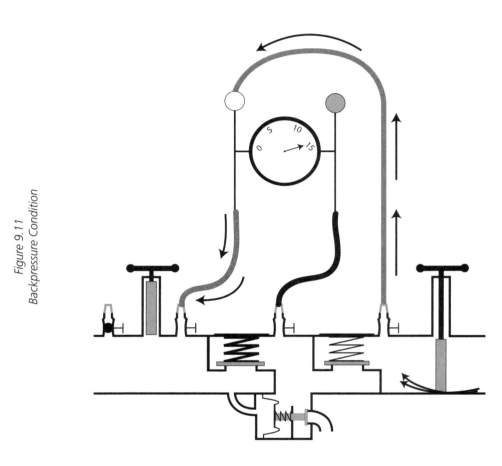

Figure 9.11
Backpressure Condition

To determine if there is a backpressure condition, close the No. 2 test cock (see Fig. 9.12), then observe the reading:

- If the differential pressure reading remains steady or drops to the low end of the scale, backpressure is not present. This false indication of backpressure may have been caused by pressure fluctuations. Open No. 2 test cock. Return to step e, Test No. 2.
- If the differential pressure reading increases then backpressure is present, close the No. 4 test cock. The following information can be recorded:
  - No. 2 Check Valve is holding tight against a backpressure evidenced by the relief valve not discharging upon arrival.
  - No. 1 Check Valve is holding tight, and the apparent differential pressure reading, Test No. 1, step i, can be recorded as the actual static differential pressure across check valve No. 1. At the time the apparent reading was observed, there was no flow going through the No. 1 Check Valve. Record No. 2 shutoff valve as leaks with backpressure. Proceed to Test No. 3, step b.

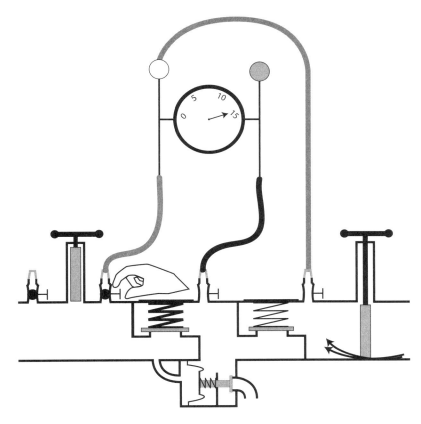

Figure 9.12
Backflow Condition due to Backpressure

9.2.2.3.7 Diagnostics Summary

NOTE: Many problems can be corrected by cleaning the internal components of the backflow prevention assemblies. Carefully observe condition of components.

Problem	May be caused by
Relief valve discharges continuously	1. Faulty No. 1 check valve 2. Faulty No. 2 check valve with backpressure condition 3. Faulty relief valve 4. Plugged relief valve sensing line(s) 5. Leak through RV diaphragm
Relief valve discharges intermittently	1. Properly working assembly with large pressure fluctuations 2. No. 1 check valve reading is close to RV opening point, with small line pressure fluctuation. 3. Water hammer 4. No. 1 check valve leak with intermittent flow through assembly.
Relief Valve discharges after No. 2 shutoff valve is closed (Test No. 1)	1. Normally indicates faulty No. 1 check valve • Dirty or damaged disc • Dirty or damaged seat 2. Leak through relief valve diaphragm
Relief valve would not open; differential on the field test kit would not drop (Test No. 1)	1. Leaky No. 2 shutoff valve with flow through the assembly
Relief valve would not open, differential drops to zero (Test No. 1)	1. Relief valve stuck closed due to corrosion or scale 2. Relief valve sensing line(s) plugged
Relief valve opens too high (with sufficiently high No. 1 check reading)	1. Faulty relief valve • Dirty or damaged disc • Dirty or damaged seat
No. 1 check reading too low; less than 5.0 (Tests No. 1 and No. 3)	1. Dirty or damaged disc 2. Dirty or damaged seat 3. Guide members hanging up 4. Damaged Spring
Leaky No. 2 Check (Test No. 2)	1. Dirty or damaged disc 2. Dirty or damaged seat 3. Guide members hanging up

NOTE: Lubricants should only be used to assist with the reassembly of components, and must be non-toxic.

9.2.3 Reduced Pressure Principle Backflow Prevention Assembly (RP) Using Three Needle Valve Field Test Kit

9.2.3.1 Equipment Required:

a. Differential pressure field test kit with minimum range 0-15 psid (0.1 or 0.2 psid graduations)
b. Temporary bypass hose (comparable to the size of the test cocks)
c. Four brass 1/4" IPS x 45° SAE flare connectors
d. Brass adapter fittings for each test cock size: 1/8" x 1/4", 1/4" x 1/2", 1/4" x 3/4"

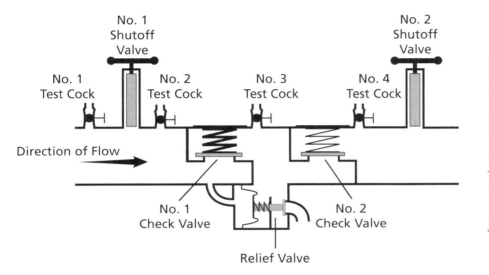

Figure 9.13
Reduced Pressure Principle Assembly

9.2.3.2 Field Test Procedure

Test No. 1   Relief Valve Opening Point

Purpose: To test the operation of the differential pressure relief valve.

Requirement: The differential pressure relief valve must operate to maintain the zone between the two check valves at least 2 psi less than the supply pressure.

> **NOTE**: It is important that during this test the tester does not cause the relief valve to discharge before step j below. (See diagnostics 9.2.3.3.1: Exercising the Differential Pressure Relief Valve.)
>
> **NOTE**: If the differential pressure relief valve is discharging continuously, see diagnostics 9.2.3.3.2 Continuously Discharging Differential Pressure Relief Valve.

Steps:
Follow all preliminary steps detailed in 9.1

a. Bleed water through test cocks to eliminate foreign material. Open No. 4 test cock to establish flow through the unit, then open test cocks No. 3, No. 2 (open No. 2 test cock slowly), and No. 1. Then close test cocks No. 1, No. 2, No. 3., and No. 4. Be careful not to activate the differential pressure relief valve while bleeding the test cocks.

b. Install appropriate fittings to test cocks.

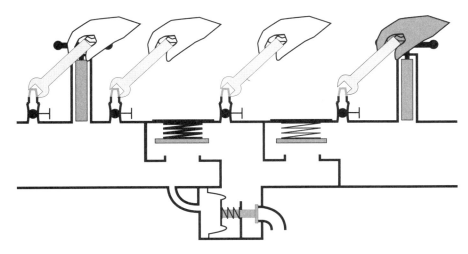

NOTE: Some assemblies are equipped with the adapter fittings which fit the field test kit hoses as an integral part of the test cocks.

c. Attach hose from the high side of the field test kit to the No. 2 test cock.

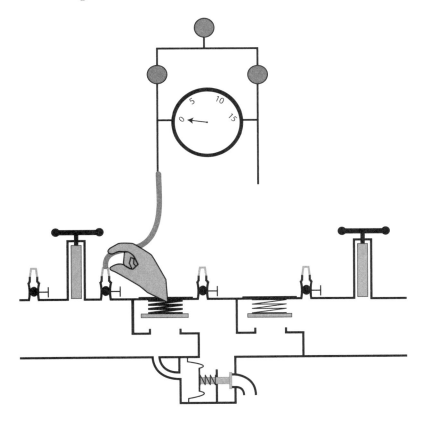

NOTE: Some types of test cocks are opened when the hose from the field test kit are attached. If assembly contains this type of test cock, see Appendix A.2.1 for hose attachment procedure.

d. Attach hose from the low side of the field test kit to the No. 3 test cock.

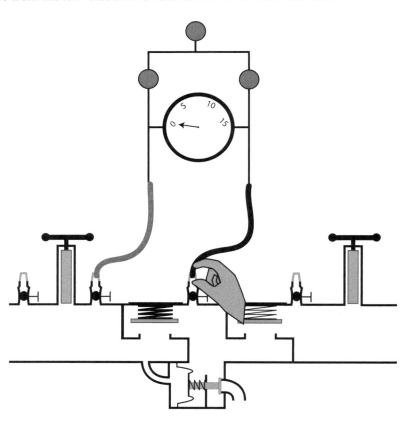

Field Test Procedures

e. Slowly open test cock No. 3 fully. Bleed all air from the low side of the field test kit by opening the bypass needle valve approximately one turn, then open the low side bleed needle valve.

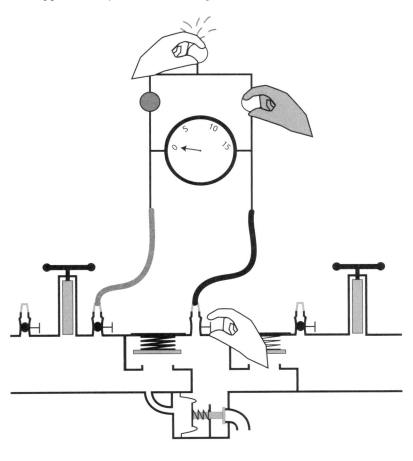

9 Field Test Procedures

f. Maintain the low side bleed needle valve in the open position and slowly open test cock No. 2 fully to pressurize the field test kit.

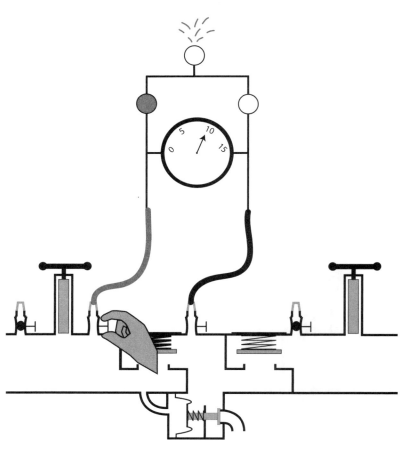

g. Open the high side bleed needle valve to bleed all air from the high side of the field test kit.

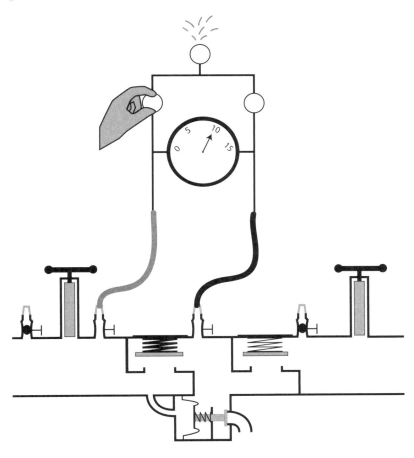

h. Close No. 2 shutoff valve.

i. Close the high side bleed needle valve. After the field test kit reading has reached the upper end of the scale, slowly close the low side bleed needle valve. Close bypass needle valve.

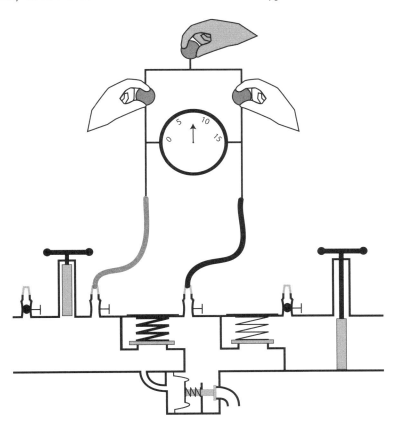

- If the reading remains above the differential pressure relief valve opening point (i.e., relief valve does not discharge), then observe the reading. This is the apparent differential pressure across the No. 1 check valve. During Tests No. 1, No. 2, and No. 3 of this procedure the field test kit is on-line showing the differential pressure across the No. 1 check valve. Proceed to step j.
- If the reading drops to the low end of the scale and the differential pressure relief valve discharges continuously, then the No. 1 check valve is leaking and should be recorded as such. Tests No. 1, No. 2 and No. 3 may not be completed. (See diagnostics 9.2.3.3.3: Leaking No. 1 Check Valve.)

j. Open the high side bleed needle valve approximately one turn, and then slowly open the low side bleed needle valve no more than one-quarter turn to bypass water from the No. 2 test cock to the No. 3 test cock. Observe the differential pressure reading as it slowly drops to the relief valve opening point. Record this opening point value when the first discharge of water is detected.

- If the low side bleed needle valve is opened one-quarter turn and the reading does not decrease to the relief valve opening point, then see diagnostics 9.2.3.3.4: Instructions for Leaking No. 2 Shutoff Valve.
- If the differential pressure reading drops to zero, and the differential pressure relief valve does not discharge any water, record that the differential pressure relief valve did not open. Proceed to Step k.

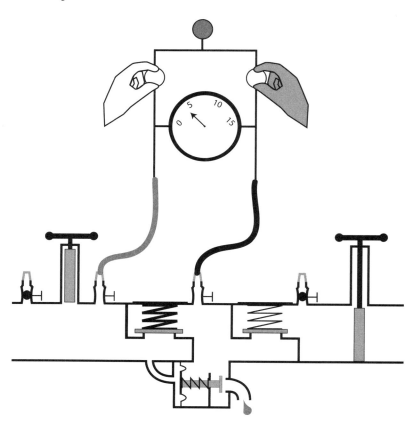

k. Close the low side bleed needle valve.

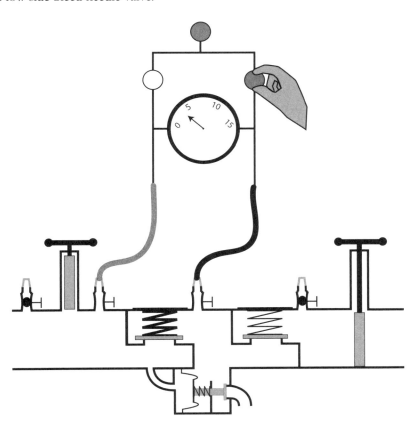

Test No. 2   Tightness of No. 2 Check Valve

Purpose: To test the No. 2 check valve for tightness against backpressure.

Requirement: The No. 2 check valve shall be tight against backpressure.

Steps:
a. Maintain the No. 2 shutoff valve in the closed position and the high side needle valve in the open position (from Test No. 1).

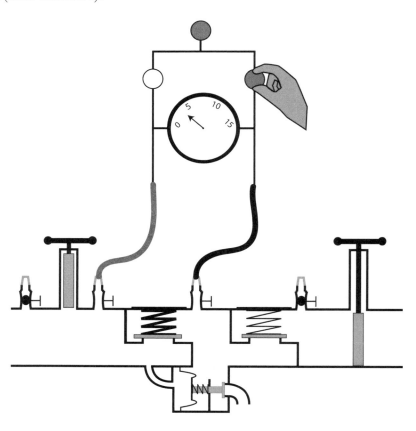

Field Test Procedures

b. Attach bypass hose to bypass needle valve. Bleed all of the air through the bypass hose by opening the bypass needle valve. Close the bypass needle valve.

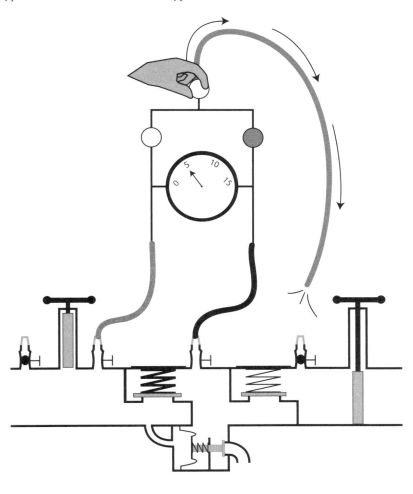

c. Attach the bypass hose from the field test kit to the No. 4 test cock, then fully open the No. 4 test cock.

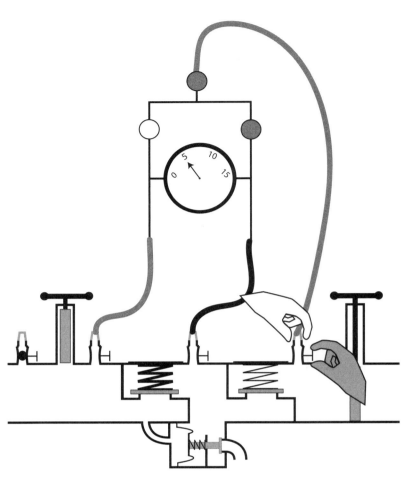

d. Loosen the low side hose on No. 3 test cock. Once the reading reaches a value above the apparent No. 1 check valve differential pressure reading, slowly tighten the low side hose on No. 3 test cock.

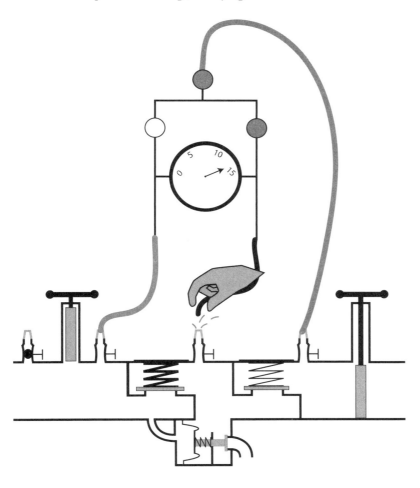

e. Open the bypass needle valve.

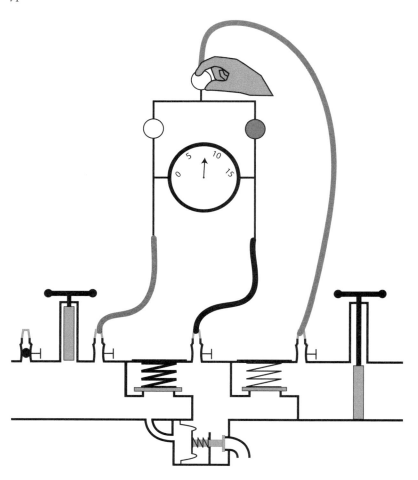

- If the differential pressure reading remains steady then the No. 2 check valve is recorded as "closed tight." Go to Test No. 3.
- If the differential pressure reading decreases, but stabilizes above the relief valve opening point, the No. 2 check valve can be recorded as "closed tight." See the diagnostics 9.2.3.3.5: Disc Compression: Second Check Valve. Go to Test No. 3.
- If the differential pressure reading falls to the relief valve opening point, loosen the low side hose at No. 3 test cock until the reading reaches a value above the apparent No. 1 check valve reading, then slowly tighten the low side hose at No. 3 test cock.
  - If the differential pressure reading settles above the relief valve opening point (relief valve does not open), record the No. 2 check valve as "closed tight," and proceed to Test No. 3. (See diagnostics 9.2.3.3.5: Disc Compression – Second Check Valve.)
  - If the differential pressure reading falls to the relief valve opening point again, then the No. 2 check valve is recorded as "leaking," and Test No. 3 below cannot be completed. Go to Test No. 3, step b.
- If the differential pressure reading falls to zero, and the relief valve did not open (as recorded in Test 1, step j), loosen the low side hose on No. 3 test cock until the reading reaches a value above the apparent No. 1 check valve reading, then slowly tighten the low side hose on No. 3 test cock.
  - If the differential pressure reading settles at a value above zero (0.0 psid) and relief valve does not open, record the No. 2 check valve as "closed tight," and proceed to Test No. 3. (See diagnostics 9.2.3.3.5: Disc Compression – Second Check Valve.)

- If the differential pressure reading falls to 0.0 psid and relief valve does not open again, then the No. 2 check valve is recorded as leaking and Test No. 3 below cannot be completed. Go to Test No. 3, step b.
- If the differential pressure reading increases, see diagnostics 9.2.3.3.6: Backpressure Condition.

Test No. 3   Tightness of No. 1 Check Valve

Purpose: To determine the tightness of No. 1 check valve, and to record the static differential pressure across No. 1 check valve.

Requirement: The static differential pressure across check valve No. 1 shall be greater than the relief valve opening point (Test No. 1), and at least 5.0 psid.

Steps:
a. With the bypass hose connected to No. 4 test cock as in step e of Test No. 2 (high side bleed needle valve and bypass needle valve remaining open), loosen the low side hose on No. 3 test cock until the reading exceeds the apparent No. 1 check valve differential pressure. Slowly tighten the low side hose on No. 3 test cock. After the reading settles, the differential pressure reading indicated (reading is not falling on the field test kit) is the actual static (i.e., no flow) differential pressure across check valve No. 1 and is to be recorded as such.

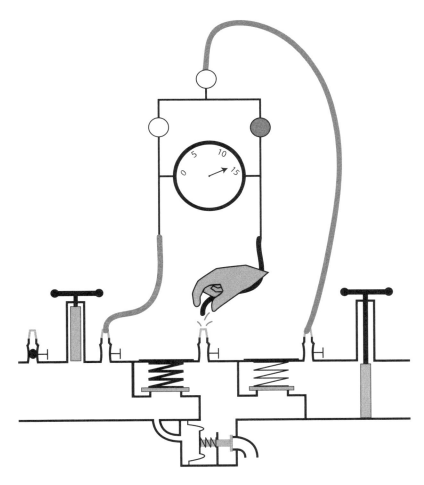

b. Close all test cocks, slowly open shutoff valve No. 2, and remove all test equipment and fittings.

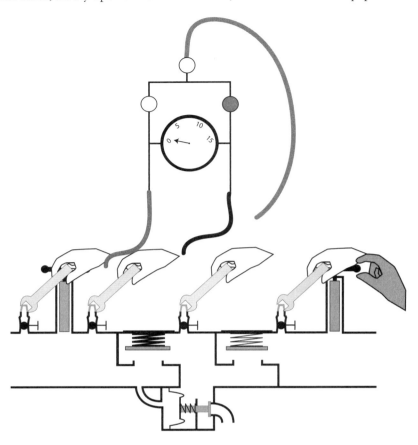

9.2.3.3 Diagnostics

9.2.3.3.1 Exercising the Differential Pressure Relief Valve

It is one of the objectives of the field test procedure to determine the opening point value of the differential pressure relief valve, the first time that it opens. If the relief valve is activated (caused to open) before step j of Test No. 1, then a misleading value may be recorded. By causing the relief valve to open prematurely, the exercising of the moving components in the relief valve will generally produce higher relief valve opening point values. If the initial opening point would have been below the minimum 2.0 psid, but the tester activates the relief valve and then records the opening point value as greater than 2.0 psid, the tester may inadvertently pass an assembly which is not functioning properly.

In normal field operation, the differential pressure relief valve may not get exercised prior to the occurrence of a backflow condition. Therefore, the corresponding field test must be performed to evaluate the unit under the same conditions.

9.2.3.3.2 Continuously Discharging Differential Pressure Relief Valve.

If the differential pressure relief valve is observed to be continuously discharging upon arrival, perform steps a through i of Test No. 1.

- If the relief valve stops discharging continuously then a backpressure condition is present with a leaking No. 2 check valve. Perform Test No. 1, step j. Record No. 2 check valve as leaking, and the apparent differential pressure reading, Test No. 1 step i, as the actual static differential pressure across check valve No. 1. Go to Test No. 3, step b.

- If relief valve continues discharging after step i.
  - If No. 1 check valve reading is 5.0 psid or greater this indicates a leaking relief valve. Record relief valve as leaking. Appropriate repair must be made before testing the check valves. Go to Test No. 3, step b.
  - If No. 1 check valve reading is less than 5.0 psid this indicates a failing. No. 1 check valve. See diagnostics 9.2.3.3.3: Leaking No. 1 Check Valve.

9.2.3.3.3 Leaking No. 1 Check Valve

If during step i of Test No. 1, the No. 1 check valve is found to be leaking, this will not allow the tester to proceed with the field test procedure. Since the relief valve is already discharging the relief valve opening point can not be accurately determined. However, the indicated differential pressure reading with the relief valve discharging is an approximate value of the relief valve opening point.

The No. 2 check valve can not be evaluated since the field test kit reading is already at the relief valve opening point. Go to Test No. 3, step b.

The following should be recorded:

- No. 1 Check Valve: Leak
- No. 2 Check Valve: (No test result)
- Differential Pressure Relief Valve: (No test result)
- Overall Assembly Status: Fail

If the same condition occurs after repairing the No. 1 check valve, then there may be a leak in the differential pressure relief valve diaphragm.

9.2.3.3.4 Leaking No. 2 Shutoff Valve (See Fig. 9.14)

If the differential pressure reading on the field test kit during Test No. 1, step j, does not decrease to the RV opening point, and the low side bleed needle valve is opened one-quarter turn, it is likely that the No. 2 shutoff valve is leaking. Close high side and low side bleed needle valves. The No. 2 shutoff valve should be re-opened and closed in an effort to get a better seal. This may particularly occur when testing units which do not have resilient seated shutoff valves. Make sure that hose fittings are drip-tight. Return to Test No. 1, step j. If this does not correct the leak in the No. 2 shutoff valve, then continue below.

> **NOTE**: In tests No. 1, No. 2, and No. 3 above, a leaking No. 2 shutoff valve may affect the accuracy of the recorded values. A small leak in the No. 2 shutoff valve can be tolerated as long as the hose capacity is enough to satisfy the leak of the No. 2 shutoff valve. In Test No. 1, step j, if the low side bleed needle valve must be opened more than one-quarter turn then the above procedure may provide invalid data. If the volume of the leak of the No. 2 shutoff requires more than one-quarter turn of the needle valve, then fittings should be placed in the No. 1 and No. 4 test cocks to accommodate an additional temporary bypass hose from the No. 1 test cock to the No. 4 test cock. A 1/2"⌀ or 3/4"⌀ hose may be needed to satisfy the leak in larger assemblies.

Close low side bleed needle valve (high side control needle valve remains open). Attach temporary bypass hose to No. 1 test cock and bleed all air from the temporary bypass hose by opening No. 1 test cock, then close. Attach the other end of the temporary bypass hose to No. 4 test cock. Fully open No. 1 test cock to pressurize the temporary bypass hose. Slowly open No. 4 test cock to the fully open position, observing the reading on the field test kit.

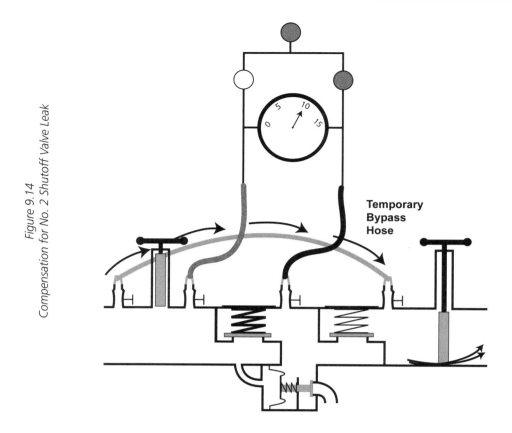

*Figure 9.14
Compensation for No. 2 Shutoff Valve Leak*

- If the differential pressure reading stabilizes above the relief valve opening point, re-open the low side bleed needle valve no more than one-quarter turn.

  - If the differential pressure reading decreases, observe the differential pressure reading as it slowly drops to the relief valve opening point. Record the opening point value when the first discharge of water is detected. Proceed to 9.2.3.3.4.a below.

  - If the differential pressure reading drops to zero, and the differential pressure relief valve does not discharge any water, record that the differential pressure relief valve did not open. Proceed to 9.2.3.3.4.a below.

    9.2.3.3.4.a: Close the low side bleed needle valve. Bleed water through test cock No. 3 by loosening the low side hose on the No. 3 test cock, once the reading reaches a value above the apparent No. 1 check valve reading, tighten the low side hose on the No. 3 test cock. The No. 2 check valve is recorded as "closed tight", and the reading indicated is the actual static differential pressure across the No. 1 check valve and recorded as such. The No. 2 shutoff valve should be recorded as leaking. Proceed to Test No. 3, step b.

  - If the reading on the field test kit still will not drop to the relief valve opening point, then the leak through the No. 2 shutoff valve is greater than the capacity of the temporary bypass hose, and should be recorded as such. Repair or replacement of the No. 2 shutoff valve is necessary before an accurate test can be completed. Close low side bleed needle valve. Proceed to Test No. 3, step b.

- If the differential pressure reading decreases to the low end of the scale and the differential pressure relief valve discharges continuously, record the opening point value when the first discharge of water is detected. Bleed water through test cock No. 3 by loosening the hose on test cock No. 3, once the reading reaches a value above the apparent No. 1 check valve reading, tighten the hose on test cock No. 3.

- If the differential pressure reading remains above the relief valve opening point, then the No. 2 check valve is recorded as "closed tight", and the reading indicated is the actual static pressure drop across the No. 1 check valve and recorded as such. The No. 2 shutoff valve should be recorded as leaking. Proceed to Test No. 3, step b.

- If the differential pressure reading decreases to the relief valve opening point again, then either the No. 1 check valve and/or No. 2 check valve are leaking. It is not possible to determine which check valve is leaking. This concludes the field test, proceed to Test No. 3, Step b; appropriate repairs must be made. The No. 2 shutoff valve should be recorded as leaking.

9.2.3.3.5 Disc Compression: Second Check Valve

As high side pressure from the No. 2 test cock is being transferred to the No. 4 test cock (i.e., backside of the No. 2 check valve) during step e of Test No. 2, the differential pressure reading on the field test kit may drop. This lowering of the No. 1 check valve reading can be caused by the backpressure created behind the No. 2 check valve. This backpressure will cause the No. 2 check valve seat to imbed more deeply into the elastomer disc. See figure 9.15

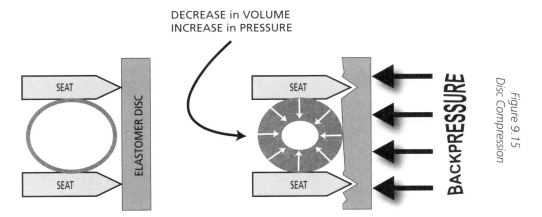

Figure 9.15
Disc Compression

This decreases the volume between the two check valves (i.e., the zone of reduced pressure) and, in turn, increases the pressure in the zone. An increase of the zone pressure will lower the differential pressure across the No. 1 check valve. To eliminate this false reading on the field test kit, the excess pressure built up in the zone must be bled off. This is done by loosening the low side hose on test cock No. 3 (see step a, Test No. 3).

The amount of disc compression will be dependent upon the hardness of the elastomer disc, the profile of the seat, or the design of check valve. With some assemblies the amount of disc compression will cause the differential pressure reading to lower to the relief valve opening point (Test 2, step e), and give a false impression that the No. 2 check valve is leaking.

9.2.3.3.6 Backpressure Condition

As the bypass needle valve is opened during step e of Test No. 2, the differential pressure reading on the field test kit may increase. This may be caused by a pressure downstream of the No. 2 check valve which is greater than the line pressure at the No. 2 test cock (i.e., backpressure condition). The backpressure condition would pass through the bypass hose from the No. 4 test cock and raise the pressure on the high side of the field test kit, and therefore increase the reading on the field test kit (see Fig. 9.16). This backpressure will cause a backflow condition through the field test kit if the field test kit is left online. For the backpressure condition to reach the No. 4 test cock, the No. 2 shutoff valve would have to be leaking.

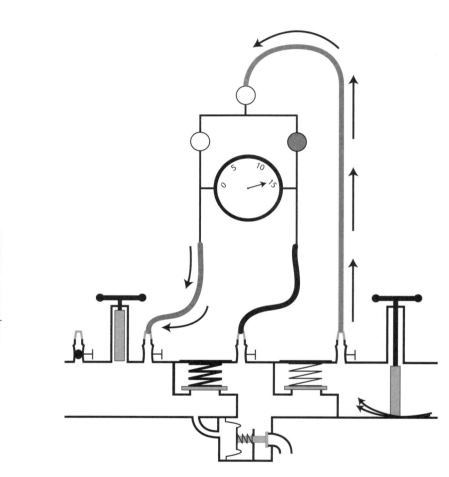

Figure 9.16
Backpressure Condition

To determine if there is a backpressure condition, close the No. 2 test cock (see Fig. 9.17), then observe the reading:

- If the differential pressure reading remains steady or drops to the low end of the scale, backpressure is not present. This false indication of backpressure may have been caused by pressure fluctuations. Open No. 2 test cock. Return to step e, Test No. 2.
- If the differential pressure reading increases then backpressure is present, close the No. 4 test cock. The following information can be recorded:
  - No. 2 Check Valve is holding tight against a backpressure evidenced by the relief valve not discharging upon arrival.
  - No. 1 Check Valve is holding tight, and the apparent differential pressure reading, Test No. 1, step i, can be recorded as the actual static differential pressure across check valve No. 1. At the time the apparent reading was observed, there was no flow going through the No. 1 Check Valve. Record No. 2 shutoff valve as leaks with backpressure. Proceed to Test No. 3, step b.

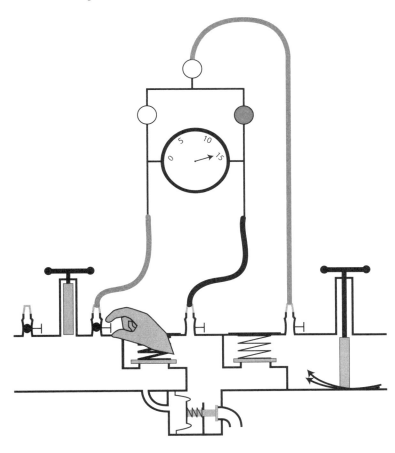

Figure 9.17
Backpressure Condition

9.2.3.3.7 Diagnostics Summary

NOTE: Many problems can be corrected by cleaning the internal components of the backflow prevention assemblies. Carefully observe condition of components.

Problem	May be caused by
Relief valve discharges continuously	1. Faulty No. 1 check valve 2. Faulty No. 2 check valve with backpressure condition 3. Faulty relief valve 4. Plugged relief valve sensing line(s) 5. Leak through RV diaphragm
Relief valve discharges intermittently	1. Properly working assembly with large pressure fluctuations 2. No. 1 check valve reading is close to RV opening point, with small line pressure fluctuation. 3. Water hammer 4. No. 1 check valve leak with intermittent flow though assembly
Relief Valve discharges after No. 2 shutoff valve is closed (Test No. 1)	1. Normally indicates faulty No. 1 check valve • Dirty or damaged disc • Dirty or damaged seat 2. Leak through relief valve diaphragm
Relief valve would not open; differential on the field test kit would not drop (Test No. 1)	1. Leaky No. 2 shutoff valve with flow through the assembly
Relief valve would not open, differential drops to zero (Test No. 1)	1. Relief valve stuck closed due to corrosion or scale 2. Relief valve sensing line(s) plugged
Relief valve opens too high (with sufficiently high No. 1 check reading)	1. Faulty relief valve • Dirty or damaged disc • Dirty or damaged seat
No. 1 check reading too low; less than 5.0 (Tests No. 1 and No. 3)	1. Dirty or damaged disc 2. Dirty or damaged seat 3. Guide members hanging up 4. Damaged spring
Leaky No. 2 Check (Test No. 2)	1. Dirty or damaged disc 2. Dirty or damaged seat 3. Guide members hanging up

NOTE: Lubricants should only be used to assist with the reassembly of components, and must be non-toxic.

9.3 Double Check Valve Backflow Prevention Assembly

9.3.1 Equipment Required:

- Differential pressure field test kit with minimum range 0-15 psid (0.1 or 0.2 psid graduations)
- One brass 1/4" IPS x 45° SAE flare connector
- Brass adapter fittings for each test cock size: 1/8" x 1/4", 1/4" x 1/2", 1/4" x 3/4"
- Street ell, pipe nipple or tube (for length, see 9.3.3.1)
- Bleed-off valve (see Appendix A.1.2: Bleed-Off Valve Arrangement)

NOTE: While field-testing the double check valve assembly it is important to pay attention to the elevation of the field test kit. Be sure that hoses not being used are also kept at the proper elevation. See diagnostics 9.3.3.1: Location of Field Test Kit.

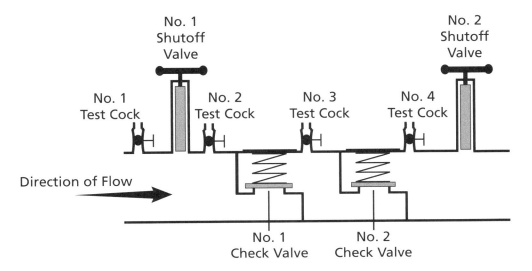

*Figure 9.18
Double Check Valve Backflow Prevention Assembly*

Field Test Kit Configurations: The illustration box below will be inserted for each step of the field test procedures where the needle valves are operated.

Two Needle Valve: Open/Close the high side bleed needle valve when indicated in the procedures.

Three Needle Valve: Fully open the bypass needle valve before starting the procedure, then open/close the high side bleed needle valve when indicated in the procedures.

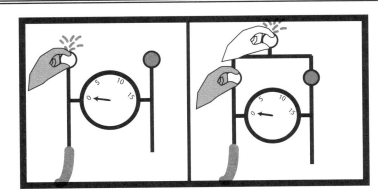

9.3.2 Field Test Procedure

Test No. 1   Tightness of Check Valve No. 1

Purpose: To determine the static differential pressure across check valve No. 1.

Requirement: The static differential pressure across check valve No. 1 must be at least 1.0 psid.

Steps:
Follow all preliminary steps in 9.1

a. Bleed water through test cocks to eliminate foreign material by opening and closing each test cock.

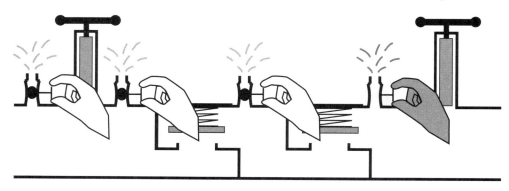

b. Install appropriate fittings.

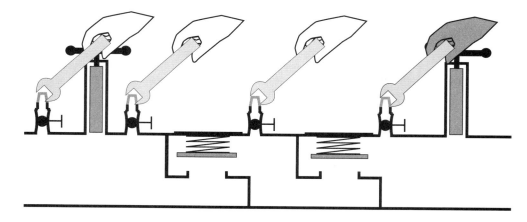

c. If test cock No. 3 is not at the highest point of the check valve body, then a vertical tube or pipe must be installed on test cock No. 3 so that it rises to the top of the check valve body. (See diagnostics 9.3.3.1: Location of Field Test Kit.)

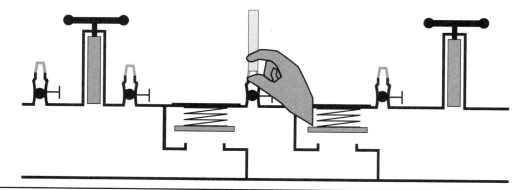

d. Attach bleed-off valve arrangement to test cock No. 2, and the hose from the high side of the field test kit to the bleed-off valve.

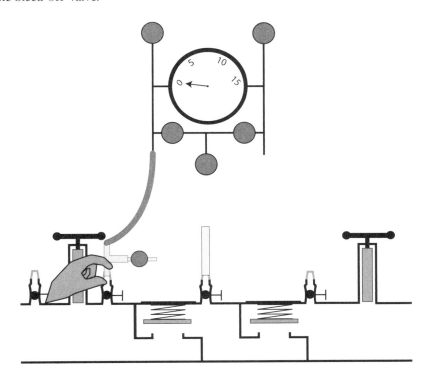

e. Open test cock No. 2 and bleed all air from the field test kit by opening the high side bleed needle valve, then close the high side bleed needle valve. Open test cock No. 3 to fill the test cock No. 3 (or tube, if attached) so that the water level is above the top of the body, then close test cock No. 3.

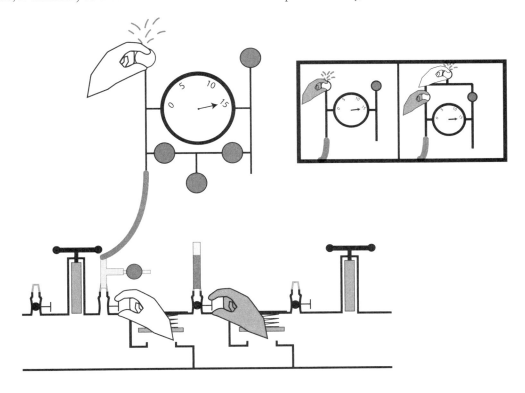

9 Field Test Procedures

279

f. Close No. 2 shutoff valve, the field test kit must be maintained at the same elevation as the water at test cock No. 3, then close No. 1 shutoff valve.

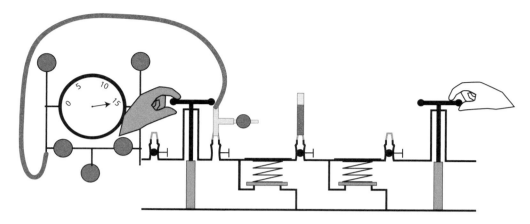

g. Slowly open test cock No. 3. After the reading stabilizes and water stops running out of test cock No. 3, or is no more than a drip, the reading indicated on the field test kit is the differential pressure across check valve No. 1 and is to be recorded as such. Proceed to step h.

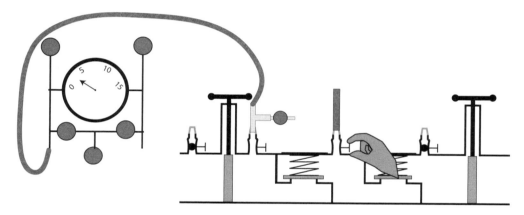

- If there is a continuous discharge of water from test cock No. 3, see diagnostics 9.3.3.2: Leaking Shutoff Valves: Water Discharges from No. 3 Test Cock.
- If the water level at test cock No. 3 recedes, see diagnostics 9.3.3.3: Leaking Shutoff Valves, Water Recedes from No. 3 Test Cock.

h. Close all test cocks, open shutoff valve No. 1, and remove all test equipment.

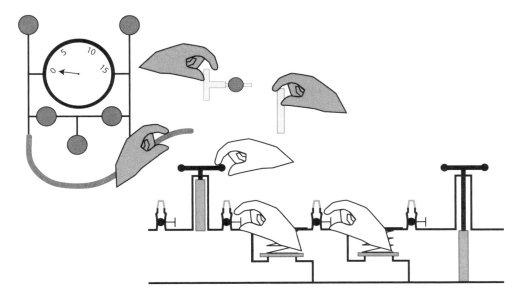

Test No. 2   Tightness of Check Valve No. 2

Purpose: To determine the static differential pressure across check valve No. 2.

Requirement: The static differential pressure across check valve No. 2 shall be at least 1.0 psid.

Steps:
a. Attach bleed-off valve arrangement to test cock No. 3 and hose from the high side of the field test kit to the bleed off valve. If test cock No. 4 is not at the highest point of the check valve body, then a vertical tube or pipe must be installed on test cock No. 4 so that it rises to the top of the check valve body. (See diagnostics 9.3.3.1: Location of Field Test Kit)

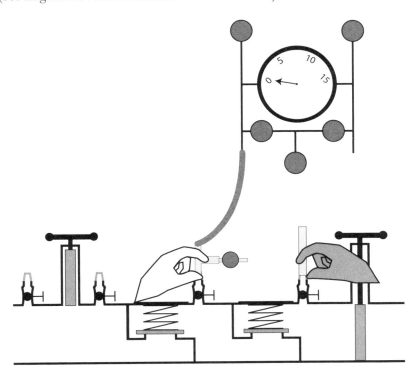

b. Open test cock No. 3, and bleed all air from the field test kit by opening the high side bleed needle valve, and then closing the high side bleed needle valve open test cock No. 4 to fill the test cock No. 4 (or tube if attached) so that the water level is above the check valve body, then close test cock No. 4.

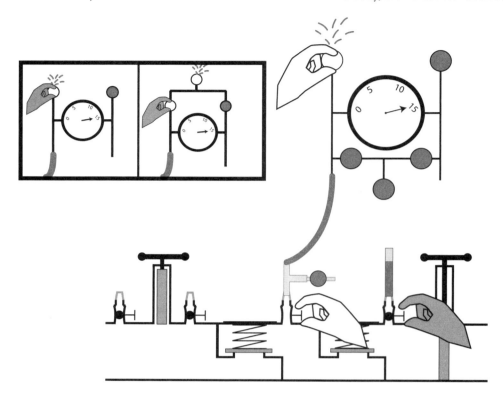

c. The field test kit must be maintained at the same elevation as the water at test cock No. 4. Close No. 1 shutoff valve.

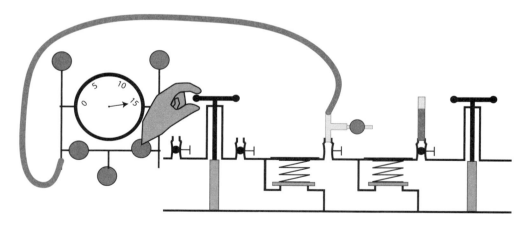

Field Test Procedures

d. Slowly open test cock No. 4. After the reading stabilizes and water stops running out of test cock No. 4, or is no more than a drip, the reading indicated on the field test kit is the differential pressure across check valve No. 2 and is to be recorded as such. Go to Step e.

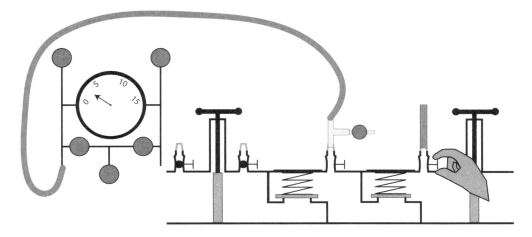

- If there is a continuous flow of water from test cock No. 4, see diagnostics 9.3.3.5: Second Check Test: Leaking Shutoff Valves.
- If the water level at No. 4 test cock recedes, see diagnostics 9.3.3.4: Leaking Shutoff Valves, Water Recedes From No. 4 Test Cock.

e. Close all test cocks, remove all test equipment.

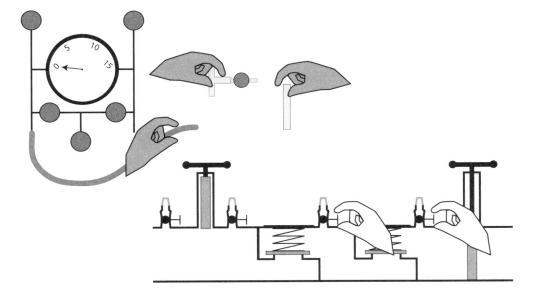

f. Remove fittings. Open shutoff valve No. 1, then slowly open shutoff valve No. 2.

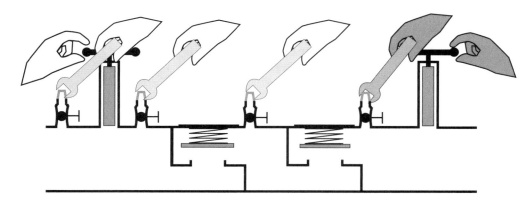

9.3.3 Diagnostics

9.3.3.1 Location of Field Test Kit

During this field test it is important to keep the field test kit and unused hoses at the appropriate elevation. A visible downstream reference point is needed for this field test. If the downstream test cock is at the highest point of the body, then this can be used as the reference point. However, if the downstream test cock is below the top of the body, then a vertical pipe or tube must be attached to the downstream test cock so that it rises above the top of the body. See figure 9.19.

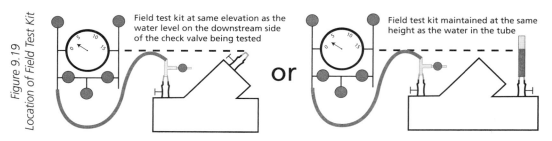

Figure 9.19 Location of Field Test Kit

To record the correct differential pressure reading the field test kit must be held at the same elevation as the water level on the downstream side of the check valve being tested. If a tube is attached to the downstream test cock, then the field test kit must be maintained at the same height as the water in the tube. If this is not done properly, then values, which are either too high or too low, may be recorded.

A common misconception is that the entire length of the high side hose must be maintained at the same level of the field test kit. The critical aspect is the elevation of the field test kit relative to the water level on the downstream side of the check valve being tested. A high side hose looping up or down in between the field test kit and the test cock will not effect the reading.

It is also important to maintain all unused hoses of the field test kit at the same elevation as the field test kit. If the low side hose of the field test kit contains water, and the hose is maintained above or below the field test kit, the water in the low side hose will create a false reading. See figure 9.20.

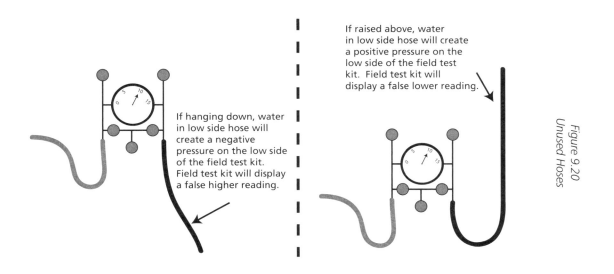

Figure 9.20
Unused Hoses

9.3.3.2 Leaking Shutoff Valves: Water Discharges from No. 3 Test Cock

T1. Verify that shutoff valves are closed. If water continues to flow from test cock No. 3, then a shutoff valve may be leaking. Observe the reading on the field test kit, but do not record at this time. Slowly open the bleed-off valve so there is a drip from test cock No. 3.

Figure 9.21
DC Diagnostics T1

- If the water continues to flow from the bleed-off valve, and the bleed-off valve can be adjusted so there is a drip from the No. 3 test cock. Proceed to step T2.
- If it is not possible to adjust the bleed-off valve so there is a drip from the No. 3. test cock, the No. 1 shutoff valve is leaking too much to continue the test, and should be recorded as such. Shutoff valve No. 1 must be repaired before the field test may be completed. If the same condition occurs after repairing the No. 1 shutoff valve, repair the No. 1 check valve, No. 2 check valve and the No. 2 shutoff valve.
- If the water does not continue to flow from the bleed-off valve with water still flowing from No. 3 test cock, proceed to T3.

T2. After adjusting the bleed-off valve so that there is a drip from the No. 3 test cock, record the reading as the differential pressure across the No. 1 check valve. This reading should be 1.0 psid or greater. Return to Test No. 1, step h. If the reading is less than 1.0 psid, the No. 1 check valve must be repaired and retested before proceeding to Test No. 2.

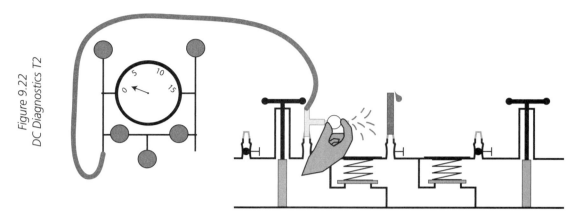

Figure 9.22
DC Diagnostics T2

T3. If water does not continue to flow from the bleed-off valve with water still flowing from the No. 3 test cock, record the observed reading from Step T1 as the static differential pressure across the No. 1 check valve. Also record the No. 2 check valve as leaking, and the No. 2 shutoff valve as leaking with backpressure. This concludes the field test, proceed to Test No. 2, step e. Appropriate repairs must be made.

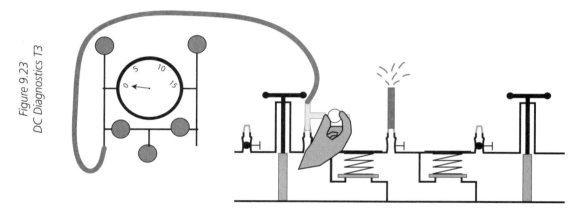

Figure 9.23
DC Diagnostics T3

9.3.3.3 Leaking Shutoff Valves: Water Recedes from No. 3 Test Cock

T4. If water recedes completely from test cock No. 3, the elevation of the field test kit must be moved to the centerline of the assembly. Record the reading on the field test kit as the differential pressure across the No. 1 check valve. The No. 2 check valve should be recorded as leaking and the No. 2 shutoff valve as leaking. Proceed to Test No. 2, step e.

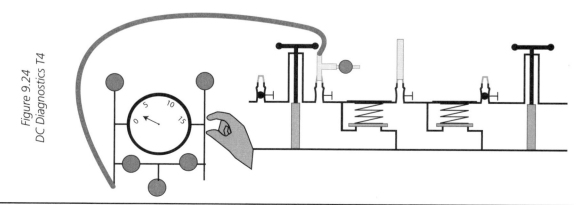

Figure 9.24
DC Diagnostics T4

9.3.3.4 Leaking Shutoff Valve – Water Recedes From No. 4 Test Cock

T5. If the water at test cock No. 4 recedes completely, there is a leaking No. 2 shutoff valve. The elevation of the field test kit must be moved to the centerline of the assembly and record the reading as the differential pressure across the No. 2 check valve. The No. 2 shutoff valve should be recorded as leaking. Proceed to Test No. 2, step e.

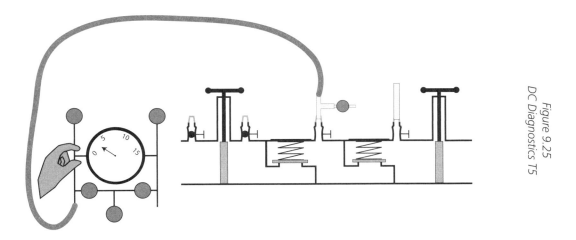

Figure 9.25
DC Diagnostics T5

9.3.3.5 Second Check Test: Leaking Shutoff Valves

T6. Verify that shutoff valves are closed. If at Test No. 2, step d water continues to flow from the No. 4 test cock, one of the shutoff valves is leaking. Observe the reading on the field test kit, but do not record it at this time. Slowly open the bleed-off valve so there is a drip from the No. 4 test cock.

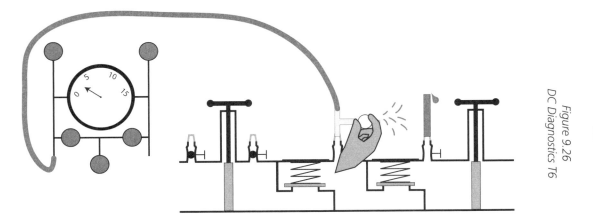

Figure 9.26
DC Diagnostics T6

- If the water does not continue to flow from the bleed-off valve proceed to step T7.
- If the water continues to flow from the bleed-off valve, and the bleed-off valve can be adjusted so there is no more than a drip from the No. 4 test cock, proceed to step T8.
- If it is not possible to adjust the bleed-off valve so there is no more than a drip from the No. 4 test cock, close bleed-off valve. Proceed to step T9.

T7. If water does not continue to flow from the bleed-off valve with water still flowing from the No. 4 test cock, record the observed reading from step T6 as the differential pressure across the No. 2 check valve. Also the No. 2 shutoff valve should be recorded as leaking with backpressure. This concludes the test, proceed to Test No. 2, step e; if necessary, appropriate repairs must be made.

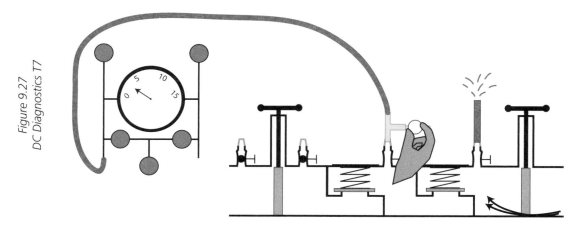

Figure 9.27
DC Diagnostics T7

T8. After adjusting the bleed-off valve so that there is no more than a drip from the No. 4 test cock record the reading on the field test kit as the differential pressure across the No. 2 check valve. Also record the No. 1 shutoff valve as leaking. Proceed to Test No. 2, step e.

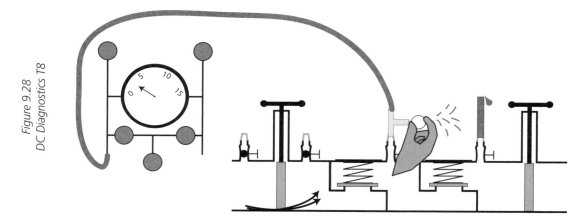

Figure 9.28
DC Diagnostics T8

T9. If it is not possible to adjust the bleed-off valve so that the water flowing from the No. 4 test cock is no more than a drip, and if check valve No. 1 was holding less than 1.0 psid in Test No. 1, the No. 1 check valve must be repaired before testing the No. 2 check valve. Then return to Test No. 1, step a. If No. 1 check valve held 1.0 psid or more, proceed to step T10.

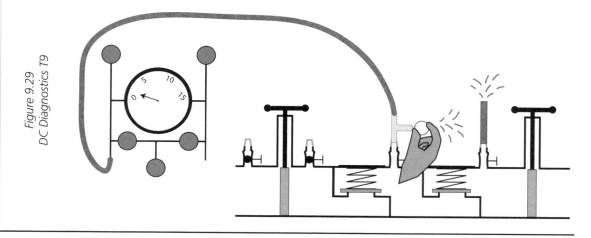

Figure 9.29
DC Diagnostics T9

T10. If check valve No. 1 was holding 1.0 psid or more in Test No. 1, close the bleed-off valve and close No. 4 test cock. Open the No. 2 test cock, then open the No. 4 test cock. Record the reading on the field test kit as the differential pressure across the No. 2 check valve. Also record No. 1 shutoff valve as leaking, and No. 2 shutoff valve as leaking with backpressure. Proceed to Test No. 2, step e.

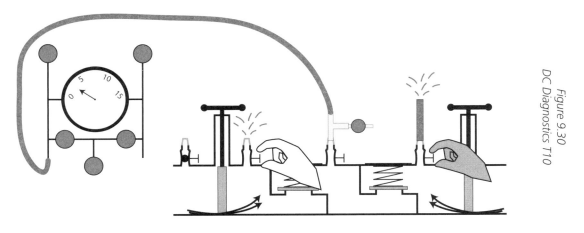

Figure 9.30
DC Diagnostics T10

9.3.3.6 Diagnostics Summary

NOTE: Many problems can be corrected by cleaning the internal components of the backflow prevention assemblies. Carefully observe condition of components.

Problem	**May be caused by**
Check valve reading below 1.0 psid	1. Dirty or damaged check disc
	2. Dirty or damaged check seat
	3. Damaged spring
Check valve reading above 5.0 psid	1. Incorrect spring, possible RP 1st check

NOTE: Lubricants shall only be used to assist with the reassembly of components, and shall be non-toxic.

9.4 Pressure Vacuum Breaker Assembly (PVB)

9.4.1 Equipment Required:
- Differential pressure field test kit with minimum range 0-15 psid (0.1 or 0.2 psid graduations)
- Two brass 1/4" IPS x 45° SAE flare connectors
- Brass adapter fittings for each test cock size: 1/8" x 1/4"
- Bleed-off valve (see Appendix A.1.2: Bleed-off Valve Arrangement)

NOTE: While field-testing the PVB it is important to pay attention to the elevation of the field test kit. Be sure that hoses not being used are also kept at the proper elevation. See diagnostics 9.4.3.1: Location of Field Test Kit.

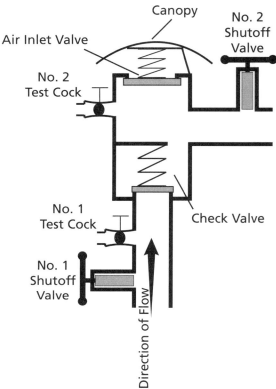

Figure 9.31
Pressure Vacuum Breaker Assembly

Field Test Kit Configurations: The illustration box below will be inserted for each step of the field test procedures where the needle valves are operated.

Two Needle Valve: Open/Close the high side bleed needle valve when indicated in the procedures.

Three Needle Valve: Fully open the bypass needle valve before starting the procedure, then open/close the high side bleed needle valve when indicated in the procedures.

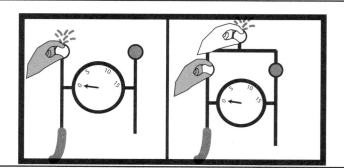

While field-testing the PVB it is important to pay attention to the elevation of the field test kit. Be sure that hoses not being used are also kept at the proper elevation.

9.4.2 Field Test Procedure

Test No. 1   Air Inlet Valve Opening Point

Purpose: To determine the pressure in the body when the air inlet valve opens.

Requirement: The air inlet valve shall open when the pressure in the body is at least 1.0 psi above atmospheric pressure. And, the air inlet valve shall be fully open when the water drains from the body.

> **NOTE**: If the air inlet valve is discharging continuously upon arrival, see diagnostics 9.4.3.2: Leaking Air Inlet Valve.

Steps:
Follow all preliminary steps detailed in 9.1

a. Remove air inlet valve canopy.

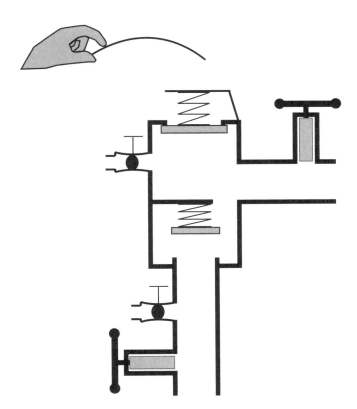

b. Bleed water through test cocks to eliminate foreign material. Open test cock No. 1, then close; open test cock No. 2, then close

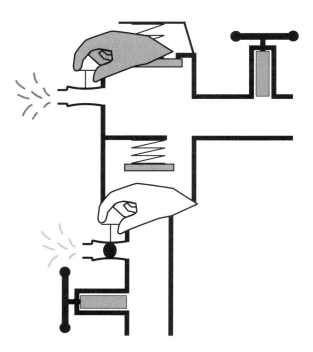

c. Install appropriate fittings to test cocks. Install bleed-off valve arrangement to test cock No. 1.

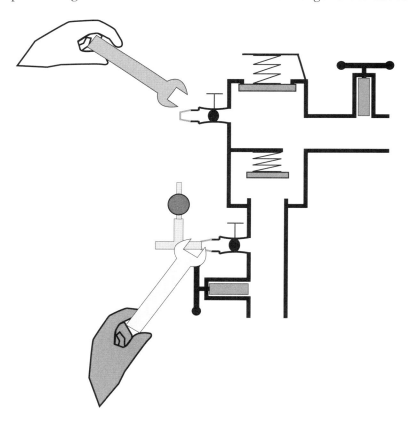

d. Attach the high side hose of the field test kit to test cock No. 2, open test cock No. 2.

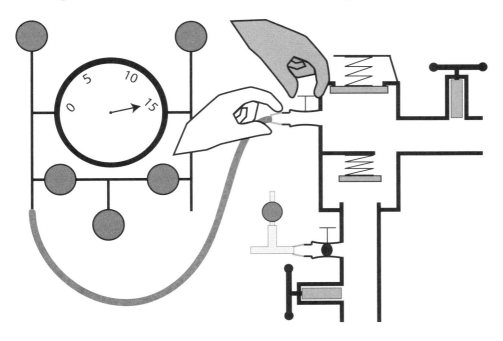

e. Bleed air from the field test kit by opening the high side bleed needle valve. Close the high side bleed needle valve.

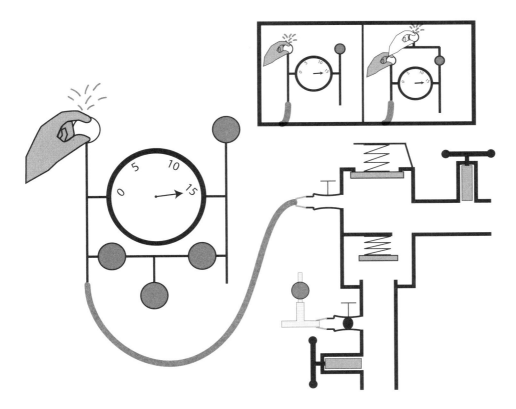

Field Test Procedures

f. Close No. 2 shutoff valve. The field test kit must be maintained at the same level as the air inlet valve being tested. (See diagnostics 9.4.3.1: Location of Field Test Kit)   Close No. 1 shutoff valve. If the reading on the test kit begins to decrease, prepare to record the reading when the air inlet opens. (see diagnostics 9.4.3.3: Leaking No. 2 Shutoff Valve, Air Inlet Valve Test).

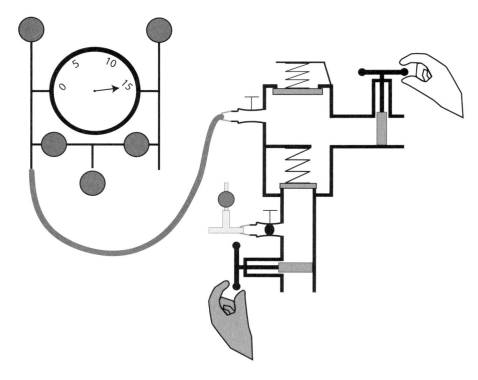

g. Slowly open the high side bleed needle valve no more than one-quarter turn, being careful not to drop the differential pressure reading on the field test kit too fast (If reading is decreased too quickly, reading can not be recorded accurately).

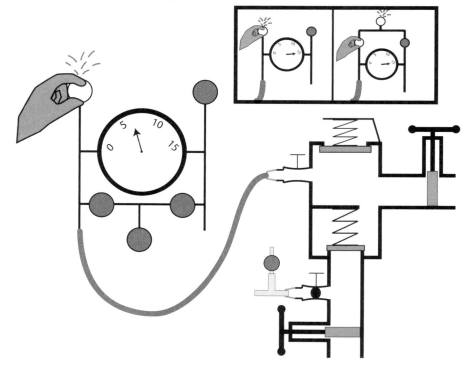

- Record the differential pressure reading on the field test kit when the air inlet valve opens. Proceed to step h.
- If the reading on the field test kit will not drop to the air inlet valve opening point with the high side bleed needle valve open one-quarter turn, see diagnostics 9.4.3.4: Leaking No. 1 Shutoff Valve, Air Inlet Valve Test.
- If the reading drops to 0.0 psid and the air inlet does not open, record that the air inlet valve did not open. Close the high side bleed needle valve and close test cock No. 2. Remove the high side hose from test cock No. 2 and proceed to Test No. 1, step j.

h. Close the high side bleed needle valve and remove the high side hose from test cock No. 2 to drain water from the body. Observe that the air inlet valve has opened to its fully open position. Record whether or not the air inlet opens to the fully open position.

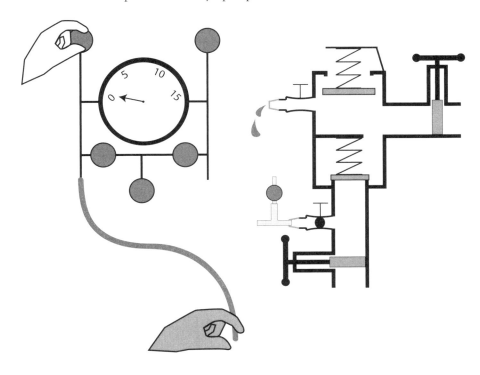

i. Close test cock No. 2.

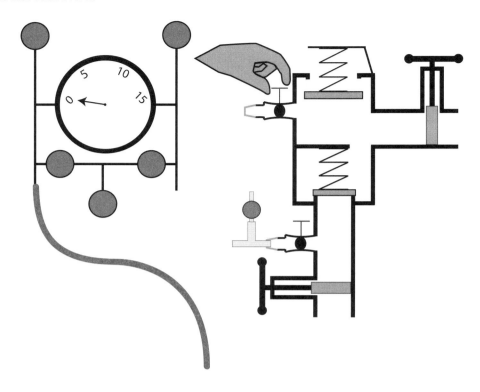

j. Open No. 1 shutoff valve.

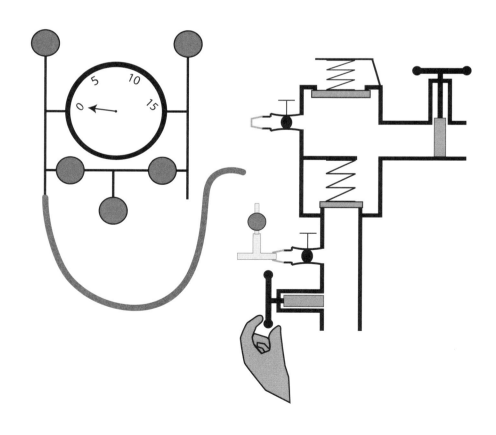

Test No. 2   Tightness of Check Valve

Purpose: To determine the static differential pressure across the check valve.

Requirement: The static differential pressure across the check valve shall be at least 1.0 psid.

Steps:
a. Attach high side hose of the field test kit to the bleed-off valve arrangement on test cock No. 1, slowly open test cock No. 1.

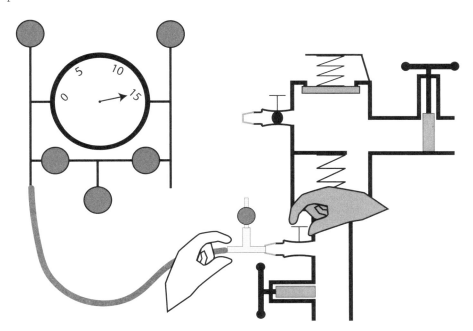

b. Bleed all air from the field test kit by opening high side bleed needle valve. Close high side bleed needle valve.

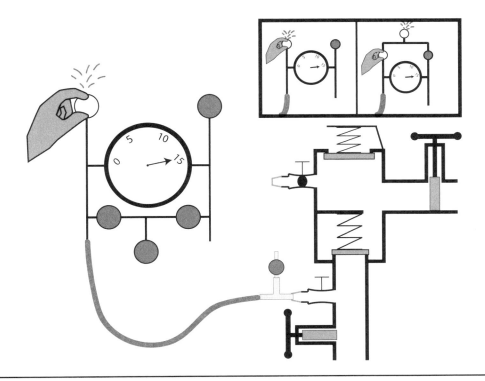

9 Field Test Procedures

297

c. The field test kit must be maintained at the same level as test cock No. 2. (See diagnostics 9.4.3.1: Location of Field Test Kit) Close No. 1 shutoff valve (No. 2 shutoff valve remains closed from Test No. 1). **NOTE**: If the reading on field test kit begins to decrease, continue to step d. This may be caused by a leaking No. 2 shutoff valve, and may not affect the outcome of the field test procedure.

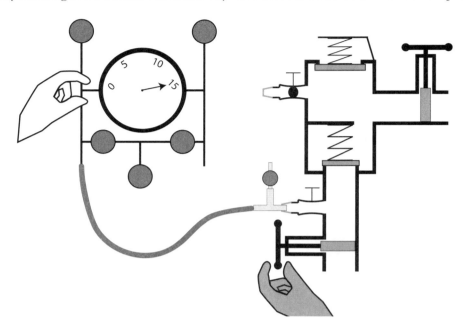

d. Open test cock No. 2. The water in the body will drain out through test cock No. 2. When this flow of water stops or is no more than a drip, and the reading indicated by the field test kit stabilizes, the reading will be the differential pressure across the check valve, and is to be recorded as such. If water continues to flow out of test cock No. 2, see diagnostics 9.4.3.5: Leaking No. 1 Shutoff Valve, Check Valve Test.

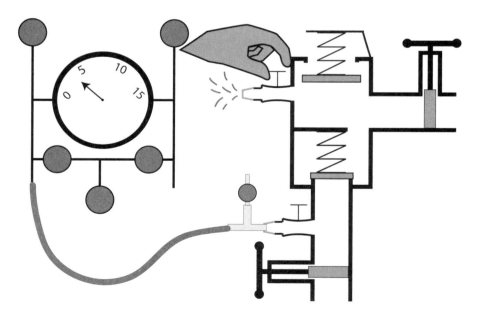

e. Close test cocks No. 1 and No. 2.

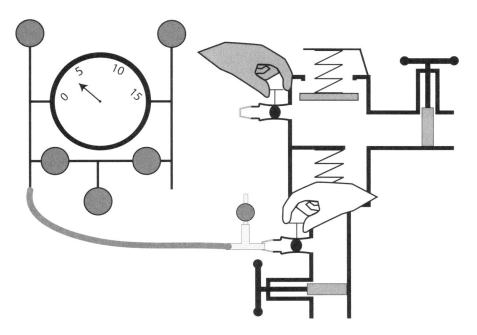

f. Remove all test equipment and fittings.

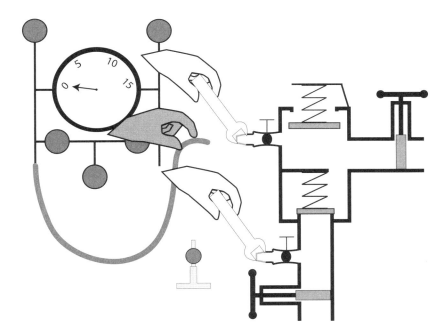

g. Open No. 1 shutoff valve, then slowly open No. 2 shutoff valve.

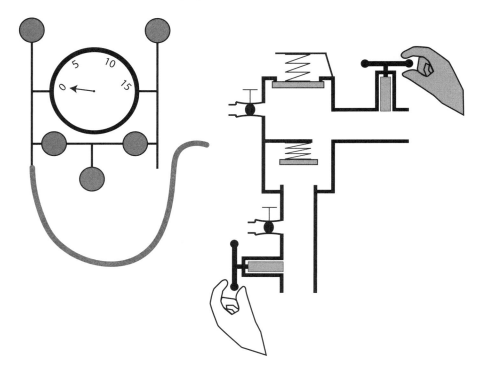

h. Replace air inlet valve canopy.

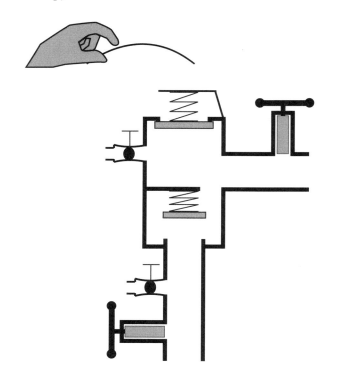

9.4.3 Diagnostics

9.4.3.1 Location of Field Test Kit

During this field test it is important to keep the field test kit and unused hoses at the appropriate elevation. To record the correct differential pressure reading during the air inlet valve test (Test No. 1), the field test kit must be held at the same elevation as the air inlet valve. See figure 9.32.

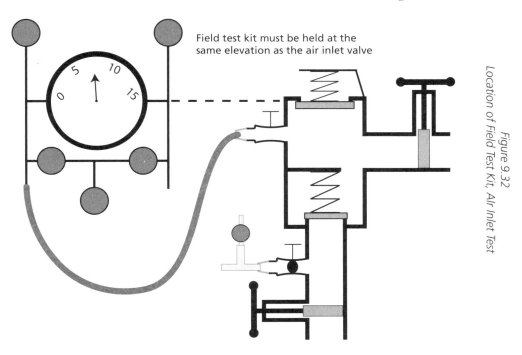

Figure 9.32
Location of Field Test Kit, Air Inlet Test

To record the correct differential pressure reading during the check valve test (Test No. 2), the field test kit must be held at the same elevation as the water level on the downstream side (i.e., Test cock No. 2) of the check valve being tested (See Fig. 9.33).

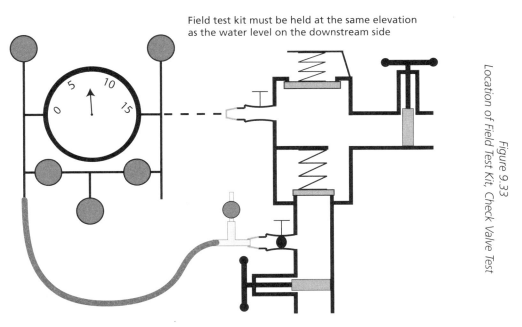

Figure 9.33
Location of Field Test Kit, Check Valve Test

A common misconception is that the entire length of the high side hose must be maintained at the same level of the field test kit. The critical aspect is the elevation of the field test kit relative to the water level on the downstream side of the component being tested. A hose looping up or down in between the field test kit and the test cock will not affect the reading.

It is also important to maintain all unused hoses of the field test kit at the same elevation as the field test kit. If the low side hose of the field test kit contains water, and the hose is maintained above or below the field test kit, the water in the low side hose will create a false reading (see Fig. 9.34).

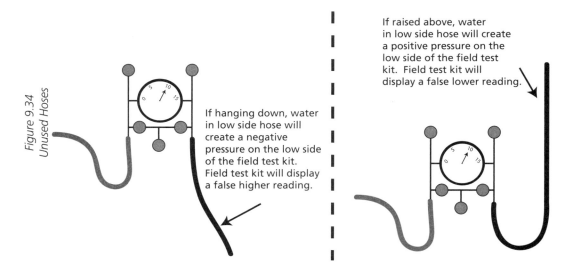

Figure 9.34
Unused Hoses

9.4.3.2 Leaking Air Inlet Valve

If the air inlet valve is discharging upon arrival, the air inlet may be fouled or damaged and can not be field tested. Record air inlet valve as leaking. The check valve may be field tested by performing only steps a through c of Test No. 1, then proceed to Test No. 2.

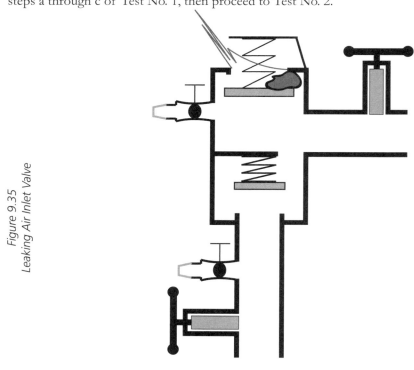

Figure 9.35
Leaking Air Inlet Valve

9.4.3.3 Leaking No. 2 Shutoff Valve: Air Inlet Valve Test

If the reading on the field test kit decreases during Test No. 1, step f, this indicates that the No. 2 shutoff valve is leaking.

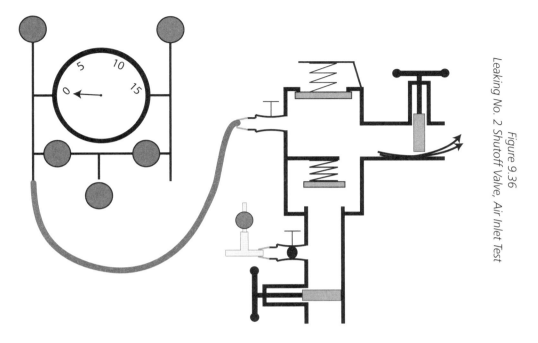

Figure 9.36
Leaking No. 2 Shutoff Valve, Air Inlet Test

- If the reading decreases slowly, prepare to observe the air inlet valve opening point.
    - If the air inlet opens, record the differential pressure reading on the field test kit when the air inlet opens. Remove the high side hose from test cock No. 2 to drain water from the body. Observe that the air inlet valve has opened to its fully open position. Record whether or not the air inlet opens to the fully open position. Proceed to Test No. 1, step i.
    - If the reading decreases to 0.0 psid and the air inlet does not open, record that the air inlet valve did not open. Close test cock No. 2. Remove high side hose from test cock No. 2. Proceed to Test No. 1, step j.
- If the reading on the field test kit decreases so quickly that the reading cannot be accurately recorded, then Test No. 1 step g can not be completed. No. 2 shutoff valve must be repaired or replaced to determine the opening point of the air inlet valve. Record the No. 2 shutoff valve as leaking. Remove the high side hose from test cock No. 2 to drain water from the body. Observe that the air inlet valve has opened to its fully open position. Record whether or not the air inlet opens to the fully open position. Proceed to Test No. 1, step i.

9.4.3.4 Leaking No. 1 Shutoff Valve: Air Inlet Valve Test

If the reading on the field test kit will not drop to the air inlet valve opening point with the high side bleed needle valve open one-quarter turn, it is likely that the No. 1 shutoff valve is leaking. Close the high side bleed needle valve. Verify that the No. 1 shutoff valve is closed. Open high side bleed needle valve one-quarter turn, and should the leak persist, the leak must be diverted so that the air inlet valve can be tested. Close the high side bleed needle valve, then open the No. 1 test cock slowly to divert the leakage from the No. 1 shutoff valve, monitoring the field test kit while this is done. Once the leakage has been diverted through the No. 1 test cock, slowly open the high side bleed needle valve no more than one-quarter turn, being careful not to drop the differential pressure reading on the field test kit too fast.

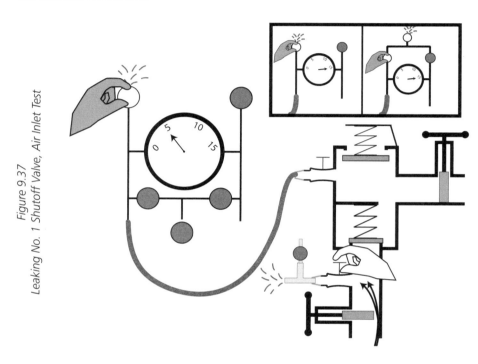

Figure 9.37
Leaking No. 1 Shutoff Valve, Air Inlet Test

- If the reading begins to decrease, record the differential pressure reading on the field test kit when the air inlet valve opens. Close the high side bleed needle valve and remove the high side hose from test cock No. 2 to drain water from the body. Observe that the air inlet valve has opened to its fully open position. Record whether or not the air inlet opens to the fully open position. Close both test cocks and proceed to Test No. 1, step j.
- If the reading drops to 0.0 psid and the air inlet does not open, record that the air inlet valve did not open. Close both test cocks. Close the high side needle valve then proceed to Test No. 1, step j.
- If the reading on the field test kit will not drop to the air inlet valve opening point with the high side bleed needle valve open one-quarter turn the No. 1 shutoff valve leak exceeds the limit of the No. 1 test cock and the field test can not be completed until the No. 1 shutoff valve is repaired or replaced. Proceed to Test No. 2, step e.

9.4.3.5 Leaking No. 1 Shutoff Valve: Check Valve Test

If water continues to flow out of test cock No. 2 during Test No. 2, this indicates that the No. 1 shutoff valve is leaking. Slowly open the bleed-off valve.

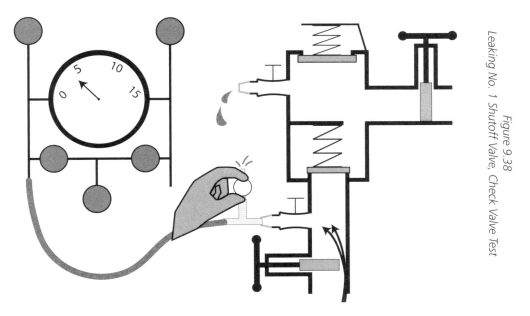

Figure 9.38
Leaking No. 1 Shutoff Valve, Check Valve Test

- If it is possible to adjust the bleed valve so that the flow of water from test cock No. 2 is no more than a drip, record the reading on the field test kit as the differential pressure across the check valve. Proceed to Test 2, step e.
- If it is not possible to adjust the bleed-off valve so there is a drip from the No. 2. test cock, the No. 1 shutoff valve is leaking too much to continue the test, and should be recorded as such. Shutoff valve No. 1 must be repaired or replaced before the field test may be completed. Proceed to Test No. 2, step e.

9.4.3.6 Diagnostics Summary

NOTE: Many problems can be corrected by cleaning the internal components. Carefully observe condition of components.

Problem	May be caused by
Air inlet valve does not open, as field test kit drops to 0.0 psid	1. Air inlet disc stuck to seat 2. Broken or missing air inlet spring 3. "Old style" pressure vacuum breaker (non-loaded air inlet valve) 4. SVB, See 9.5
Air inlet valve does not open, and reading on field test kit will not drop	1. Leaky No. 1 shutoff valve 2. Parallel installation with leaky No. 2 shut-off valve
Air inlet opens below 1.0 psid	1. Dirty or damaged air inlet disc 2. Scale buildup on seat 3. Damaged spring
Check valve reading below 1.0 psid	1. Dirty or damaged check disc 2. Dirty or damaged seat 3. Damaged spring
Water runs continuously from test cock No. 2 (Test No. 2)	1. Leaky No. 1 shutoff valve

NOTE: Lubricants should only be used to assist with the reassembly of components, and must be non-toxic.

9.5 Spill-Resistant Pressure Vacuum Breaker Assembly (SVB)

9.5.1 Equipment Required:

a. Differential pressure field test kit with minimum range 0-15 psid (0.1 or 0.2 psid graduations)
b. One brass 1/4" IPS x 45° SAE flare connector
c. Brass Adapter fittings for each test cock size: 1/8" x 1/4"
d. Bleed-off valve (see Appendix A.1.2: Bleed-off Valve Arrangement)

NOTE: While field testing the SVB it is important to pay attention to the elevation of the field test kit Be sure that hoses not being used are also kept at the proper elevation. See diagnostics 9.5.3.1, Location of Field Test Kit.

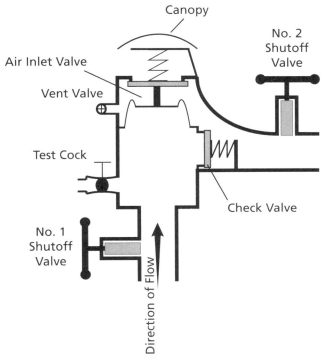

Figure 9.39
Spill-Resistant Pressure Vacuum Breaker Assembly

Field Test Kit Configurations: The illustration box below will be inserted for each step of the field test procedures where the needle valves are operated.

Two Needle Valve: Open/Close the high side bleed needle valve when indicated in the procedures.

Three Needle Valve: Fully open the bypass needle valve before starting the procedure, then open/close the high side bleed needle valve when indicated in the procedures.

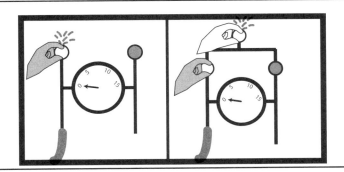

9.5.2 Field Test Procedure

Test No. 1   Tightness of Check Valve

Purpose: To determine the static differential pressure across the check valve.

Requirement: The static differential pressure across the check valve shall be at least 1.0 psid.

Steps: Follow all preliminary steps detailed in 9.1

NOTE: If the air inlet valve is discharging continuous upon arrival, see diagnostics 9.5.3.2: Leaking Air Inlet Valve.

a. Remove air inlet valve canopy.

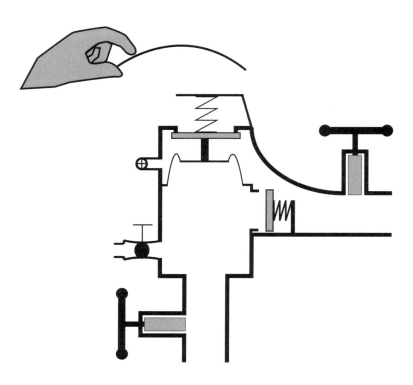

Field Test Procedures

b. Bleed water through the test cock and vent valve to eliminate foreign material.

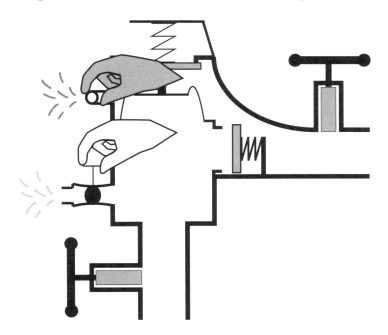

c. Install appropriate fitting to test cock. Attach bleed-off valve arrangement to test cock.

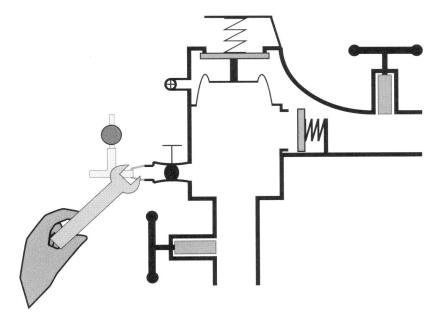

d. Attach the high side hose of the field test kit to the bleed-off valve, open the test cock, and bleed air from the field test kit by opening the high side bleed needle valve. (**NOTE**: To help determine the opening point of the air inlet valve, the area on top of the air inlet valve may be filled with the water discharging from the field test kit.) Close the high side bleed needle valve.

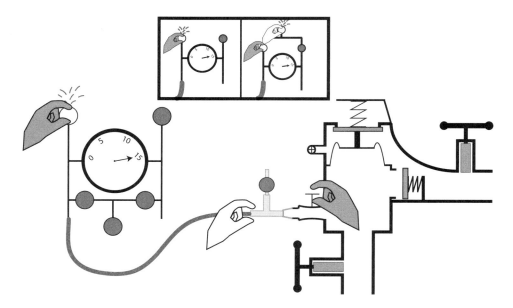

e. Close No. 2 shutoff valve. The field test kit must be maintained at the same level as the vent valve. (See diagnostics 9.5.3.1: Location of Field Test Kit) Close No. 1 shutoff valve.

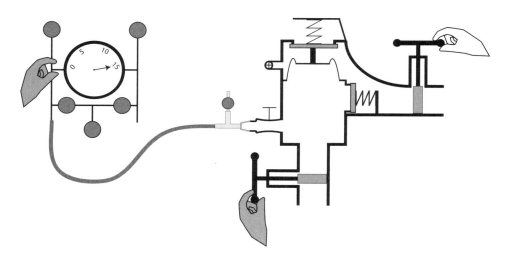

- If reading on field test kit does not decrease, continue to step f.
- If reading on the field test kit begins to decrease and the air inlet valve does not open, the No. 2 shutoff valve is leaking and should be recorded as such. Continue to step f.
- If reading on the field test kit begins to decrease and the air inlet valve opens, see diagnostics 9.5.3.3: Leaking No. 2 Shutoff Valve.
- If air inlet valve was continuously discharging upon arrival, continue to step f.

To help determine the opening point of the air inlet valve, the area on top of the air inlet valve may be filled with the water discharging from the field test kit.

f. Open vent valve to lower outlet pressure to atmospheric.

> **NOTE:** If the vent valve is a machine screw, this may be accomplished by loosening the machine screw, or totally removing the machine screw from the body. Removing the machine screw will lower outlet pressure more quickly. It is recommended to totally remove the machine screw to drain water pressure to atmosphere.

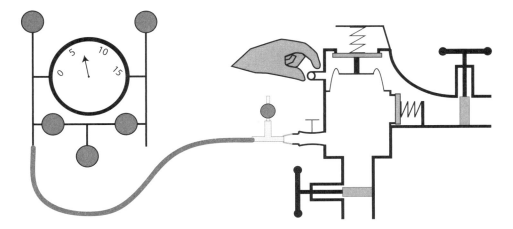

- If air inlet valve does not open, proceed to step g.
- If air inlet valve opens, observe reading and record value for air inlet valve. See diagnostics 9.5.3.4: Check Valve.

g. When the flow of water from the vent valve stops or is a no more than a drip, and the reading indicated by the field test kit stabilizes, the reading will be the differential pressure across the check valve, and recorded as such. If water continues to flow out of vent valve, see diagnostics 9.5.3.5: Leaking No. 1 Shutoff Valve.

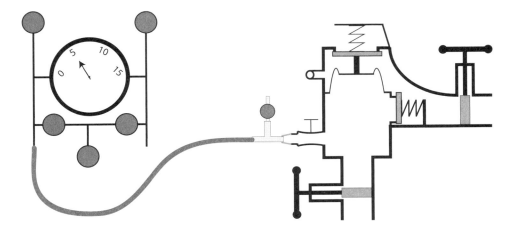

Test No. 2   Air Inlet Valve Opening Point

Purpose: To determine the operating characteristics of the air inlet valve.

Requirement: The air inlet valve shall open when the inlet pressure is at least 1.0 psi above atmospheric pressure and the outlet pressure is atmospheric pressure. The air inlet valve shall be fully open when the inlet pressure is atmospheric.

Steps:
a. The field test kit must be maintained at the same level as the vent valve. (See diagnostics 9.5.3.1: Location of Field Test Kit)

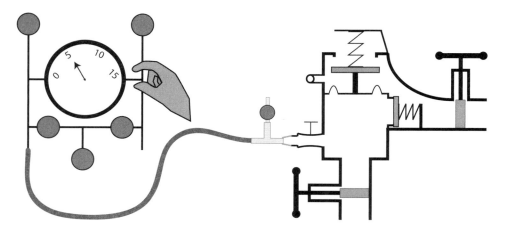

b. Slowly open the high side bleed needle valve no more than one-quarter turn, being careful not to drop the differential pressure reading on the field test kit too fast (If reading decreases too quickly, reading can not be recorded accurately).

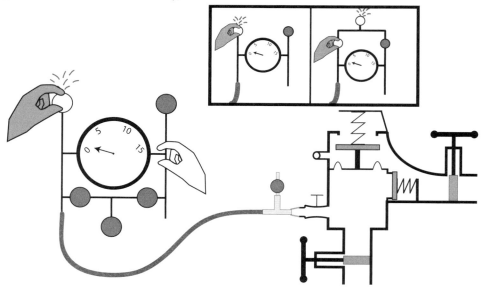

NOTE: An indicator that the air inlet valve has opened is when water starts to flow out of the vent valve.

- Record the differential pressure reading on the field test kit when the air inlet valve opens. Proceed to step c.
- If the reading drops to 0.0 psid and the air inlet does not open, record that the air inlet valve did not open. Proceed to Test No. 2, step d.

c. Close the high side bleed needle valve and remove the high side hose from the bleed-off valve to drain water from the body. Observe that the air inlet valve has opened to its fully open position. Record whether or not the air inlet opens to the fully open position.

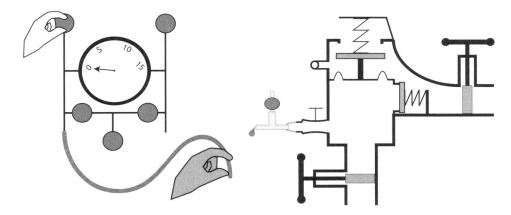

d. Close test cock and vent valve.

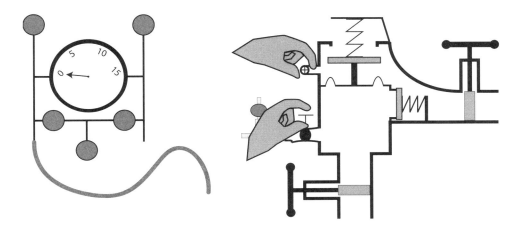

e. Remove all test equipment and fittings.

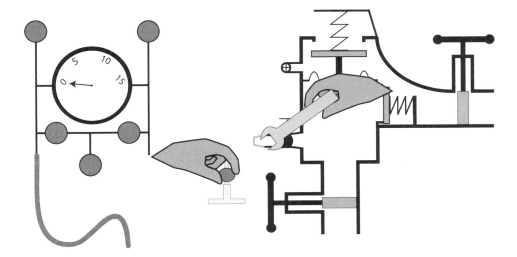

f. Open No. 1 shutoff valve, then slowly open No. 2 shutoff valve.

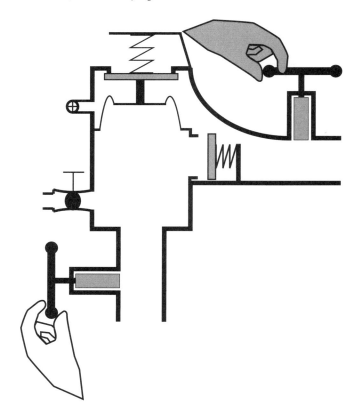

g. Replace air inlet valve canopy.

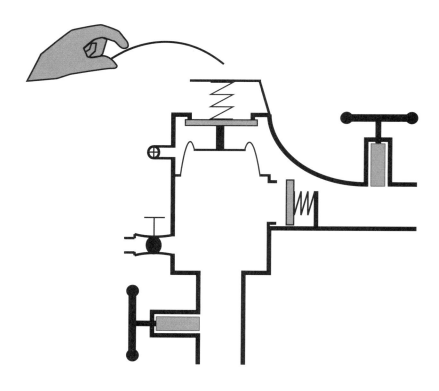

9.5.3 Diagnostics

9.5.3.1 Location of Field Test Kit

During this field test it is important to keep the field test kit and unused hoses at the appropriate elevation. To record the correct differential pressure reading during the check valve test (Test No. 1) and the air inlet valve test (Test No. 2), the field test kit must be held at the same elevation as the vent valve (i.e., water level on the downstream side of the check valve).

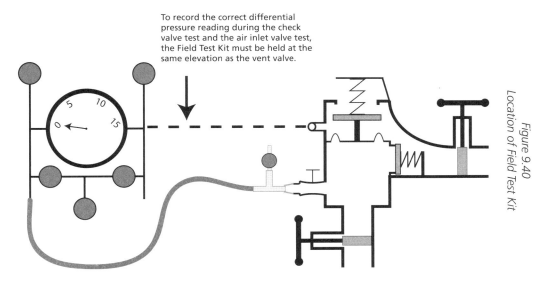

Figure 9.40
Location of Field Test Kit

A common misconception is that the entire length of the high side hose must be maintained at the same level of the field test kit. The critical aspect is the elevation of the field test kit relative to the water level on the downstream side of the component being tested. A high side hose looping up or down in between the field test kit and the test cock will not affect the reading.

It is also important to maintain all unused hoses of the field test kit at the same elevation as the field test kit. If the low side hose of the field test kit contains water, and the hose is maintained above or below the field test kit, the water in the low side hose will create a false reading.

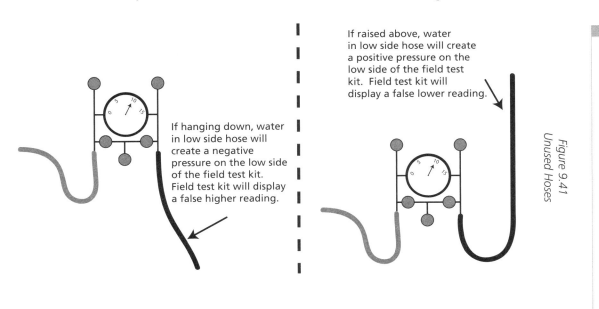

Figure 9.41
Unused Hoses

9.5.3.2 Leaking Air Inlet Valve

If the air inlet valve is discharging upon arrival, the air inlet valve may be fouled or damaged and can not be field tested. Record air inlet valve as leaking. The check valve may be field tested by performing only Test No. 1.

9.5.3.3 Leaking No. 2 Shutoff Valve and Failing/Leaking Check Valve

If the reading on the field test kit decreases and the air inlet opens during Test No. 1, step e, this indicates that the No. 2 shutoff valve is leaking and the check valve is leaking/failing or holding at a value less than the air inlet opening point.

- If the reading decreases slowly, prepare to observe the air inlet valve opening point, and record the differential reading on the field test kit when the air inlet opens. Record the No. 2 shutoff valve as leaking. Open vent valve to lower outlet pressure to atmospheric. When the differential pressure reading indicated by the field test kit settles, the reading will be the differential pressure across the check valve, and recorded as such.

   NOTE: An acceptable check valve reading (1.0 psid or greater) which is less than the air inlet valve opening point may also produce this result.

   Remove the high side hose from test cock to drain water from the body. Observe that the air inlet valve has opened to its fully open position. Record whether or not the air inlet opens to the fully open position. Proceed to Test No. 2, step d.

- If the reading on the field test kit decreases so quickly that the air inlet reading can not be accurately recorded, then Test No. 1 can not be completed. No. 2 shutoff valve must be repaired or replaced to determine the opening point of the air inlet valve. Record the No. 2 shutoff valve as leaking. Open vent valve. When the differential pressure reading indicated by the field test kit settles, the reading will be the differential pressure across the check valve, and recorded as such. Remove the high side hose from test cock to drain water from the body. Observe that the air inlet valve has opened to its fully open position. Record whether or not the air inlet opens to the fully open position. Proceed to Test No. 2, step d.

9.5.3.4 Check Valve

If the reading decreases as the vent valve is opened, prepare to observe the air inlet valve opening point, and record the differential reading on the field test kit when the air inlet opens.

- If the differential pressure reading indicated by the field test kit settles, and flow of water from the vent valve stops or is no more than a drip, the reading will be the differential pressure across the check valve and recorded as such.
- If the differential pressure reading indicated by the field test kit settles, and water continues to flow from the vent valve, slowly open the bleed-off valve until the flow of water from the vent valve is no more than a drip. Record the reading on the field test kit as the differential pressure across the check valve.

Remove the high side hose from test cock to drain water from the body. Observe that the air inlet valve has opened to its fully open position. Record whether or not the air inlet opens to the fully open position. Proceed to Test No. 2, step d.

9.5.3.5 Leaking No. 1 Shutoff Valve

If water continues to flow out of the vent valve during Test No. 1, this indicates that the No. 1 shutoff valve is leaking, and should recorded as such. Slowly open the bleed-off valve until the flow of water from the vent valve is no more than a drip (See Fig. 9.42). Record the reading on the field test kit as the differential pressure across the check valve. With the bleed-off valve open so that the flow of water from the vent valve is no more than a drip, proceed to Test No. 2, step b.

If it is not possible to adjust the bleed-off valve so there is a drip from the vent valve, the No. 1 shutoff valve is leaking too much to continue the field test, and should be recorded as such. Shutoff valve No. 1 must be repaired before the field test may be completed. Proceed to Test No. 2, step d.

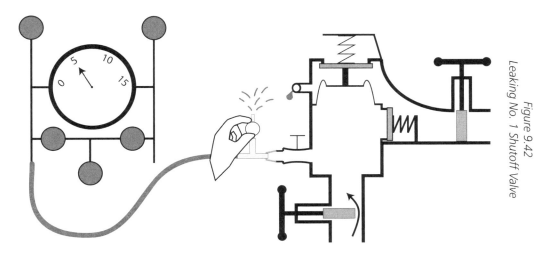

Figure 9.42
Leaking No. 1 Shutoff Valve

9.5.3.6 Diagnostics Summary

NOTE: Many problems can be corrected by cleaning the internal components. Carefully observe condition of components.

Problem	May be caused by
Air inlet valve does not open, as field test kit drops to 0.0 psid	1. Air inlet disc stuck to seat 2. Broken or missing air inlet spring 3. Parallel installation with leaky No.2 shutoff valve 4. PVB not SVB
Air inlet valve does not open, and reading on field test kit will not drop	1. Leaky No. 1 shutoff valve
Air inlet opens below 1.0 psid	1. Dirty or damaged air inlet disc
Check valve reading below 1.0 psid	1. Dirty or damaged check disc 2. Damaged seat 3. Leaking diaphragm/seal
Water runs continuously from vent valve (Test No. 1)	1. Leaky No. 1 shutoff valve

NOTE: Lubricants should only be used to assist with the reassembly of components, and must be non-toxic.

9.6 Reduced Pressure Principle-Detector Backflow Prevention Assembly (RPDA)

9.6.1 Equipment Required:

a. Differential pressure field test kit with minimum range 0-15 psid (0.1 or 0.2 psid graduations)
b. Four brass 1/4" IPS x 45° SAE flare connectors
c. Brass adapter fittings for each test cock size: 1/8" x 1/4", 1/4" x 1/2", 1/4" x 3/4"

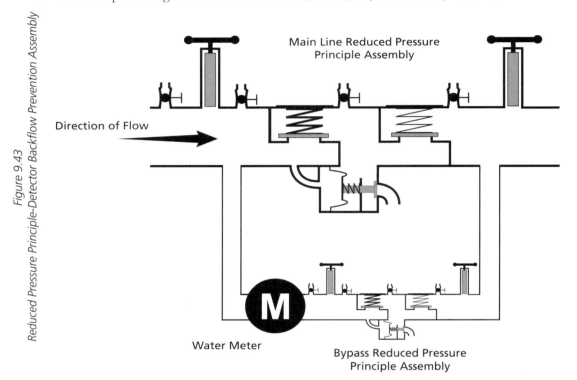

Figure 9.43 Reduced Pressure Principle-Detector Backflow Prevention Assembly

9.6.2 Field Test Procedure

Test No. 1   Test of Main-line Assembly

Purpose: To test the operation of the main-line reduced pressure principle backflow prevention assembly.

Requirement: The main-line reduced pressure principle backflow prevention assembly shall comply with field test requirements of 9.2.

> **NOTE**: Follow all preliminary steps detailed in 9.1. The tester shall request that the owner or occupant notify the authority having jurisdiction, the fire department, if required, and the alarm receiving facility before shutting down a fire sprinkler system or its water supply. The notification prior to testing shall include the purpose for the shutdown of the system, the component(s) involved, and the estimated time required.

> **NOTE**: The water meter reading should be recorded at this time and at the conclusion of the field test. This will document the water usage during the field test and aid in the determination of unauthorized water usage between periodic field tests.

The tester should request that the owner or occupant notify the authority having jurisdiction, the fire department, if required, and the alarm receiving facility before shutting down a fire sprinkler system or its water supply.

Steps:

a. Close No. 2 shutoff valve of bypass assembly

b. Perform field test procedure per 9.2 for main-line reduced pressure principle backflow prevention assembly.

c. Maintain No. 2 shutoff valve of main-line assembly in closed position.

Test No. 2   Test of Bypass Assembly

Purpose: To test the operation of the bypass reduced pressure principle backflow prevention assembly.

Requirement: The bypass reduced pressure principle backflow prevention assembly shall comply with field test requirements of 9.2.

Steps:

a. Perform field test procedure per 9.2 for bypass reduced pressure principle backflow prevention assembly.

> **NOTE**: Procedures for verifying the operation of the bypass water meter can be found in Appendix A.5.1.

b. Open all shutoff valves of RPDA.

9.7 Double Check-Detector Backflow Prevention Assembly (DCDA)

9.7.1 Equipment required:

a. Differential pressure field test kit with minimum range 0-15 psid (0.1 or 0.2 psid graduations)
b. One brass 1/4" IPS x 45° SAE flare connector
c. Brass adapter fittings for each test cock size: 1/8" x 1/4", 1/4" x 1/2", 1/4" x 3/4"
d. Street ell, pipe nipple or tube (for length see 9.3.3.1)
e. Bleed-off valve (see Appendix A.1.2: Bleed-Off Valve Arrangement)

NOTE: While field testing the double check detector assembly it is important to pay attention to the elevation of the field test kit Be sure that hoses not being used are also kept at the proper elevation . See diagnostics 9.3.3.1: Location of Field Test Kit.

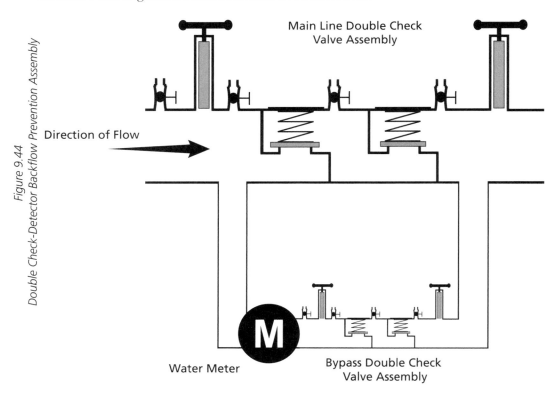

Figure 9.44 Double Check-Detector Backflow Prevention Assembly

9.7.2 Field Test Procedure

Test No. 1   Test of Bypass Assembly

Purpose: To test the operation of the bypass double check valve backflow prevention assembly.

Requirement: The bypass double check valve backflow prevention assembly shall comply with field test requirements of 9.3.

NOTE: Follow all preliminary steps detailed in 9.1 The tester should request that the owner or occupant notify the authority having jurisdiction, the fire department, if required, and the alarm receiving facility before shutting down a fire sprinkler system or its water supply. The notification prior to testing shall include the purpose for the shutdown of the system, the component(s) involved, and the estimated time required.

NOTE: The water meter reading should be recorded at this time and at the conclusion of the field test. This will cut the water usage during the field test and aid in the determination of unauthorized water usage between periodic field tests.

Steps:

a. Perform field test procedure per 9.3 for bypass double check valve backflow prevention assembly.

b. Maintain No. 2 shutoff valve of the bypass assembly in closed position.

Test No. 2   Test of Main-line Assembly

Purpose: To test the operation of the main-line double check valve backflow prevention assembly.

Requirement: The main-line double check valve backflow prevention assembly shall comply with field test requirements of 9.3.

Steps:

a. Perform field test procedure per 9.3 for main-line double check valve backflow prevention assembly.

    NOTE: Procedures for verifying the operation of the bypass water meter can be found in Appendix A.5.2.

b. Open all shutoff valves of DCDA.

9.8 Reduced Pressure Principle-Detector Backflow Prevention Assembly-Type II (RPDA-II)

9.8.1 Equipment Required:

a. Differential pressure field test kit with minimum range 0-15 psid (0.1 or 0.2 psid graduations)
b. Three brass 1/4" IPS x 45° SAE flare connectors
c. Brass adapter fittings for each test cock size: 1/8" x 1/4", 1/4" x 1/2", 1/4" x 3/4"
d. Street ell, pipe nipple or tube (for length see 9.3.3.1)
e. Bleed-off valve (see Appendix A.1.2: Bleed-Off Valve Arrangement)

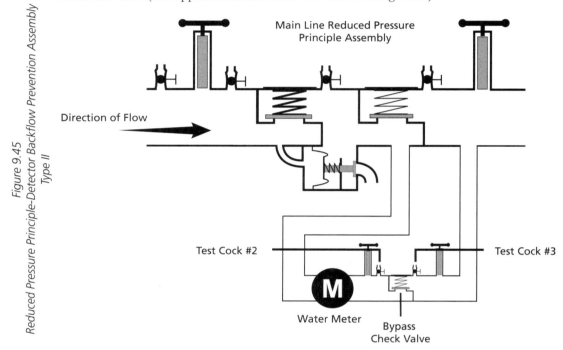

Figure 9.45
Reduced Pressure Principle-Detector Backflow Prevention Assembly Type II

9.8.2 Field Test Procedure

Test No. 1   Test of Main-line Assembly

Purpose: To test the operation of the main-line reduced pressure principle backflow prevention assembly.

Requirement: The main-line reduced pressure principle backflow prevention assembly shall comply with field test requirements of 9.2.

> **NOTE**: Follow all preliminary steps detailed in 9.1. The tester should request that the owner or occupant notify the authority having jurisdiction, the fire department, if required, and the alarm receiving facility before shutting down a fire sprinkler system or its water supply. The notification prior to testing shall include the purpose for the shutdown of the system, the component(s) involved, and the estimated time required.

> **NOTE**: The water meter reading should be recorded at this time and at the conclusion of the field test. This will cut the water usage during the field test and aid in the determination of unauthorized water usage between periodic field tests.

Steps:

a. Close No. 2 shutoff valve of bypass single check assembly.

b. Perform field test procedure per 9.2 for main-line reduced pressure principle backflow prevention assembly.

c. Maintain No. 2 shutoff valve of main-line assembly in closed position.

Test No. 2   Test of Bypass Single Check Assembly

Purpose: To test the operation of the bypass single check assembly.

Requirement: The bypass single check assembly shall comply with field test requirements of 9.3.2 Test No. 1.

Steps:

a. Perform field test procedure per 9.3.2 Test No. 1 for bypass single check assembly.

>   **NOTE**: Procedures for verifying the operation of the bypass water meter can be found in Appendix A.5.3.

b. Open all shutoff valves of the RPDA-II assembly.

9.9 Double Check-Detector Backflow Prevention Assembly-Type II (DCDA-II)

9.9.1 Equipment required:

a. Differential pressure field test kit with minimum range 0-15 psid (0.1 or 0.2 psid graduations)
b. One brass 1/4" IPS x 45° SAE flare connector
c. Brass adapter fittings for each test cock size: 1/8" x 1/4", 1/4" x 1/2", 1/4" x 3/4"
d. Street ell, pipe nipple or tube (for length see 9.3.3)
e. Bleed-off valve (see Appendix A.1.2: Bleed-Off Valve Arrangement)

NOTE: While field testing the DCDA-II it is important to pay attention to the elevation of the field test kit. Be sure that hoses not being used are also keep at the proper elevation. See diagnostics 9.3.3.1: Location of Field Test Kit.

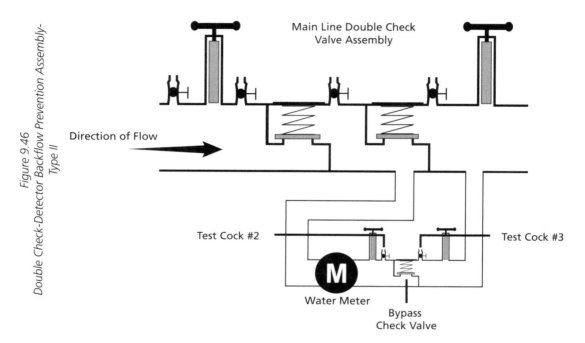

Figure 9.46 Double Check-Detector Backflow Prevention Assembly-Type II

9.9.2 Field Test Procedure

Test No. 1   Test of Bypass Single Check Assembly

Purpose: To test the operation of the bypass single check assembly.

Requirement: The bypass single check assembly shall comply with field test requirements of 9.3.2 Test No. 1.

NOTE: Follow all preliminary steps detailed in 9.1 The tester shall request that the owner or occupant notify the authority having jurisdiction, the fire department, if required, and the alarm receiving facility before shutting down a fire sprinkler system or its water supply. The notification prior to testing shall include the purpose for the shutdown of the system, the component(s) involved, and the estimated time required.

NOTE: The water meter reading should be recorded at this time and at the conclusion of the field test. This will document the water usage during the field test and aid in the determination of unauthorized water usage between periodic field tests.

Steps:

a. Perform field test procedure per 9.3.2 Test No. 1 for bypass single check assembly.

b. Maintain No. 2 shutoff valve of the bypass single check assembly closed.

Test No. 2   Test of Main-line Assembly

Purpose: To test the operation of the main-line double check valve backflow prevention assembly.

Requirement: The main-line double check valve backflow prevention assembly shall comply with field test requirements of 9.3.

Steps:
a. Perform field test procedure per 9.3 for main-line double check valve backflow prevention assembly.

> **NOTE**: Procedures for verifying the operation of the bypass water meter can be found in Appendix A.5.4.

b. Open all shutoff valves of the DCDA-II assembly.

Field Test Procedures

Chapter 10
Standards

Standards

Inside this Chapter

Backflow Prevention Assemblies 330
    General Design and Material Requirements and Laboratory Testing 330
    Reduced Pressure Principle Backflow Prevention Assemblies(RP) 349
    Double Check Valve Backflow Prevention Assemblies(DC) 361
    Pressure Vacuum Breaker Backsiphonage
        Prevention Assemblies (PVB) 368
    Atmospheric Vacuum Breaker Backsiphonage
        Prevention Assemblies(AVB) 376
    Double Check Detector Backflow Prevention Assemblies(DCDA) 381
    Reduced Pressure Principle Detector Backflow
        Prevention Assemblies(RPDA) 385
    Spill-Resistant Pressure Vacuum Breaker
        Backsiphonage Preventions Assemblies (SVB) 390
    Double Check-Detector Backflow Prevention
        Assemblies-Type II (DCDA-II) 397
    Reduced Pressure Principle-Detector Backflow Prevention
        Assemblies-Type II (RDPA-II) 402

Differential Pressure Gage Field Test Kits Used for
    Field Testing of Backflow Prevention Assemblies 407
    Scope 407
    Definitions 407
    General Design Requirements 408
    Design Requirements 409
    Material Requirements 410
    Evaluation of Design and Performance 410

10.1 Backflow Prevention Assemblies

10.1.1 General Design and Material Requirements and Laboratory Testing

10.1.1.1 General Requirements

10.1.1.1.1 Flow Characteristics and Pressure Loss Requirements

The flow characteristics and pressure loss requirements of these assemblies are of prime consideration in insuring their functional operation. In all cases the flow channels shall be streamlined to minimize pressure loss. The lowest possible pressure loss through the backflow prevention assembly is necessary to deal with low distribution system pressure and high in-plant pressure losses.

10.1.1.1.2 Rated Flow and Maximum Allowable Pressure Loss

The American Water Works Association (AWWA) has adopted values for the rated flow which must be met by displacement or compound meters in sizes $5/8 \times 3/4$ inch to 10 inch, inclusive. Those flow rates are included in Table 10-1 and Table 10-7 for the above sizes together with those for 1/4, 3/8, 1/2, 1 1/4, 2 1/2, 12, 14 and 16 inch, which have been extrapolated or interpolated. For each size of backflow prevention assembly at any flow rate up to and including the rated flow, the maximum pressure loss shall not exceed the values given in Table 10-1 for Reduced Pressure Principle Assemblies and Double Check Valve Assemblies and Table 10-7 for Pressure Vacuum Breaker Assemblies, Spill-Resistant Pressure Vacuum Breaker Assemblies and Atmospheric Vacuum Breaker Assemblies.

As a guideline for assemblies intended for fire sprinkler service, the pressure loss at the following rates of flow are now being asked for by some agencies.

4 inch (100 mm)	750 gpm (47.31 L/s)
6 inch (150 mm)	1500 gpm (94.63 L/s)
8 inch (200 mm)	3000 gpm (189.25 L/s)
10 inch (250 mm)	4900 gpm (309.11 L/s)

In these cases there has been no maximum pressure loss restrictions placed on the assemblies at these rates of flow.

Table 10-1
Rated Flow and Maximum Allowable Pressure Loss
for Various Sizes of Backflow Prevention Assemblies

Size of the Assembly (1)		Rated Flow (2)		Maximum Allowable Pressure Loss			
				Reduced Pressure Principle Assembly (3)		Double Check Valve Assembly (4)	
(inches)	(mm)	(gpm)	(L/s)	(psi)	(KPa)	(psi)	(KPa)
1/4	(6)	1*	(0.06)	24	(165.4)	10	(68.9)
3/8	(10)	3*	(0.19)	22	(151.6)	10	(68.9)
1/2	(15)	7.5*	(0.47)	22	(151.6)	10	(68.9)
5/8×3/4 ***	(16x20)	20	(1.26)	20	(137.8)	10	(68.9)
3/4	(20)	30	(1.89)	20	(137.8)	10	(68.9)
1	(25)	50	(3.15)	18	(124.1)	10	(68.9)
1 1/4	(32)	75**	(4.73)	18	(124.1)	10	(68.9)
1 1/2	(40)	100	(6.31)	16	(110.3)	10	(68.9)
2	(50)	160	(10.10)	16	(110.3)	10	(68.9)
2 1/2	(65)	225**	(14.20)	16	(110.3)	10	(68.9)
3	(80)	320	(20.19)	15	(103.4)	10	(68.9)
4	(100)	500	(31.55)	14	(96.5)	10	(68.9)
6	(150)	1000	(63.10)	14	(96.5)	10	(68.9)
8	(200)	1600	(100.96)	14	(96.5)	10	(68.9)
10	(250)	2300	(145.13)	14	(96.5)	10	(68.9)
12	(300)	3000	(189.30)	13	(89.6)	10	(68.9)
14	(350)	3700	(233.47)	13	(89.6)	10	(68.9)
16	(400)	4400	(277.64)	13	(89.6)	10	(68.9)

* Extrapolated
**Interpolated
*** 3/4 inch inlet and outlet connections

NOTE: The pressure losses shown in Column (3) and Column (4) represent the maximum permissible pressure loss at any flow rate up to and including the rated flow shown in Column (2).

10.1.1.1.3 Standard Sizes

The following standard sizes have been adopted for backflow prevention assemblies: 1/4 inch (6 mm), 3/8 inch (10 mm), 1/2 inch (15 mm), 5/8×3/4 inch (16×20 mm), 3/4 inch (20 mm), 1 inch (25 mm), 1 1/4 inch (32 mm), 1 1/2 inch (40 mm), 2 inch, (50 mm) 2 1/2 inch (65 mm), 3 inch (80 mm), 4 inch (100 mm), 6 inch (150 mm), 8 inch (200 mm), 10 inch (250 mm), 12 inch (300 mm), 14 inch (350 mm) and 16 inch (400 mm). All assemblies designed and constructed in sizes other than those aforementioned shall be given separate consideration.

The inlet and outlet of the assembly shall be threaded in accordance with ANSI/ASME B1.20.1 for taper pipe connections; or ANSI B16.24 for bronze flanges; or ANSI B16.1 for iron flanges; or ANSI/AWWA C606 for grooved and shoulder joints or flared connections per SAE J513. Metric equivalents (International System of Units [SI]) have been provided wherever needed per ASTM Designation E380.

10.1.1.1.4 Markings

Size, model and serial number markings on backflow prevention assemblies for sizes 5/8×3/4 inch (16×20 mm) through 16 inch (400 mm) shall be with letters or numbers at least 1/4 inch (6 mm) in height [Note: For fractions (i.e., 1/4, 1/2, 5/8) the total height of the fraction shall be permitted to be 1/4 inch.] For backflow prevention assemblies sized 1/4, 3/8 and 1/2 inch (6, 10, 15 mm), the size, model and serial number markings shall be with letters or numbers a minimum of 1/8 inch (3 mm) in height. All lettering cast on a body shall be with letters or numbers a minimum of 1/4 inch (6 mm) in height. All markings shall be easily read and shall be either cast or stamped on the body; or stamped, engraved or etched on a durable nameplate permanently affixed to the body of the assembly and shall be located either; a) on both sides (i.e., quadrants adjacent to upper quadrant) of the body or b) on a top surface of the body (i.e., upper quadrant). The nameplate shall be either brass or stainless steel and permanently affixed with stainless steel escutcheon pins. In affixing a nameplate with escutcheon pins or stamping data in the metal of the assembly, caution shall be exercised so as not to produce an area of stress concentration.

The markings on a DCDA, RPDA, DCDA-II and RPDA-II, which identify the backflow prevention assembly or single check valve assembly in the bypass, shall be with letters or numbers at least 1/8 inch (3 mm) in height.

The markings on a pressure vacuum breaker (PVB and SVB) shall be as noted above except that all markings shall be located either, a) on a top surface of the body (i.e., upper quadrant) or b) on the side of the body. The removable bonnet or cover shall not be used for identification purposes.

Markings shall be permanent and not easily defaced. The markings shall include those shown in Table 10-2.

Name, model and size markings on atmospheric vacuum breakers shall be with letters or numbers a minimum of 1/8 inch (3 mm) in height. All lettering cast on a body shall be a minimum of 1/4 inch (6 mm) in height. The markings on an atmospheric vacuum breaker shall be as noted above except that all markings shall be located either, a) on a top surface of the assembly or b) on the side of the assembly.

All markings shall be in English units (i.e., PSI, °F, inches). Metric equivalents (i.e., KPa, °C, mm) shall be permitted in addition to the English units at the manufacturer's discretion.

Table 10-2
Markings for Backflow Prevention Assemblies

Make: Name or Trademark
Type of assembly (i.e., reduced pressure principle backflow prevention assembly or double check valve backflow prevention assembly, double check detector assembly-type II, pressure vacuum breaker assembly, spill-resistant pressure vacuum breaker assembly, double check detector assembly, double check detector assembly-type II, reduced pressure principle detector assembly, reduced pressure principle detector assembly-type II or acceptable abbreviation RP, DC, PVB, SVB, DCDA, DCDA-II, RPDA, RPDA-II respectively.)

Size
Model (full model designation)
Direction of flow (shown by an arrow)
Unit serial number
Maximum rated working water pressure (MWWP)
Maximum rated working water temperature (MWWT)

For the DCDA and RPDA, the following additional information is required on the mainline assembly:

Make, Model and Size of backflow prevention assembly in bypass.

For the DCDA-II and RPDA-II, the following additional information is required on the mainline assembly:

Make, Model and Size of single check valve assembly in bypass.

10.1.1.1.5 Hydrostatic Test for Structural Integrity

a. All assemblies shall be pressure tested according to their designed operating pressure for use on cold water service (maximum 110°F [44.3°C]). Normal testing shall be accomplished at a minimum of 150 psi (1034 KPa) line pressure. The hydrostatic test pressure shall be twice the maximum rated working water pressure of the assembly.
b. The entire assembly shall be subjected to the hydrostatic test both in the normal direction of flow and in the reverse direction of flow of all closed barriers with the opposite side of the barrier open to the atmosphere. There shall be no leakage across any barrier. And, the hydrostatic test pressure shall be maintained for 10 minutes for each test.
c. No damage or permanent deformation of any parts of the assembly or impairment of operation shall occur resulting from the full hydrostatic test pressure.

10.1.1.1.6 General Statement of Policy Regarding Assembled Assemblies

All assemblies, which consist of independent units assembled for the purpose of preventing backflow, shall comply with the material, operational and other specifications as required for backflow prevention assemblies. In order to ensure proper installations, all backflow prevention assemblies shall be delivered for installation completely assembled by the original manufacturer with all components as approved. Resilient seated shutoff valves and test cocks are considered integral parts of the assembly.

All replaceable parts of assemblies of the same size and model shall be interchangeable with the original parts.

10.1.1.1.7 Modifications

Modifications to backflow prevention assemblies will invalidate the Foundation for Cross-Connection Control and Hydraulic Research's approval. Assemblies must be installed and maintained in the configuration(s) and orientation(s) in which they were evaluated and approved. Shutoff valves may be replaced only with shutoff valves, which are approved for each size and model of assembly. Detector assemblies (i.e., DCDA, DCDA-II, RPDA, RPDA-II) are permitted to have the water meters replaced only with the acceptable water meters approved with each size and model of assembly. The bypass assemblies on detector assemblies may only be replaced with the specific assemblies approved with each size and model of assembly.

10.1.1.2 Design Requirements

10.1.1.2.1 Policy Regarding Design

In the design of any backflow prevention assembly, prime consideration shall be given to the construction of a trouble-free unit. The water way shall be as free as possible from obstructions and pockets, which could interfere with the free flow of water. All moving parts shall be designed to operate up to the rated flow without chatter or vibration. The moving parts shall have adequate clearance to prevent binding and galling. The moving parts shall also have adequate clearance to prevent the assembly from becoming inoperative by being thrown out of balance, by being distorted, by having one part interfere with another or by becoming encrusted with lime, rust or scale deposits.

All parts shall conform, with allowable tolerances, to the manufacturer's specifications and shall be free from defects. All flanged joints shall be faced true and machined at right angles to their respective axes; while threaded joints must be concentric and accurately cut. All joints shall be watertight where subject to water pressure. All ferrous parts receiving a bronze or other mounting shall be finished to fit. Such handwork as required in finishing shall produce a neat, workmanlike, well-fitting and smoothly operating product. All replaceable parts of assemblies of the same size and model shall be interchangeable with the original parts.

Assemblies intended for cold water service shall be designed and tested so as to function satisfactorily up to a minimum temperature of 140°F (60°C) at the maximum rated working water pressure.

Assemblies intended for elevated temperature service above 140°F (60°C) shall be designed and tested so as to function satisfactorily over the rated temperature and pressure ranges. Any changes of design, materials or coating, require full disclosure by the manufacturer to the approving laboratory for approval by the approving laboratory prior to implementation.

10.1.1.2.2 Accessibility of Internal Parts

A backflow prevention assembly shall be provided with one or more openings through which the internal wetted parts may be removed, repaired or inspected without having to remove the body of the assembly from the line, using the assembly's shutoff valves to isolate the assembly. The body of the backflow preventer is that portion of the assembly that rigidly connects the No. 1 and No. 2 shutoff valves. This portion of the backflow prevention assembly is not necessarily a pressure-containing component.

10.1.1.2.3 Interdependence of Components

In the double check valve assembly there shall be no mechanical linkage between the two check valves. Each check valve shall be free to operate independently through its entire movement. The movement of either the first or second check valve through its full limit of travel shall not affect the operation of the other check valve.

In the reduced pressure principle backflow prevention assembly each of the check valves shall likewise be free of any coupling mechanical linkage and shall be free to operate independently through its entire movement. Further, the relief valve shall be mechanically independent of both check valves yet hydraulically dependent upon the pressure differential across the first check valve. The movement of either the first check valve, second check valve or differential pressure relief valve through their respective full limits of travel shall not affect the operation of the other components.

In the pressure vacuum breaker (PVB and SVB) the check valve shall be free to operate independently from the air inlet valve. The movement of either the check valve or the air inlet valve through their full limits of travel shall not affect the operation of the other.

The operation of a backflow prevention assembly's shutoff valves through their full limit of travel shall not affect the operation of the internal components (i.e., check valve(s) or air inlet valve).

10.1.1.2.4 Differential Pressure Relief Valve

The differential pressure relief valve of a reduced pressure principle backflow prevention assembly shall be located so that its valve seat(s) and port(s) to the atmosphere are below the lowest point of the first check valve so as to preclude backsiphonage. The water passageway from the differential pressure relief valve seat(s) to the discharge port(s) to the atmosphere shall not cause any backsiphonage air to pass through a pocket of water that is in or been discharged from the differential pressure relief valve.

For designs with a separate differential pressure relief valve body, the relief valve shall not be attached to the body of the assembly with standard tapered pipe thread, pipe flange or grooved joint sizes as detailed in 10.1.1.1.3. To reduce the possibility of unauthorized field modifications, the shape of the connection or boss on the body shall be such that no piping, hose connection, flange or plate can easily be attached to it.

The differential pressure relief valve discharge port(s) to the atmosphere shall not be of a size that can be threaded for NPT threads or adapted to tubing internally or externally. Further, the shape of the port or its boss shall be such that no piping, hose connection, flange or plate can easily be attached to it.

If a drain funnel is available as an accessory to an assembly, the drain funnel shall provide adequate backflow protection between the discharge port(s) of the relief valve and the drain funnel. (See 10.1.2.2.3.9) The outlet of the drain funnel shall be provided with a standard plumbing connection. The drain funnel shall be attached to the assembly in such a way to maintain the adequate backflow protection between the discharge port(s) of the relief valve and the drain funnel.

10.1.1.2.5 Design of Waterway

a. Area of Waterway: Backflow prevention assemblies shall be so designed that the minimum waterway area normal to the direction of flow shall not adversely affect the overall pressure loss.
b. Obstructions: The waterways of the assembly shall be designed so as to minimize cavitation and eliminate all cavities that could entrap foreign materials.
c. Turbulence: The turbulence within or created by the assembly shall not be excessive for any flow rate up to the rated flow conditions.

In the pressure vacuum breaker (PVB and SVB) the check valve shall be free to operate independently from the air inlet valve.

10.1.1.2.6 Clearance

a. Between guide stems and guides, valve stems and guides, hinge pins and bushings and other similar parts the clearances shall be adequate to prevent sticking or binding.
b. Binding or clogging of parts might occur between the body of a valve or hinge arm, clapper, counterweight, poppet valve or other similar free-moving part. Wherever such binding or clogging of parts might occur which would prevent the assembly from being free to operate normally, there shall be provided minimum clearances as follows: for ferrous-body assemblies of all sizes, 1/2 inch (12.7 mm) clearance; for bronze-body assemblies in sizes up to and including 1 inch (25 mm), 1/8 inch (3.1 mm) clearance; for bronze assemblies over 1 inch (25 mm), 1/4 inch (6.35 mm) clearance.

10.1.1.2.7 Body and Bonnet

In addition, a notch shall be permitted in the bonnet or a lug on the bonnet and body by means of which the bonnet may be loosened from the body without damaging the seal or sealing surfaces.

10.1.1.2.8 Tapping and Threading

Valve bodies shall be provided with an adequate boss at each location where a tapped hole is required. Female pipe threaded connections shall be tapped into bosses in the body and shall be so constructed that it shall not be possible to thread a pipe into them far enough to interfere with internal components.

All bolts and cap screws used internally and/or externally are to be of one standard, either English or SI.

10.1.1.2.9 Test Cocks

Double check valve assemblies (Fig. 10.5) and reduced pressure principle backflow prevention assemblies (Fig 10.1) shall be equipped with resilient seated test cocks located as follows:

a. On the upstream side of the No. 1 shutoff valve.
   Exception: Not required on 1/4 inch (6 mm) or 3/8 inch (10 mm) reduced pressure principle backflow prevention assemblies.
b. Between the No. 1 shutoff valve and the No. 1 check valve
c. Between the check valves
d. Between the No. 2 check valve and the No. 2 shutoff valve.

Pressure vacuum breaker backsiphonage prevention assemblies (Fig. 10.6) shall be equipped with resilient seated test cocks located as follows:

a. Between the No. 1 shutoff valve and the check valve
b. Between the check valve and the air inlet valve.

Spill-resistant pressure vacuum breaker backsiphonage prevention assemblies (Fig. 10.10) shall be equipped with a resilient seated test cock and a resilient seated vent valve located as follows:

a. Test cock: Between the No. 1 shutoff valve and the check valve.
b. Vent valve: Between the check valve and the air inlet valve.

The sizes of these test cocks shall be as given in Table 10-3. The body port into which the test cock is threaded shall be rated size for the test cock and the waterway through the test cock itself shall be at minimum rated size. All 1/4 inch, 1/2 inch and 3/4 inch test cocks shall have IPS female ends on the discharge side in accordance with ANSI standards for the standard pipe sizes mentioned in Table 10-3. The 1/8 inch test cock shall be permitted to be either IPS female threads or male 1/4 inch flared type connection per SAE J513.

The water passageway shall be at least equal in area to the rated size of the unit and shall not adversely affect the overall pressure loss.

The operating stem of a ball valve type test cock shall be designed so that it is inserted from inside the body (blow out proof stem). The operator on quarter turn valves must clearly show if the valve is open or closed, even if the handle is removed.

The test cocks shall withstand all hydraulic testing of the assembly, in the fully open, partially open and closed positions, without leakage, damage or permanent deformation.

Table 10-3
Sizes of Test Cocks

Size of Assembly (inches)	Minimum Equivalent Orifice Size of Test Cock (inches)
1 (25mm) and below	1/8 ⌀ (3.2 mm)
1 1/4 (32mm) to 2 (50mm) inclusive	1/4 ⌀ (6.4 mm)
2 1/2 (65mm) to 4 (100mm) inclusive	1/2 ⌀ (12.7 mm)
6 (150mm) and larger	3/4 ⌀ (19.1 mm)

10.1.1.2.10 Control Piping and Diaphragms
All control piping or passageways shall be of corrosion-resistant material or protected with a suitable protective coating in accordance with 10.1.1.3.14. The pipes or passageways shall be sized sufficiently large to prevent clogging. The pipe or passageways shall be located in a manner so as to prevent entrapment of foreign materials or air. Furthermore, all external control piping shall be placed in such a manner that it is not readily damaged when the assembly is located in unprotected areas.

When diaphragms or bellows are used as barriers in control piping which bypasses one or more check valves in the backflow prevention assembly, such a diaphragm or bellows shall be installed in such a manner that its failure shall produce visible evidence of such failure.

10.1.1.2.11 Air Release
Provision shall be made for bleeding trapped air from the highest point of the assembly when the normal flow of water will not displace it.

10.1.1.2.12 Valve Seats
The DC, RP, DCDA, RPDA, DCDA-II, RPDA-II, PVB and SVB backflow prevention assemblies shall be fitted with replaceable valve seats on the check valves, differential pressure relief valves and air inlet valves, which shall be provided with a means of in situ insertion and removal without damage or permanent deformation.

10.1.1.2.13 Alignment
The clapper or poppet elastomer disc, clamping ring and the securing bolt, stud or cap screw each shall be concentric to an axis, which is normal to the face of the elastomer disc. A means shall be provided to prevent clamping the elastomer disc so tight as to deform its face.

Provision shall be made to prevent the clapper from tipping and catching under the seat ring or any faulty action, which would prevent true alignment. All parts shall be constructed and supported in such a manner as to preclude distortion or misalignment.

10.1.1.2.14 Chatter, Vibration

Furthermore, all external control piping shall be placed in such a manner that it is not readily damaged when the assembly is located in unprotected areas.

All moving parts shall be designed to operate from static to the rated flow condition in a positive manner without chatter or vibration.

10.1.1.2.15 Material Selection
When a relative motion is to be allowed between mating parts the materials shall be sufficiently different so that there will be no scuffing or galling.

10.1.1.2.16 Lubrication of Components
When proper operation of an assembly is dependent upon a lubricant, the manufacturer shall provide documentation (i.e., assembly installation and maintenance literature).

10.1.1.2.17 Shutoff Valves
The shutoff valves used as an integral component of approved backflow prevention assemblies are typically separated into two categories. The 2 inch (50 mm) and smaller assemblies generally use bronze bodied shutoff valves, whereas the 2 1/2 inch (65 mm) and larger assemblies generally use ferrous-bodied shutoff valves. But this is not a restriction of use.

The specifications for resilient seated gate valves on 3 inch (80 mm) through 12 inch (300 mm) units are taken from the American Water Works Association (AWWA) Standard #C509 with an additional requirement for the use of a polymerized coating on all ferrous surfaces. The 2 1/2 inch (65 mm), 14 inch (350 mm) and 16 inch (400 mm) shutoff valves are to be constructed per the applicable requirements of AWWA #C509 Standard, even though they are not specifically listed in this standard.

The specifications for resilient seated butterfly valves on 3 inch (80 mm) through 16 inch (400 mm) units are taken from the American Water Works Association (AWWA) Standard #C504 with an additional requirement for the use of a polymerized coating on all ferrous surfaces. The 2 1/2 inch (65 mm) shutoff valve is to be constructed per the applicable requirements of AWWA #C504 Standard, even though it is not specifically listed in this standard.

All ferrous bodies shall be coated with a holiday-free polymerized coating of suitable thickness to comply with AWWA Standard #C550. The protective coating shall be composed of materials deemed acceptable in the Food and Drug Administration document, "Resinous and Polymeric Coatings" (Title 21 of the Federal Regulations on food additives, Sec.175.300). The manufacturer shall provide documentation showing suitability for potable water use.

The AWWA #C550 Standard is listed as a minimum requirement.

Each shutoff valve shall be marked with the following:

    a. Manufacturer's or private labeler's name or identifying symbol.
    b. Nominal size of valve.
    c. Model number must be cast, molded or affixed onto the body or bonnet of the valve.
    d. Working pressure.

Below are the requirements for the resilient seated shutoff valves which do not come under the AWWA Standard #C509.

The inlet of the upstream shutoff valve and the outlet of the downstream shutoff valve body shall be threaded, flanged or grooved in accordance with the ANSI Standards for the standard pipe sizes mentioned in 10.1.1.1.3. The valve body shall be provided with an adequate boss at each location where a tapped hole is required for the No. 1 test cock.

Each shutoff valve shall be marked with the following:

a. Manufacturer's or private labeler's name or identifying symbol.
b. Nominal size of valve.
c. Model number must be cast, molded or affixed onto the body of the valve.
d. Working pressure.

For threaded valve bodies, the hole in the wall of the No. 1 shutoff valve for the No. 1 test cock shall not enter into the body where it can be covered by the inlet piping. A thread stop must be provided so that the inlet piping does not obstruct the test cock sensing hole. The sensing hole must be fully sized according to the test cock sizes in Table 10-3.

The operating stem in a ball valve shall be designed so that it is inserted from inside the body and is seated with an adjustable stem-packing nut on the outside. This blow-out-proof type of stem allows in-line service of the stem packing. The operating stem and handle on quarter turn shutoff valves must clearly show if the valve is open or closed, even if the handle is removed. Stops must be provided to limit the valve actuation to 90 degrees. The handle must be in line with the piping for the open position and at 90 degrees to the piping in the closed position. The operating handle must be suitably protected from corrosion.

The water passageway shall be at least equal in area to the rated size of the unit and shall not adversely affect the overall pressure loss. The water passageway shall be permitted to be less than the rated size of the unit providing that the connections between the shutoff valves and the check valve body(s) are not standard taper pipe connections, flange or groove per 10.1.1.1.3 and comply with all performance requirements of 10.1.2.

Each valve shall be capable of withstanding all pressures from atmospheric up to a hydrostatic pressure of twice the working pressure of the assembly for 10 minutes: in the fully open, partially open and closed positions, in both directions, with no leakage, damage or permanent deformation.

10.1.1.2.18 Manifold Assembly

A manifold assembly shall be comprised of two or more backflow prevention assemblies (double check valve assemblies or reduced pressure principle backflow prevention assemblies) in parallel with a single inlet and outlet connection per 10.1.1.1.3. The size of the manifold assembly shall be determined by the inlet and outlet connections. The manifold assembly shall comply with the following requirements of this chapter.

10.1.1.2.18.1
For each size of manifold assembly at any flow rate up to and including the rated flow, the maximum allowable pressure loss shall not exceed the values given in Table 10-1.

10.1.1.2.18.2
The manifold assembly shall comply with Hydrostatic Test requirements per 10.1.1.1.5.

10.1.1.2.18.3
Each of the individual backflow prevention assemblies contained in the manifold assembly shall be approved backflow prevention assemblies.

10.1.1.2.18.4
Each of the individual backflow prevention assemblies contained in the manifold assembly shall be of the same manufacturer, model and size.

10.1.1.2.18.5
The manifold adaptor fittings on both the inlet and outlet of the manifold assembly are considered integral components and shall meet all of the material specifications in 10.1.1.3. The manifold adaptor fittings shall be marked per the marking requirements in 10.1.1.1.4 with the manifold assembly model designation to identify the adaptor fittings as integral components of the manifold assembly.

Each of the individual backflow prevention assemblies contained in the manifold assembly shall be flow tested at the rated flow of the manifold assembly. See s 10.1.2.2.3.2 or 10.1.2.3.3.2.

10.1.1.2.19 Repair Tools
Tools not commercially available for the repair/maintenance of a backflow prevention assembly shall be made readily available from the backflow prevention assembly manufacturer.

10.1.1.3 Material Requirements

10.1.1.3.1 Statement of Policy

10.1.1.3.1.1 General
The following material specifications and current ASTM (American Society for Testing Materials) and/or ANSI (American National Standards Institute) designations shall be adhered to at all times except where a manufacturer desires to use an equivalent or better material. In such cases the substitute specifications shall be submitted for approval prior to the inclusion of such materials in an assembly. In the subsequent list of materials no attempt has been made to bar the use of alloys, rubbers, plastics or other materials which may be adaptable and which will give at least the equivalent trouble free service. All materials shall be non-toxic. The manufacturer shall provide documentation showing suitability for potable water use.

10.1.1.3.1.2 Dissimilar Metals
In the construction of backflow prevention assemblies a minimum of dissimilar metals shall be used. In all cases where it is impossible to use similar metals, steps shall be taken, insofar as it is practicable, to prevent the formation of galvanic electrolytic couples. To this end, when two dissimilar metals must come close to each other, the metals chosen shall be as nearly as possibly electrolytically similar and shall be insulated wherever possible.

10.1.1.3.1.3 Corrosion
Ferrous metals contained in backflow prevention assemblies shall be protected to resist corrosion.

10.1.1.3.2 Bodies and Covers
Materials to be used in construction of these parts of the assembly shall be either valve bronze that conforms to ASTM Designation B61 or B62 or B584 UNS number C84400 or other ASTM designated bronze alloys that contain at least 79% copper and less than 15% zinc or gray iron that conforms to ASTM Designation A126, Class B or Class C; or ductile iron that conforms to ASTM Designation A536, Grade 65-45-12; or stainless steel that conforms to ASTM Designation A276 or A296 either UNS No. S30400, S30500, S31600 or steel pipe and flanges that are suitably protected against corrosion (see 10.1.1.3.12 Protective Coatings); or engineered plastic.

10.1.1.3.3 Seat Rings
The seat rings shall be constructed of valve bronze that conforms to ASTM Designation B61 or B62 or B584 UNS number C84400 or other ASTM Designated bronze alloys that contain at least 79% copper and less than 15% zinc; or stainless steel that conforms to ASTM Designation A276 either UNS No. S30400, S30500 or S31600; or engineered plastic.

10.1.1.3.4 Check Valves
The clapper, poppet or similar check valve shall be constructed of valve bronze that conforms to ASTM Designation B61 or B62 or B584 UNS number C84400 or other ASTM Designated bronze alloys that contain at least 79% copper and less than 15% zinc; or stainless steel that conforms to ASTM Designation A276 either UNS No. S30400, S30500 or S31600; or engineered plastic.

10.1.1.3.5 Elastomer Discs

a. Check valves, differential pressure relief valves and air inlet valve discs shall be composed of natural, synthetic or thermal plastic elastomers.
b. For elastomer discs, the material must be capable of withstanding the rated temperature and pressure of the assembly in which it is being used with no permanent change in characteristics per performance evaluation in 10.1.2. It is suggested that the manufacturer add a means of identification to the elastomer discs to identify them as original manufactured parts.
c. The manufacturer shall supply the ASTM Designation D2000 for all elastomers used in the assembly and material properties such as tensile strength, tear resistance and compression set for identification purposes.

10.1.1.3.6 Swing Arms
The swing arm shall be made of valve bronze (that conforms to ASTM Designation B61, B62 or B584; UNS number C84400 or other ASTM Designated bronze alloys that contain at least 79% copper and less than 15% zinc), stainless steel (that conforms to ASTM Designation A276; UNS numbers S30400, S30500 or S31600) or engineered plastic.

10.1.1.3.7 Swing Pins and Guide Stems
The swing pin or guide stem shall be made either of phosphor bronze (that conforms to ASTM Designation B139, UNS numbers C51000, C52100 or C52400), stainless steel (that conforms to ASTM Designation A276, UNS numbers S30400, S30500 or S31600) or engineered plastic.

10.1.1.3.8 Bushings
Bushings shall have corrosion resistance at least equal to ASTM Designation B584, UNS number C84400 or other ASTM Designated bronze alloys that contain at least 79% copper and less than 15% zinc.

10.1.1.3.9 Counterbalances
The counterbalance shall be weighted with a corrosion resistant material and/or protected from corrosion with a suitable protective coating in accordance with 10.1.1.3.12.

10.1.1.3.10 Springs
All springs shall be made of either stainless steel (that conforms to ASTM Designation A313), phosphor bronze (that conforms to ASTM Designation B159, UNS numbers C52100 or C52400) or equivalent.

10.1.1.3.11 Diaphragms
The diaphragm material shall be of a natural rubber, synthetic or thermoplastic elastomer.

10.1.1.3.12 Protective Coatings
All ferrous bodies and parts shall be coated with a holiday-free polymerized coating per AWWA #C550. The use of synthetic protective coatings on ferrous bodies and parts shall be subject to the individual approval not only of the specific use but of the coating and process of application. The protective coating shall be composed of materials deemed acceptable in the Food and Drug Administration document, "Resinous and Polymeric Coatings" (Title 21 of the Federal Regulations on food additives, Sec.175.300). The manufacturer shall provide documentation showing suitability for potable water use.

Machining of a protective coating after the coating has been applied in order to facilitate the assembling of the assembly or to provide dimensional control is to be avoided. Extra care is to be taken with areas around ports or threaded surfaces such that focal points for corrosion are eliminated.

10.1.1.3.13 Studs, Bolts and Cap Screws

Such parts for internal use, bonnet plates or closure plates shall be made of naval bronze rod (that conforms to ASTM Designation B21) or stainless steel (that conforms to ASTM Designation F593, UNS numbers S30400, S30500 or S31600). Bolts for line flanges may be cadmium or zinc plated steel.

10.1.1.3.14 Control Piping

Control piping shall have corrosion resistance at least equal to ASTM Designation B584 or UNS number C84400.

10.1.1.3.15 Test Cocks

The test cocks on backflow prevention assemblies are required so that the assembly may be field tested periodically, in situ. Any leak through a closed test cock constitutes an undesirable nuisance. Hence, a test cock that will not shut off tightly or that is easily damaged by frequent use is to be avoided. The test cock shall be resilient seated and have a full flow characteristic.

Materials to be used in construction of the body and internal wetted components of the test cock shall be either valve bronze (that conforms to ASTM Designation B61, B62 or B584, UNS number C84400 or other ASTM Designated bronze alloys that contain at least 79% copper and less than 15% zinc), stainless steel (that conforms to ASTM Designation A276, UNS numbers S30400, S30500 or S31600) or suitable engineered plastic.

In ball valves, the solid ball shall be stainless steel (that conforms to ASTM Designation A276, UNS numbers S30400, S30500 or S31600) or hard chromium plated brass (per Federal Specification QQ-C-320B: Chromium Plating [Electrodeposited]).

In plug valves, the plug shall be stainless steel (that conforms to ASTM Designation A276, UNS numbers S30400, S30500 or S31600), valve bronze (that conforms to ASTM Designation B61, B62 or B584 UNS number C84400 or other ASTM Designated bronze alloys that contain at least 79% copper and less than 15% zinc) or hard chromium plated brass (per Federal Specification QQ-C-320B–Chromium Plating [Electrodeposite]).

Material to be used for the resilient seals shall be of a suitable elastomer or polymer and capable of withstanding the action of line fluids and operation under long term service at rated conditions.

An approved backflow prevention assembly shall include test cocks as detailed above and in 10.1.1.2.9.

10.1.1.3.16 Shutoff Valves

The shutoff valves on backflow prevention assemblies are required so that the assembly may be field tested and/or maintained periodically, in situ. Any internal leakage of a closed shutoff valve may eliminate the possibility of accurate testing. Hence, a shutoff valve that will not shut off tightly or that is easily damaged by frequent use is to be avoided. The shutoff valves shall be resilient seated and shall have a full flow characteristic.
Below are the requirements for the resilient seated shutoff valves which do not come under the AWWA C509 Standard as detailed in 10.1.1.2.17.

Check valves, differential pressure relief valves and air inlet valve discs shall be composed of natural, synthetic elastomers or thermal plastic elastomers.

Materials to be used in construction of the body of the shutoff valve shall be valve bronze (that conforms to ASTM Designation B61, B62 or B584; UNS number C84400 or other ASTM Designated bronze alloys that contain at least 79% copper and less than 15% zinc); gray iron (that conforms to ASTM Designation A126, Class B or Class C); ductile iron (that conforms to ASTM Designation A536 Grade 65-45-12); stainless steel (that conforms to ASTM Designation A276, UNS numbers S30400, S30500 or S31600); or engineered plastic.

Materials to be used in construction of the internal wetted components of the shutoff valve shall be either valve bronze (that conforms to ASTM Designation B61, B62 or B584; UNS number C84400 or other ASTM Designated bronze alloys that contain at least 79% copper and less than 15% zinc); stainless steel (that conforms to ASTM Designation A276, UNS numbers S30400, S30500 or S31600); or engineered plastic.

In ball valves, the solid ball shall be stainless steel (that conforms to ASTM Designation A276, UNS numbers S30400, S30500 or S31600); or hard chromium plated brass (per Federal Specification QQ-C-320B–Chromium Plating [Electrodeposited]).

In plug valves, the plug shall be stainless steel (that conforms to ASTM Designation A276, UNS numbers S30400, S30500 or S31600); valve bronze (that conforms to ASTM Designation B61, B62 or B584; UNS number C84400 or other ASTM Designated bronze alloys that contain at least 79% copper and less than 15% zinc); or hard chromium plated brass (per Federal Specification QQ-C-320B–Chromium Plating [Electrodeposited]).

Material to be used for the resilient seals shall be of a suitable elastomer or polymer and capable of withstanding the action of line fluids and operation under long term service at rated conditions.

An approved backflow prevention assembly shall use shutoff valves as detailed above and in 10.1.1.2.17.

10.1.1.3.17 Lubricants
The lubricant shall be non-toxic, shall not support the growth of bacteria and shall have no deteriorating effects on any component. The manufacturer must supply documentation stating that lubricant is suitable for contact with potable water.

10.1.2 Evaluation of Design and Performance

10.1.2.1 General
The standards set forth in this Manual supersede the specifications of previous editions of this manual as well as those of Paper No. 5 and USCEC Report No. 48-101.

10.1.2.1.1 Drawings and Requirements
A full set of working drawings and specifications of all materials shall be furnished with each make, model and size of assembly that is submitted for evaluation, including shutoff valves and test cocks. The Foundation for Cross-Connection Control and Hydraulic Research's (Foundation) Request for Evaluation Form must accompany each submittal. See Appendix B for a sample of this form and submittal information.

10.1.2.1.2 Assemblies Required for Evaluation

a. Laboratory Evaluation: One or more assemblies, selected at random from the manufacturer's stock, of each size and model shall undergo a complete laboratory evaluation of its design and operating characteristics under the supervision of the Foundation for Cross-Connection Control and Hydraulic Research. For the sizes up to and including 2 inch (50 mm), the manufacturer shall submit a minimum of three assemblies. For the sizes 2½ inch (65 mm) and larger, the manufacturer shall submit a minimum of one assembly.

The manufacturer shall identify repair/maintenance tools necessary to remove or install all components of the assembly. Tools not commercially available, per 10.1.1.2.19, shall be supplied with the assembly(s) submitted.

The laboratory evaluation shall be performed in the orientation for which each assembly is submitted. Horizontal, vertical (flow up and/or down) and axial rotation orientations shall be identified by the manufacturer at the time of submittal.

For the initial submittal of DC's or DCDA's in the horizontal and vertical up orientations, the laboratory evaluation shall include 10.1.2.3.3.8 in each orientation.

Following the successful completion of the laboratory evaluation and before the assembly(s) can be granted approval; a production quality assembly(s) shall be submitted for review and/or evaluation. The Foundation for Cross-Connection Control and Hydraulic Research, at its own discretion retains the right of determining the extent of re-evaluation required on the assembly(s) containing production components.

b. Field Evaluation: Upon successful completion of the laboratory evaluation, a minimum of three assemblies, selected at random from the manufacturer's stock, of each size and model shall, simultaneously and successfully, complete a twelve consecutive month field evaluation under the supervision of the Foundation for Cross-Connection Control and Hydraulic Research.

The field evaluation shall be performed in each of the orientation(s) successfully completing the laboratory evaluation.

10.1.2.1.3 Selection of Field Locations

The manufacturer shall be responsible for locating and installing all assemblies in acceptable field evaluation sites; however, the Foundation for Cross-Connection Control and Hydraulic Research shall be permitted to reject any field evaluation site submitted by the manufacturer. In general, the following conditions will govern site selection but these conditions will not constitute the only site selection criteria:

a. When a common body design is used for two or more line sizes, at least two of the three field sites shall be of the largest size.
b. Reduced pressure principle assemblies shall not normally be installed where conditions would require the use of this type of assembly.
c. The sites shall provide as wide a range of water conditions (i.e., varying line pressures and flow rates) and types of use as practical. A minimum of one site shall have flow rates reaching the range of 50-100% of rated flow.
d. No more than one site per model and size shall be permitted on a non-flowing static service (e.g., fire sprinkler system). The one non-flowing static service may be installed in a service one size smaller than the size of the assembly.

> For the purposes of the field evaluation in vertical orientation(s), the manufacturer shall be permitted to install up to two sites per size on a non-flowing static service provided that:
> 1. The assembly is currently Approved by the Foundation for Cross-Connection Control and Hydraulic Research in the horizontal orientation; or,
> 2. The assembly is currently undergoing the field evaluation in the horizontal orientation.

e. For the initial submittal of DC & DCDA's in the horizontal (H) and vertical flowing upward (VU) orientations, the field evaluation will require three assemblies in acceptable field sites including:

The test cocks on backflow prevention assemblies are required so that the assembly may be tested periodically, in situ. Any leakage of a closed test cock constitutes an undesirable nuisance.

1. One flowing horizontal orientation and
2. One flowing vertical up (VU) orientation and
3. Third site at discretion of the manufacturer

f. Parallel installations for field evaluation are not normally recognized; but, if such a site is accepted the two assemblies shall be considered to be only one of the required sites.
g. All sites shall be freely accessible during normal working hours to the Foundation's field evaluation personnel.
h. The owner of the property (or the manager), the water agency and/or health agency having cognizance must all supply written acknowledgment prior to the installation of the assembly(s) to the Foundation for Cross-Connection Control and Hydraulic Research that the site will be an acceptable field evaluation location. (See Appendix B for Letter of Acknowledgement for Field Evaluation Sites).
i. Each field site shall be suitably protected from freezing conditions and vandalism.
j. The manufacturer shall submit the Field Site Application Form (See Appendix B for a sample of this form) for each proposed field site prior to the installation of the assembly.

10.1.2.1.4 Period of Field Evaluation

Once the manufacturer submits the Field Site Application Form for each of the specific field sites, and it has been reviewed and accepted by the Foundation for Cross-Connection Control and Hydraulic Research, the initial in situ field evaluation test will be conducted, if the laboratory evaluation has been successfully completed. When a minimum of three assemblies for each size and model have successfully complied with the initial field test by the Foundation for Cross-Connection Control and Hydraulic Research, the twelve month field evaluation period begins. The initial field test shall include each of the following:

a. Recording of field test data per Chapter 9 of this manual
b. Recording of static differential pressure reading across each check valve
c. Recording of inlet line pressure
d. Disassembly and physical inspection of all components, reassembly
e. Recording of field test data per Chapter 9 of this manual
f. Procedures and recording requirements detailed in Appendix B.2.

A minimum of three assemblies undergoing field evaluation for each size and model of assembly shall provide simultaneous trouble-free operation for a minimum period of twelve consecutive months in conformance with the field evaluation Standards as set forth herein for each type of assembly. Assemblies at each of the field test sites shall be field tested on a nominal thirty day schedule. When more than three assemblies of each size and model are undergoing the field evaluation, and two assemblies fail to comply with the performance requirements in a similar manner, this shall be cause for rejection. Before a specific size and model of assembly shall be released from the field evaluation it shall be inspected to determine that all of the working parts continue to conform to the design and performance standards. The concluding field test at the end of the field evaluation shall include each of the following:

a. Recording of field test data per Chapter 9 of this manual
b. Recording of static differential pressure reading across each check valve
c. Recording of inlet line pressure
d. Disassembly and physical inspection of all components, reassembly
e. Recording of field test data per Chapter 9 of this manual
f. Procedures and recording requirements detailed in Appendix B.2.

A full set of working drawings and specifications of all materials shall be furnished with each make, model and size of assembly that is submitted for evaluation, including shutoff valves and test cocks.

Should a make, model and size assembly fail to comply with the standards set forth in this Manual, the following offices, having cognizance of the field test locations, in addition to the manufacturer, shall be immediately notified:

a. health agency
b. plumbing authority
c. water supplier

Should a model and size assembly not complete the field evaluation and not gain approval, then the manufacturer shall be responsible for replacing the field test assemblies in all field evaluation sites with currently approved assemblies.

During the entire field evaluation program only the Foundation for Cross-Connection Control and Hydraulic Research shall be permitted to test, repair or inspect an assembly undergoing field evaluation. The Foundation for Cross-Connection Control and Hydraulic Research shall be permitted to reject a field site should any testing, repair or inspection of an assembly(s) be performed by unauthorized personnel.

10.1.2.1.5 Approval

In addition to the satisfactory completion of the laboratory and field evaluation, the manufacturer shall supply to the Foundation for Cross-Connection Control and Hydraulic Research the sales literature and installation/maintenance literature for the model and size assembly being evaluated under these Standards. Before consideration for approval these materials shall be reviewed for accuracy, including the following:

a. Installation detailing, including orientation and clearances
b. General operating parameters of the assembly MWWP, MWWT, Pressure loss vs. flow rate curves
c. Repair/Maintenance Procedures/Tool requirements per 10.1.1.2.19
d. Field test procedures.

Following the successful completion of the laboratory evaluation and before the assembly(s) can be granted approval, a production quality assembly(s) shall be submitted for review and/or evaluation The Foundation for Cross-Connection Control and Hydraulic Research, at its own discretion, retains the right of determining the extent of re-evaluation required on the assembly(s) containing production components. Upon the completion of a satisfactory laboratory and field evaluation and acceptable review of the manufacturer's literature, the manufacturer shall supply to the Foundation for Cross-Connection Control and Hydraulic Research the serial number of the first assembly manufactured in the approved configuration. The Foundation for Cross-Connection Control and Hydraulic Research shall then grant an approval for the specific size and model evaluated under these standards. This approval shall be valid for a period of no more than three years and may be rescinded for cause before that time. The Foundation for Cross-Connection Control and Hydraulic Research shall issue a list of all assemblies that are currently approved by said agency.

Should only original design replacement parts (see 10.1.2.1.9) be available from the manufacturer and not the entire assembly, the manufacturer shall notify the Foundation for Cross-Connection Control and Hydraulic Research. The particular assembly shall be designated on the list of approved assemblies as only having spare parts available. When the original manufacturer ceases to maintain a supply of original design replacement parts for an approved model and size of assembly that particular model and size shall have its approval rescinded.

The manufacturer shall be responsible for locating and installing all assemblies in acceptable field evaluation sites...

10.1.2.1.6 Renewal of Approval

Continuing verification of compliance with these standards and field performance shall be accomplished at a maximum of once every three years to the satisfaction of the Foundation for Cross-Connection Control and Hydraulic Research. The Foundation for Cross-Connection Control and Hydraulic Research, at its own discretion, retains the right of determining the extent of re-evaluation required before renewal is granted. The manufacturer's current sales, installation and maintenance literature shall be submitted for review at this time. Acceptable re-evaluation, literature, past performance of the assembly under field operating conditions and spare parts availability shall be considered before the renewing of an approval. Failure to meet these requirements shall result in the automatic rescinding of the approval for that size and model of assembly. The latest listing of approved assemblies shall reflect the current status of a particular model and size assembly.

10.1.2.1.7 Change of Design, Materials or Operation

The Foundation for Cross-Connection Control and Hydraulic Research shall be notified by the manufacturer in writing of any proposed change in design, materials or operation of an assembly approved by the Foundation for Cross-Connection Control and Hydraulic Research. The Foundation for Cross-Connection Control and Hydraulic Research, at its own discretion, shall be permitted to require another laboratory and/or field evaluation. Failure to notify the Foundation for Cross-Connection Control and Hydraulic Research of any changes of design, materials or operation prior to implementation shall be cause for the rescinding of the approval for the size and model of assembly involved.

10.1.2.1.8 Prior Approval

a. On the publication date of this manual all makes, models and sizes of assemblies having been previously approved shall henceforth be designated by reference to the specifications under which each assembly was approved. The list of approved assemblies will show compliance with either Paper No. 5[1], USCEC 48-101[2] or Editions 1, 2, 3, 4, 5, 6, 6 (Revised), 7, 8 or 9 of this manual[3].

b. Backflow prevention assemblies approved under the above paragraph (10.1.2.1.8a) shall be approved for installation as set forth herein.

c. Continuing verification of compliance with the standard under which a size and model of assembly was originally approved shall be accomplished at least once every three years from the date of the most recent renewal to the satisfaction of the original approving agency. The Foundation for Cross-Connection Control and Hydraulic Research, at its own discretion retains the right of determining the extent of re-evaluation required before renewal is granted. Acceptable current literature and past performance of the assembly under field operation conditions shall be considered before the renewing of an approval. Failure to meet these requirements shall result in the automatic rescinding of the approval for that make, model and size of assembly.

d. A Certificate of Full Approval or Approval granted under conditions of prior approval (10.1.2.1.8a) or a Certificate of Approval granted under the conditions of this 10th Edition of the Manual shall be permitted to be rescinded at any time by the Foundation for Cross-Connection Control and Hydraulic Research for cause.

A minimum of three assemblies undergoing field evaluation for each size and model of assembly shall provide simultaneous trouble-free operation for a minimum period of twelve (12) consecutive months...

---

1 Paper No.5:"Objectives, General Testing Procedure, Specifications, Results of Tests;" Foundation for Cross-Connection Control Research, University of Southern California; April, 1948.
2 USCEC Report 48-101 "Definitions and Specification of Double Check Valve Assemblies and Reduced Pressure Principle Backflow Prevention Devices;" Foundation for Cross-Connection Control Research, University of Southern California; January 30, 1959.
3 Manual of Cross-Connection Control: Foundation for Cross-Connection Control and Hydraulic Research, University of Southern California; 1st Ed., 1960; 2nd Ed., 1965; 3rd Ed., 1966; 4th Ed., 1969; 5th Ed., 1974; 6th Ed., 1979; Revised 6th Ed., 1982; 7th Ed., 1985, 8th Ed., 1988, 9th Ed., 1993.

10.1.2.1.9 Replacement Parts

The replacement parts of a backflow prevention assembly shall include all control piping, internal components and access covers for internal components of the assembly, excluding the primary assembly body(s), i.e., main pressure containing vessel(s).

10.1.2.1.10 Evaluation Policies

The Foundation at its own discretion retains the right to issue evaluation policies updating or clarifying evaluation procedures as needed. Current details may be found at www.usc.edu/fcchr.

When the original manufacturer ceases to maintain a supply of original design replacement parts for an approved model and size of assembly that particular model and size shall have its Approval rescinded.

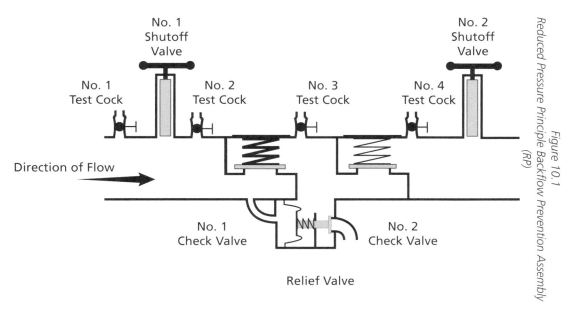

Figure 10.1 Reduced Pressure Principle Backflow Prevention Assembly (RP)

10.1.2.2 Design, Operational and Evaluation Standard for Reduced Pressure Principle Backflow Prevention Assemblies (RP)

10.1.2.2.1 Design and Operational Requirements

a. This assembly shall include two approved, independently operating check valves with an automatically operating, mechanically independent, hydraulically dependent pressure differential relief valve located between the two check valves. The assembly shall include a tightly closing resilient seated shutoff valve on each end of the body and each assembly shall be fitted with four properly located resilient seated test cocks (see figure 10.1). During normal flow and at the cessation of normal flow, the pressure in the zone, (i.e., the zone between these two check valves) shall be at least 2 psi (13.78 KPa) less than the upstream (or supply) pressure. The pressure drop across the first check valve at static (no flow) conditions shall be essentially a constant value for all line pressures from 20 psi (137.8 KPa) through the maximum working water pressure (MWWP), but not less than 150 psi (1034 KPa).
b. With no flow from the upstream side, when the pressure on the supply side drops to 2 psi (13.78 KPa) above the zone pressure, the relief valve shall discharge water to maintain the zone at 2 psi (13.78 KPa) below the supply pressure. When the pressure on the upstream side of the first check valve drops below 2 psi (13.78 KPa) above the zone pressure the relief valve shall be open.
c. Under a backflow condition, when the upstream pressure is 2 psi (13.78 KPa) up to maximum working water pressure, the relief valve shall discharge from the zone to atmosphere the quantities of water flowing back given in Table 10-4 and the pressure in the zone shall be at least 1/2 psi (3.44 KPa) below the upstream pressure.
d. The second check valve shall be internally loaded and shall at all times be drip tight in the normal direction of flow with the inlet pressure at least 1 psi (6.89 KPa) and the outlet under atmospheric pressure.
e. When the upstream pressure is atmospheric the differential pressure relief valve shall discharge water from the zone to atmosphere with the rate of discharge corresponding to the data shown in Table 10-4; and the pressure in the zone shall not exceed 1 1/2 psi (10.34 KPa).
f. The differential pressure relief valve shall open and close positively and quietly. Further, it shall not spit water under fluctuating line conditions when the upstream pressure is 3.0 psi (20.68 KPa) or more above the differential pressure required to open the relief valve.

Failure to notify the Foundation... of any changes of design, materials or operation prior to implementation shall be cause for the rescinding of the approval for the size and model of assembly involved.

g. The differential pressure relief valve shall be located so that the valve seat(s) and the discharge port(s) are below the lowest portion of the No. 1 check valve so as to preclude backsiphonage.
h. The differential pressure relief valve and the No. 1 check valve shall be designed so that their characteristic opening points shall not drift unduly as a function of the line pressure.
i. The differential pressure relief valve shall not discharge while flowing up to and including 200% of the rated flow of the assembly. This includes both increasing and decreasing rates of flow.
j. For assemblies intended for fire sprinkler service, the differential pressure relief valve shall not discharge while flowing up to and including the rates of flow detailed in 10.1.1.1.2.
k. The differential pressure relief valve shall not discharge water when the test cocks are fully opened one at a time and there shall be a continuous flow of water from each test cock.

Table 10-4
Minimum Flow Rates
of Differential Pressure Relief Valve Opening

Size of Assembly		Minimum Flow Rate Through Relief Valve	
(inches)	(mm)	(gpm)	(L/s)
1/4	6	0.5	0.03
3/8	10	1	0.06
1/2	15	3	0.19
5/8×3/4	16×20	5	0.32
3/4 -1	20 - 25	5	0.32
1 1/4 - 1 1/2	32 - 40	10	0.63
2	50	20	1.26
2 1/2	65	20	1.26
3	80	30	1.89
4	100	40	2.52
6	150	40	2.52
8	200	60	3.78
10	250	60	3.78
12	300	75	4.73
14	350	90	5.68
16	400	100	6.31

10.1.2.2.2 Laboratory Evaluation

One assembly shall be inspected and evaluated in the following order. Unless otherwise noted, tests shall be performed at ambient temperature.

a. Conformance to the general, design and material requirements outlined in 10.1.
b. Conformance to the operational requirements outlined in 10.1.2.
c. Conformance to the working drawings and materials specifications.

d. Hydrostatic tests: the test assembly shall be subjected to the conditions of 10.1.1.1.5.
e. Pressure loss characteristics for flow rates up to the rated conditions.
f. Differential pressure relief valve opening point test.
g. Sensitivity of differential pressure relief valve to opening of test cocks.
h. Static closing point of check valve No. 1.
i. Static closing point of check valve No. 2.
j. Interdependence of Components
k. Differential pressure relief valve discharge capacity.
l. The assembly shall be evaluated for effectiveness when simultaneous backsiphonage and backpressure conditions are applied.
m. Differential pressure relief valve drain funnel backsiphonage test.
n. Thermal Test: the assembly shall be evaluated at the greater of 140°F (60°C) or the maximum working water temperature (MWWT), maximum working water pressure (MWWP) and specified flow rate.
o. Life cycle test.

10.1.2.2.3 Evaluation Procedure

10.1.2.2.3.1
Purpose: To determine the capability of the assembly to withstand the required hydrostatic test pressure.

Requirement: All components of the assembly shall withstand a hydrostatic pressure of twice the maximum working water pressure (MWWP), as stated by the manufacturer, for a period of 10 minutes without any damage, permanent deformation or impairment of operation.

Steps:

a. With both shutoff valves closed and all of the air bled from the assembly, raise pressure from atmospheric to a hydrostatic pressure of twice the maximum working water pressure supplied through the No. 2 test cock and maintain for a period of 10 minutes. Close No. 2 test cock and isolate assembly from pressure and open No. 4 test cock to relieve pressure, then close No. 4 test cock. Any evidence of leakage up to and including the hydrostatic pressure shall be cause for rejection.
b. Raise pressure from atmospheric to the required hydrostatic pressure supplied through test cock No. 4 with test cock No. 3 open to atmosphere for a period of 10 minutes. Close No. 4 test cock and isolate assembly from pressure, open No. 4 test cock to relieve pressure. Any evidence of the assembly leaking when subjected to pressures up to and including the hydrostatic pressure shall be cause for rejection.
c. Raise pressure from atmospheric to the required hydrostatic pressure supplied through test cock No. 3 with test cock No. 2 open to atmosphere for a period of 10 minutes. Close No. 3 test cock and isolate assembly from pressure, then open No. 4 test cock to relieve pressure. Any evidence of the assembly leaking when subjected to pressures up to and including the hydrostatic pressure shall be cause for rejection.
NOTE: A reverse hydrostatic pressure cannot be imposed on the No. 1 check valve without sealing the differential pressure relief valve, which is an abnormal condition. The backpressure test on the No. 1 check valve must be performed with the differential pressure relief valve isolated from the assembly or mechanically held in the closed position.
d. The assembly shall be disassembled and inspected for any damage, permanent deformation or impairment of operation. Evidence of such shall be cause for rejection.

10.1.2.2.3.2

Purpose: To determine the overall pressure loss of the complete assembly as a function of the rate of flow.

Requirement: At any rate of flow from static (0.0 gpm [0.0 L/s]) up to and including the rated flow, the overall pressure loss shall not exceed the values shown in Table 10-1. The assembly shall withstand 200% of the rated flow without any damage, permanent deformation or impairment of operation.

Steps:

a. Install the assembly in a suitable hydraulic test line having a flow capacity of twice the rated flow (see Table 10-1) and an available line pressure of at least 150 psi (1034 KPa).
b. Install in the supply line at a suitable location upstream from the assembly a piezometer ring; and, install a similar piezometer ring at a suitable location downstream from the assembly (located as indicated in ANSI/ISA Standard S75.02). Then install a mercury manometer or other suitable means of measuring pressure differentials between the two-piezometer rings to measure the observed overall pressure loss at various rates of flow.
c. Record differential pressure between piezometer rings at static condition. Increase flow rate and record steady state differential pressure at a sufficient number of increments to fully define the flow curve characteristics. Data shall be taken for both increasing and decreasing flow conditions. The flow rate shall then be increased to 200% of the rated flow for 5 minutes, then returned to static.
Manifold Assembly (10.1.1.2.18) Only: Close all shutoff valves so that only one of the individual assemblies of the manifold assembly is open. Gradually increase the flow of water through the individual assembly until the rated flow of the manifold assembly is reached, then return to static condition. Repeat flow test for each individual assembly contained in the manifold assembly.
d. Remove the assembly from the line and couple the downstream piping directly to the supply piping. Then observe data for a friction head correction curve for the pipe fittings required to adapt the test assembly to the supply line between the existing piezometer rings. The data of this correction curve is then subtracted from the observed overall pressure loss data of the assembly from the upstream face of the upstream shutoff valve to the downstream face of the downstream shutoff valve.
e. Exceeding the maximum allowable pressure loss listed in Table 10-1 for any rate of flow up to and including rated flow or discharge from the differential pressure relief valve, shall be cause for rejection. The differential pressure relief valve shall not discharge while flowing up to and including 200% of rated flow of the assembly, including both increasing and decreasing rates of flow. The assembly shall be disassembled and inspected for any damage, permanent deformation or impairment of operation. Evidence of such shall be cause for rejection.

10.1.2.2.3.3

Purpose: To test the operation of the differential pressure relief valve.

Requirement: The differential pressure relief valve shall operate to maintain the zone between the two check valves at least 2 psi (13.78 KPa) less than the supply pressure for all line pressures from 20 psi (137.8 KPa) up to the maximum working water pressure, but not less than 150 psi (1034.1 KPa) line pressure.

Steps:

a. Connect the high side hose of the differential pressure gage (or manometer) to test cock No. 2.

b. Connect the low side hose of the differential pressure gage (or manometer) to test cock No. 3.
c. Open test cocks No. 2 and No. 3 and bleed air from the gage (or manometer).
d. Close the No. 2 shutoff valve.
e. Slowly (maximum rate of 0.2 psid per second) equalize the pressure between the high side and low side hoses, noting the gage (or manometer) reading at the initial opening of the differential pressure relief valve.
f. Repeat step e for each 10 psi (68.9 KPa) increment between 20 psi (137.8 KPa) and the maximum working water pressure (MWWP), but not less than 150 psi (1034.1 KPa) line pressure.
g. Failure of the differential pressure relief valve to open at or before 2.0 psid (13.78 KPa) is reached, for all pressures designated in step f, shall be cause for rejection.

10.1.2.2.3.4

Purpose: To determine if the differential pressure relief valve shall discharge when the test cocks are opened one at a time and there is a continuous flow of water from each of the test cocks.

Requirement: The differential pressure relief valve shall not discharge water when the test cocks are fully opened one at a time and there is a continuous flow of water from each test cock when opened one at a time.

Steps:

a. Install the assembly in a suitable hydraulic test line, which is capable of maintaining an inlet pressure of at least 150 psi (1034 KPa) during the following steps.
b. Close the No. 2 shutoff valve while maintaining the No. 1 shutoff valve fully open.
c. Slowly open (4 seconds ± 1 second) test cock No. 1 until fully open. Then slowly close (4 seconds ± 1 second) test cock No. 1. Open No. 2 shutoff valve to re-establish the normal pressure gradient through the assembly.
d. Repeat step c with test cocks No. 2, No. 3 and No. 4.
e. Evidence of discharge from the differential pressure relief valve shall be cause for rejection. A non-continuous flow of water from each test cock shall be cause for rejection.

10.1.2.2.3.5

Purpose: To determine the static pressure drop across check valve No. 1.

Requirement: The static pressure drop across check valve No. 1 shall be at least 3.0 psi (20.7 KPa) greater than the pressure differential between inlet line pressure and the zone required to open the differential pressure relief valve, for all line pressures from 20 psi (137.8 KPa) up to maximum working water pressure (MWWP), but not less than 150 psi (1034.1 KPa).

Steps:

a. With a differential pressure gage or manometer connected as in 10.1.2.2.3.3, flow sufficient amounts of water through the No. 2 shutoff valve of the assembly to re-establish the normal pressure gradient across check valve No. 1. Record the static pressure differential across check valve No. 1.
b. Repeat step a for each 10 psi (68.9 KPa) increment between 20 psi (137.8 KPa) and the maximum working water pressure (MWWP), but not less than 150 psi (1034.1 KPa) line pressure.

c. Failure of the first check valve to maintain a static pressure differential of at least 3.0 psi (20.7 KPa) greater than the differential pressure relief valve opening point at the corresponding line pressure, shall be cause for rejection.

10.1.2.2.3.6

Purpose: To determine the static pressure drop across check valve No. 2.

Requirement: No. 2 check valve shall be drip-tight in the normal direction of flow with the inlet pressure at least 1 psi (6.89 KPa) and the outlet pressure at atmospheric.

Steps:

a. Install a vertical transparent tube at least six feet (1.83 m) long at test cock No. 3. If test cock No. 4 is not at the highest point of the check valve body, then another vertical transparent tube must be installed on test cock No. 4 so that it just rises above the top of the check valve body. See figure 10.2.

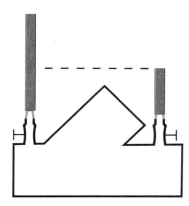

Figure 10.2
Location of vertical transparent tubes

b. Open test cock No. 3 and fill the tube with water, then close test cock No. 3. If a tube is attached to test cock No. 4, open test cock No. 4 to fill the tube so that the water level is above the highest point of the check valve body, then close test cock No. 4.
c. Close the No. 2 shutoff valve, then close the No. 1 shutoff valve.
d. Open test cock No. 4, then open test cock No. 3.
e. Water from the tube on test cock No. 3 will flow through check valve No. 2 until closure of check valve No. 2 is attained. When no further fall of the water in the tube on test cock No. 3 is observed, record the height of the water in the tube on test cock No. 3 above the water level at test cock No. 4. The difference of height of water in the transparent tube(s) must be 27 3/4 inches (704.85 mm) or greater.
f. Failure to maintain a minimum level of 27 3/4 inches (704.85 mm) shall be cause for rejection.

10.1.2.2.3.7

Purpose: To determine that each check valve, differential pressure relief valve and shutoff valve shall be free to operate independently through their entire respective movements.

Requirement: With the assembly in its intended orientation(s), the movement of either the first check valve, second check valve, differential pressure relief valve or shutoff valves through their respective full limits of travel shall not affect the operation of the other components.

Steps:

  a. Visually inspect the movement of the first check valve through its full limit of travel to determine if it contacts and affects the operation of the second check valve or the differential pressure relief valve. (The proper operation of the second check valve shall be confirmed by performing test 10.1.2.2.3.6 to determine static pressure drop across second check valve.) If necessary, remove or disconnect the loading on the first check valve to allow for free movement of the first check valve.
  b. Visually inspect the movement of the second check valve through its full limit of travel to determine if it contacts and affects the operation of the first check valve or the differential pressure relief valve. (The proper operation of the first check valve shall be confirmed by performing test 10.1.2.2.3.5 to determine static pressure drop across first check valve.) If necessary, remove or disconnect the loading on the second check valve to allow for free movement of the second check valve.
  c. Visually inspect the movement of the differential pressure relief valve through its full limit of travel to determine if it contacts and affects the operation of either the first or second check valve.
  d. Visually inspect the movement of each of the shutoff valves through their full limit of travel to determine if they contact and affect the operation of either the first or second check valve or differential pressure relief valve.
  e. Failure in steps a, c or d of the second check valve to maintain a minimum level of 27 3/4 inches (704.85 mm) shall be cause for rejection. Failure in steps b, c or d for the first check valve to maintain a static pressure differential of at least 3.0 psi (20.7 KPa) greater than the differential pressure relief valve opening point shall be cause for rejection. Failure in steps a, b or c of the differential pressure relief valve to move through its full range of motion shall be cause for rejection.

10.1.2.2.3.8

Purpose: To determine the discharge capacity of the differential pressure relief valve.

Requirement: For the backpressure requirements specified by 10.1.2.2.1.c and 10.1.2.2.1.e, the differential pressure relief valve discharge capacity shall be a minimum of the rates shown in Table 10-4. These rates shall be achieved with and without the manufacturer's drain funnel attached to the assembly.

Steps:

  a. Either cause line pressure to be piped into the downstream piping or install the assembly so that the supply line feeds directly into the No. 2 shutoff valve (i.e., assembly installed backwards)
  b. Remove the No. 2 check valve moving member (i.e., poppet or clapper).
  c. Locate a weigh tank directly beneath the relief valve discharge port to measure the rate of discharge from the relief valve or provide other suitable means to measure the rate of flow.
  d. Install a vertical transparent tube on the No. 3 test cock so that the backpressure in the zone can be observed.
  e. By means of a control valve located a minimum of five pipe diameters from the No. 2 shutoff valve (maintain the No. 2 shutoff valve in the fully open position), control the flow into the zone so that pressure in the zone is maintained at 1-1/2 psi (10.34 KPa). The inlet of the No. 1 shutoff valve and No. 2 test cock shall be open to atmosphere.
  f. Record the discharge rate of the differential pressure relief valve while the zone pressure is maintained as in step e. Shut off the control valve. Failure of the differential pressure relief valve to discharge the required flow rate shall be cause for rejection

g. Install a tee on the No. 2 test cock and a bypass hose from the supply line to the No. 2 test cock; and also, place a differential pressure gage or manometer between the No. 2 and No. 3 test cocks. Pressurize the bypass hose to the maximum working water pressure or at least a minimum of 150 psi (1034 KPa) and bleed all of the air out of the hoses and gage, as well as the assembly being tested.

h. Open the control valve on the supply line until the pressure in the zone is maintained at 1/2 psi (3.44 KPa) below the line pressure or the control valve is fully open. Determine the discharge rate of flow by means of a weigh tank or other suitable means. Failure of the differential pressure relief valve to discharge the required flow rate shall be cause for rejection

i. If a drain funnel is provided by the manufacturer, repeat steps e through h with the manufacturer's drain funnel attached to the assembly. Failure of the differential pressure relief valve to discharge the required flow rate with the drain funnel attached to the assembly shall be cause for rejection. (All water need not flow through the drain funnel.) Any damage or permanent deformation of the drain funnel during steps e through h shall be cause for rejection.

10.1.2.2.3.9

Purpose: To determine if simultaneous backsiphonage and backpressure conditions coupled with leaking No.1 and No.2 check valves will permit the carry over of water from the downstream piping into the upstream piping.

Requirement: There shall be no backsiphonage of water from the downstream piping through No.1 and No.2 check valves under conditions of up to 25 inches of mercury vacuum (84.5 KPa vacuum) in the supply piping and a backpressure conditions of 1 psi (6.89 KPa). See figure 10.3

Steps:

a. Remove the moving members (i.e., poppet or clapper) from the No.1 and the No.2 check valves. Maintain the No. 1 and No. 2 shutoff valves in the fully open position.

b. Install upstream and downstream of the assembly under test, orifices equivalent to the table below. Orifices shall comply with ASME/ANSI MFC-14M-2003[4] and the downstream orifice shall be located a minimum of five pipe diameters from the No. 2 shutoff valve. The resulting flow of water through the downstream orifice shall be normal to the pipeline. See figure 10.3. The assembly shall be subjected to a downstream condition of 1 psi (6.89 KPa).

c. With the backpressure established as in step b above, the assembly shall then be subjected to an upstream condition of 25 inches of mercury vacuum (84.5 KPa vacuum)

d. The vacuum pressure is to be slowly imposed (20 seconds ± 5 seconds) and maintained for 5 minutes. Any amount of water collecting in the upstream transparent tube shall be cause for rejection.

e. The 25 inches of mercury vacuum (84.5 KPa vacuum) is to be shock applied (by means of a quick opening valve) for five cycles and any amount of water collecting in the upstream transparent tube shall be cause for rejection.

f. The assembly shall be inspected for any damage due to the above vacuum tests. Any damage or permanent deformation shall be cause for rejection.

g. If a drain funnel is provided, tests e through f shall be repeated with the manufacturer's drain funnel attached to the assembly. Evidence of any amount of water collecting in the upstream transparent tube shall be cause for rejection.

---

[4] ASME Measurement of Fluid Flow Using Small Bore Precision Orifice Meters, 2003.

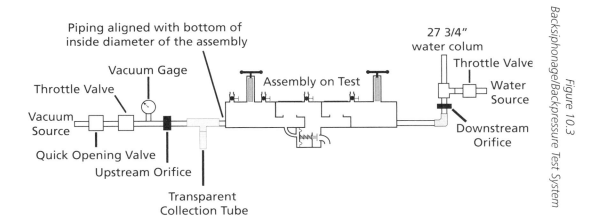

*Figure 10.3
Backsiphonage/Backpressure Test System*

*Table 10-5
Size of Orifices for Backpressure/Backsiphonage Test
10.1.2.2.3.9*

Size of Assembly		Upstream Orifice (Backsiphonage)		Downstream Orifice (Backpressure)	
(inches)	(mm)	(inches)	(mm)	(inches)	(mm)
1/2 and below	12	0.032	0.81	0.032	0.81
5/8×3/4	16×20	0.032	0.81	0.040	1.01
3/4	20	0.032	0.81	0.040	1.01
1	25	0.040	1.01	0.048	1.22
1 1/4	32	0.048	1.22	0.056	1.42
1 1/2	40	0.056	1.42	0.064	1.62
2	50	0.064	1.62	0.080	2.03
2 1/2	65	0.080	2.03	0.096	2.44
3	80	0.096	2.44	0.112	2.84
4	100	0.144*	2.84	0.144	3.66
6	150	0.144*	3.66	0.210	5.33
8	200	0.210*	5.33	0.275	6.98
10	250	0.210*	6.98	0.340	8.64
12	300	0.340	8.64	0.400	10.16
14	350	0.400	10.16	0.460	11.68
16	400	0.460	11.68	0.525	13.34

*NOTE: Since the relief discharge rates are the same for the 4 and 6 inch, 8 and 10 inch assemblies, the backsiphonage orifice will be the same for these pairings.

10.1.2.2.3.10

Purpose: To determine if there is adequate backflow protection between the drain funnel and the differential pressure relief valve discharge port(s).

Requirement: There shall be no backsiphonage of water from the drain funnel back into the differential pressure relief valve discharge port(s) under vacuum conditions.

Steps:

a. Attach inlet of assembly to vacuum source.
b. Hold in fully open position or remove the No. 1 check valve moving member (i.e., poppet or clapper).
c. Attach drain funnel to assembly per manufacturer's installation requirements. Plug outlet of drain funnel, then fill drain funnel to overflow with water.
d. Apply 25 inches of mercury vacuum (84.5 KPa vacuum) to inlet of assembly by means of a quick opening valve.
e. Evidence of water in the drain funnel carrying over into the relief valve discharge port(s) shall be cause for rejection.

10.1.2.2.3.11

Purpose: To determine if all components of the assembly shall operate properly under rated temperature and pressure conditions and flow rates as specified in Table 10-6

Requirements: All components of the assembly shall operate at and withstand a thermal test at the greater of 140°F (60°C) or at the maximum working water temperature (MWWT), maximum working water pressure (MWWP) and the flow rate as specified in Table 10-6, for a period of 100 hours without any leakage, damage, permanent deformation or impairment of operation.

Steps:

a. Install the assembly into a suitable test line, which can simultaneously generate the following minimum parameters:
   Maximum working water pressure (minimum of 150 psi [1034 KPa])
   Maximum working water temperature {minimum of 140°F (60°C)}
   Flow requirements per Table 10-6
b. The dimensions and durometer hardness of all elastomer components shall be inspected and recorded.
c. Set controls to the rated temperature and pressure conditions of the assembly, as well as rate of flow as specified in Table 10-6.
d. The temperature and pressure conditions shall be continuously monitored and recorded.
e. During and after 100 hours at rated temperature and pressure the assembly shall be tested to determine if the assembly operates satisfactorily per 10.1.2.2.3.3, 10.1.2.2.3.5 and 10.1.2.2.3.6. Failure to comply with these tests shall be cause for rejection.
f. The assembly shall then be disassembled and inspected for any damage or permanent deformation. Evidence of such shall be cause for rejection.
g. Once the assembly returns to ambient temperature conditions, the assembly shall then be tested per 10.1.2.2.3.1. Failure to comply with this test shall be cause for rejection.

Table 10-6
Thermal Test Minimum Flow Rates

Size of Assembly		Minimum Flow Rate	
(inches)	(mm)	(gpm)	(L/s)
1/4	(6)	1	(0.06)
3/8	(10)	3	(0.19)
1/2	(15)	7.5	(0.47)
5/8×3/4	(16×20)	20	(1.26)
3/4 - 1	(20 - 25)	30	(1.89)
1-1/4 - 2	(32 - 50)	40	(2.52)
2-1/2 - 16	(65 - 400)	50	(3.15)

10.1.2.2.3.12

Purpose: To determine if the assembly sustains any damage, permanent deformation or impairment of operation following the specified cycles.

Requirement: The assembly shall withstand the specified cycles without leakage, damage, permanent deformation or impairment of operation.

Steps:

a. Install assembly into suitable test line as shown in figure 10.4 that can generate the following parameter
    Rated flow of assembly under test (per Table 10-1)
    Line pressure of 60 psi ± 10 psi (414 KPa ± 69 KPa)
    Temperature of 110°F +30/-0°F (43°C. +17/-0°C)
    Backpressure of 150 psi ± 10 psi (1034 KPa ± 69 KPa)
    Control valves capable of closing/opening within 5 seconds.*
    *(Resulting rise of pressure shall not exceed MWWP of the assembly.)
b. With Valve #3 (Valve #3A open and Valve #3B closed) and Valve #4 (see figure 10.3) in the closed position and Valve #1 and Valve #2 in the open position and test cock No. 3 partially open, establish flow through the assembly equal to 25% of the rated flow per Table 10-1.
c. Maintain the flow through the assembly for a minimum of 6 seconds. Close Valve #1.
d. After 5 seconds, close Valve #2
e. After 3 seconds, open Valve #4.
f. After 3 seconds, open Valve #3. Maintain Valve #3 open for 6 seconds. Close Valve #3.
g. After 2 seconds, close Valve #4.
h. After 3 seconds, open Valve #2.
i. After 2 seconds, open Valve #1.
j. Repeat steps c through i 1250 times. The assembly shall be tested to determine if the assembly complies with 10.1.2.2.3.3 and 10.1.2.2.3.5 at line pressure only and 10.1.2.2.3.6.
k. Raise flow through the assembly to 50% of rated flow. Repeat steps c through i 1250 times. The assembly shall be tested to determine if the assembly complies with 10.1.2.2.3.3 and 10.1.2.2.3.5 at line pressure only and 10.1.2.2.3.6.

   1. With the pressure between the two check valves at atmospheric pressure (i.e., test cock No. 3 open), raise pressure at test cock No. 4 from atmospheric pressure to 1 psi (6.89KPa) and maintain for 10 minutes.
   2. Raise pressure at test cock No. 4 to 150 psi (1034 KPa) for 10 minutes. Lower pressure at test cock No. 4 to atmospheric pressure. Close test cocks No. 3 & No. 4.

3. Raise pressure at the inlet of the assembly to 20 psi (138 KPa) and maintain for 10 minutes.
4. Raise pressure at the inlet of the assembly to maximum working water pressure (MWWP) and maintain for 10 minutes.
5. Leakage during any of the above tests, k1- k4, shall be cause for rejection

l. Raise flow through the assembly to 75% of rated flow. Repeat steps c through i 1250 times. The assembly shall be tested to determine if the assembly complies with 10.1.2.2.3.3 and 10.1.2.2.3.5 at line pressure only and 10.1.2.2.3.6.
m. Raise flow through the assembly to 100% of rated flow. Repeat steps c through i 1250 times.
n. After 5000 total cycles the assembly shall be tested to determine if the assembly complies with 10.1.2.2.3.3 and 10.1.2.2.3.5 at line pressure only and 10.1.2.2.3.6. Failure to comply with these tests shall be cause for rejection. Perform steps k1 through k5.
o. The assembly shall then be disassembled and inspected for any damage or permanent deformation to any component(s), other than the elastomer discs. Evidence of such shall be cause for rejection.

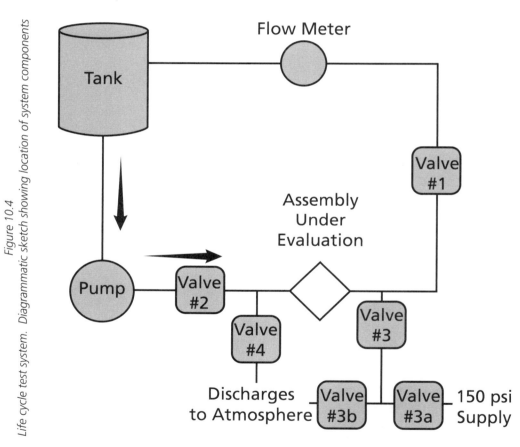

Figure 10.4
Life cycle test system. Diagrammatic sketch showing location of system components

10.1.2.2.3.13

Purpose: To determine the overall compliance of the assembly with the Standard as set forth in Chapter 10 of this manual.

Requirement: All sections of the Standard of this manual that are not specifically covered by the above tests shall be satisfactorily met.

Steps:
a. For each pertinent section of the Standard inspect, test or otherwise be assured that the assembly meets the minimum conditions of this Standard.

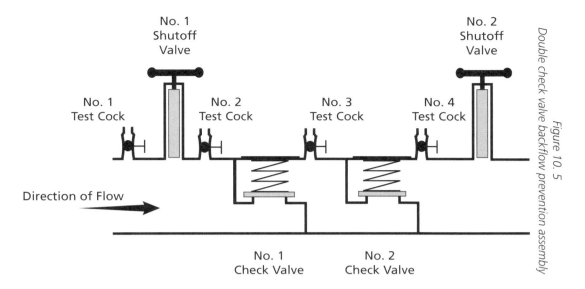

Figure 10.5 Double check valve backflow prevention assembly

10.1.2.3 Design, Operational and Evaluation Standard for Double Check Valve Backflow Prevention Assemblies (DC)

10.1.2.3.1 Design and Operational Requirements

a. This assembly shall include two independently acting approved check valves mounted between two tightly closing resilient seated shutoff valves and the necessary four resilient seated test cocks. See figure 10.5.
b. Each check valve shall be internally loaded and shall at all times be drip-tight in the normal direction of flow with the inlet pressure at least 1 psi (6.89 KPa) and the outlet under atmospheric pressure.
c. Each check valve shall permit no leakage in a direction reverse to the normal flow under all conditions of a pressure differential.

10.1.2.3.2 Laboratory Evaluation

One assembly shall be inspected and evaluated in the following order. Unless otherwise noted, tests shall be performed at ambient temperature.

a. Conformance to the general design and material requirements outlined in 10.1.
b. Conformance to the operational requirements outlined in 10.1.2.
c. Conformance to the working drawings and materials specifications.
d. Hydrostatic tests: the test assembly shall be subjected to the conditions of 10.1.1.1.5.
e. Pressure loss characteristics for flow rates up to the rated conditions.
f. Continuous flow of water from test cocks
g. Static closing point of check valve No. 1.
h. Static closing point of check valve No. 2.
i. Interdependence of components
j. Thermal Test: the assembly shall be evaluated at the greater of 140°F (60°C) or the maximum working water temperature (MWWT), maximum working water pressure (MWWP) and specified flow rate.
k. Life cycle test.

10.1.2.3.3 Evaluation Procedure

10.1.2.3.3.1

Purpose: To determine the capability of the assembly to withstand the required hydrostatic test pressure.

Requirement: All components of the assembly shall withstand a hydrostatic pressure of twice the maximum working water pressure (MWWP), as stated by the manufacturer, for a period of ten minutes without any damage, permanent deformation or impairment of operation.

Steps:

a. With both shutoff valves closed and all of the air bled from the assembly, raise the pressure from atmospheric to a hydrostatic pressure of twice the maximum working water pressure supplied through the No. 2 test cock for a period of 10 minutes. Close test cock No. 2, then open test cock No. 4 to relieve pressure. Close test cock No. 4. Any evidence of leakage up to and including hydrostatic pressure shall be cause for rejection.
b. Raise pressure from atmospheric to the required hydrostatic pressure supplied through test cock No. 3 with test cock No. 2 open to atmosphere for a period of 10 minutes. Close test cock No. 3, then open test cock No. 4 to relieve pressure. Any evidence of leakage up to and including hydrostatic pressure shall be cause for rejection.
c. Raise pressure from atmospheric to the required hydrostatic pressure supplied through test cock No. 4 with test cock No. 3 open to atmosphere for a period of ten minutes. Any evidence of leakage shall be cause for rejection.
d. The assembly shall be disassembled and inspected for any damage, permanent deformation or impairment of operation. Evidence of such shall be cause for rejection.

10.1.2.3.3.2

Purpose: To determine the overall pressure loss of the complete assembly as a function of the rate of flow.

Requirement: At any rate from static (0.0 gpm [0.0 L/s]) up to and including the rated flow, the overall pressure loss shall not exceed the values shown in Table 10-1. The assembly shall withstand 200% of rated flow without damage, permanent deformation or impairment of operation.

Steps:

a. Install the assembly in a suitable hydraulic test line having a flow capacity of twice the rated flow conditions (see Table 10-1) and an available line pressure of at least 150 psi (1034 KPa).
b. Install in the supply line at a suitable location upstream from the assembly a piezometer ring; install a similar piezometer ring at a suitable location downstream from the assembly (located as indicated in ANSI/ISA S75.02). Then install a mercury manometer or other suitable means of measuring pressure differentials between the two piezometer rings to measure the observed overall pressure loss at various rates of flow.
c. Record differential pressure between piezometer rings at static condition. Increase flow rate and record steady state differential pressure at a sufficient number of increments to fully define the flow curve characteristics. Data shall be taken for both increasing and decreasing flow conditions. The flow rate shall then be increased to 200% of the rated flow for 5 minutes and then returned to static.
Manifold Assembly (10.1.1.2.18) Only: Close all shutoff valves so that only one of the individual backflow prevention assemblies of the manifold assembly is open. Gradually increase the flow of water through the individual assembly until the rated flow of the manifold assembly is reached, then return to static condition. Repeat flow test for each individual assembly contained in the manifold assembly.

d. Remove the assembly from the line and couple the downstream piping directly to the supply piping. Then observe data for a friction head correction curve for the pipe fittings required to adapt the test assembly to the supply line between the existing piezometer rings. The data of this correction curve are then subtracted from the observed overall pressure loss data of the assembly from the upstream face of the upstream shutoff valve to the downstream face of the downstream shutoff valve.
e. Exceeding the maximum allowable pressure loss listed in Table 10-1 for any rate of flow up to and including rated flow, shall be cause for rejection. The assembly shall be disassembled and inspected for any damage, permanent deformation or impairment of operation. Evidence of such shall be cause for rejection.

10.1.2.3.3.3

Purpose: To determine if there is a continuous flow of water from each of the test cocks..

Requirement: There shall be a continuous flow of water for each test cock when opened one at a time.

Steps:

a. Install the assembly in a suitable hydraulic test line, which is capable of maintaining an inlet pressure of at least 150 psi (1034 KPa) during the following steps.
b. Close the No. 2 shutoff valve while maintaining the No. 1 shutoff valve fully open.
c. Slowly open (4 seconds ± 1 second) test cock No. 1 until fully open. Then slowly close (4 seconds ± 1 second) test cock No. 1.
d. Repeat step c with test cocks No. 2, No. 3 and No. 4.
e. A non-continuous flow of water from a test cock shall be cause for rejection.

10.1.2.3.3.4

Purpose: To determine the static pressure drop across check valve No. 1.

Requirement: The No. 1 check valve shall be drip-tight in the normal direction of flow with the inlet pressure at least 1 psi (6.89 KPa) and the outlet pressure at atmospheric.

a. Install a vertical transparent tube at least six feet (1.83 m) long at test cock No. 2. If test cock No. 3 is not at the highest point of the check valve body, then another vertical transparent tube must be installed on test cock No. 3 so that it just rises above the top of the check valve body. (See figure 10.2.)
b. Open test cock No. 2 and fill the tube with water, then close test cock No. 2. If a tube is attached to test cock No. 3, open test cock No. 3 to fill the tube, then close test cock No. 3.
c. Close the No. 2 shutoff valve, then close the No. 1 shutoff valve.
d. Open test cock No. 3, then open test cock No.2.
e. Water from the tube on test cock No. 2 will flow through check valve No. 1 until closure of check valve No. 1 is attained. When no further fall of the water in the tube on test cock No. 2 is observed, record the height of the water in the tube on test cock No. 2 above the water level at test cock No. 3. The difference of height of water in the transparent tube(s) must be 27 3/4 inches (704.85 mm) or greater.
f. Failure to maintain a minimum level of 27 3/4 inches (704.85 mm) shall be cause for rejection.

10.1.2.3.3.5

Purpose: To determine the static pressure drop across check valve No. 2.

Requirement: The No. 2 check valve shall be drip-tight in the normal direction of flow with the inlet pressure at least 1 psi (6.89 KPa) and the outlet pressure at atmospheric.

Steps:

a. Install a vertical transparent tube at least six feet (1.83 m) long at test cock No. 3. If test cock No. 4 is not at the highest point of the check valve body, then another vertical transparent tube must be installed on test cock No. 4 so that it just rises above the top of the check valve body. (See figure 10.2.)
b. Open test cock No. 3 and fill the tube with water, then close test cock No. 3. If a tube is attached to test cock No. 4, open test cock No. 4 to fill the tube, then close test cock No. 4.
c. Close the No. 2 shutoff valve, then close the No. 1 shutoff valve.
d. Open test cock No. 4, then open test cock No. 3.
e. Water from the tube on test cock No. 3 will flow through check valve No. 2 until closure of check valve No. 2 is attained. When no further fall of the water in the tube on test cock No. 3 is observed, record the height of the water in the tube on test cock No. 3 above the water level at test cock No. 4. The difference of height of water in the transparent tube(s) must be 27 3/4 inches (704.85 mm) or greater.
f. Failure to maintain a minimum level of 27 3/4 inches (704.85 mm) shall be cause for rejection.

10.1.2.3.3.6

Purpose: To determine that each check valve shall be free to operate independently through its entire movement.

Requirement: With the assembly in its intended orientation(s), the movement of either the first or second check valve through its full limit of travel shall not affect the operation of the other check valve.

Steps:

a. Visually inspect the movement of the first check valve through its full limit of travel to determine if it contacts and affects the operation of the second check valve. (The proper operation of the second check valve shall be confirmed by performing test 10.1.2.3.3.5 to determine static pressure drop across second check valve.) If necessary, remove or disconnect the loading on the first check valve to allow for free movement of the first check valve.
b. Visually inspect the movement of the second check valve through its full limit of travel to determine if it contacts and affects the operation of the first check valve. (The proper operation of the first check valve shall be confirmed by performing test 10.1.2.3.3.4 to determine static pressure drop across first check valve.) If necessary, remove or disconnect the loading on the second check valve to allow for free movement of the second check valve.
c. Visually inspect the movement of each of the shutoff valves through their full limit of travel to determine if they contact and affect the operation of either the first or second check valve.
d. Failure to maintain a minimum level of 27 3/4 inches (704.85 mm) shall be cause for rejection.

10.1.2.3.3.7

Purpose: To determine if all components of the assembly shall operate properly under rated temperature and pressure conditions and flow rates as specified in Table 10-6.

Requirements: All components of the assembly shall operate at and withstand a thermal test at 140°F (60°C) or at the maximum working water temperature (MWWT) whichever is greater, maximum working water pressure (MWWP) and flow rate as specified in Table 10-6, for a period of 100 hours without any damage, permanent deformation or impairment of operation.

Steps:

a. Install the assembly into a suitable test line that can generate the following minimum parameters:

    Maximum working water pressure (minimum of 150 psi [1034 KPa])
    Maximum working water temperature (minimum of 140°F [60°C])
    Flow requirements per Table 10-6

b. The dimensions and durometer hardness of all elastomer components shall be inspected and recorded.
c. Set controls to the rated temperature and pressure conditions of the assembly, as well as rate of flow as specified in Table 10-6.
d. The temperature and pressure conditions shall be continuously monitored and recorded.
e. During and after 100 hours at rated temperature and pressure the assembly shall be tested to determine if the assembly operates satisfactorily per 10.1.2.3.3.4 and 10.1.2.3.3.5. Failure to comply with these tests shall be cause for rejection.
f. The assembly shall then be disassembled and inspected for any internal damage or permanent deformation. Evidence of such shall be cause for rejection.
g. Once the assembly returns to ambient temperature conditions, the assembly shall then be tested per 10.1.2.3.3.1. Failure to comply with this test shall be cause for rejection.

10.1.2.3.3.8

Purpose: To determine if the assembly sustains any damage, permanent deformation or impairment of operation following the specified cycles.

Requirement: The assembly shall withstand the specified cycle without leakage, damage, permanent deformation or impairment of operation.

Steps:

a. Install assembly into suitable test line, as shown in figure 10.4, that can generate the following parameters:

    Rated flow of assembly under test (per Table 10-1)
    Line pressure of 60 psi ± 10 psi (414 KPa ± 69 KPa)
    Temperature of 110°F +30/-0°F (43°C. +17/-0°C)
    Backpressure of 150 psi ± 10 psi (1034 KPa ± 69 KPa)
    Control valves capable of closing/opening within 5 seconds.*

    *(Resulting rise of pressure shall not exceed MWWP of the assembly.)

b. With Valve #3 (Valve #3A open and Valve #3B closed) and Valve #4 (see figure 10.4) in the closed position and Valve #1 and Valve #2 in the open position and test cock No. 3 partially open, establish flow through the assembly equal to 25% of the maximum rated flow per Table 10-1.
c. Maintain the flow through the assembly for a minimum of 6 seconds. Close Valve #1.
d. After 5 seconds, close Valve #2.
e. After 3 seconds, open Valve #4.
f. After 3 seconds, open Valve #3. Maintain Valve #3 open for 6 seconds. Close Valve #3.
g. After 2 seconds, close Valve #4.
h. After 3 seconds, open Valve #2.
i. After 2 seconds, open Valve #1.
j. Repeat steps c through i 1250 times. The assembly shall be tested to determine if the assembly complies with 10.1.2.3.3.4 and 10.1.2.3.3.5.
k. Raise flow through the assembly to 50% of rated flow. Repeat steps c through i 1250 times. The assembly shall be tested to determine if the assembly complies with 10.1.2.3.3.4 and 10.1.2.3.3.5.

1. With the pressure between the two check valves at atmospheric pressure (i.e., test cock No. 3 open), raise pressure at test cock No. 4 from atmospheric pressure to 1 psi (6.89 KPa) and maintain for 10 minutes.
2. Raise pressure at test cock No. 4 to 150 psi (1034 KPa) for 10 minutes. Lower pressure at test cock No. 4 to atmospheric pressure. Close test cocks No. 4.
3. With the pressure in front of No. 1 check valve atmospheric pressure (i.e., test cock No. 2 open), raise pressure at test cock No. 3 from atmospheric pressure to 1 psi (6.89 KPa) and maintain for 10 minutes.
4. Raise pressure at test cock No. 3 to 150 psi (1034 KPa) for 10 minutes. Lower pressure at test cocks No. 3 and No. 4 to atmospheric pressure. Close test cock No. 4 and maintain test cock No. 3 in the partially open position.
5. Leakage during any of the above tests, k1- k4, shall be cause for rejection

l. Raise flow through the assembly to 75% of rated flow. Repeat steps c through i 1250 times. The assembly shall be tested to determine if the assembly complies with 10.1.2.3.3.4 and 10.1.2.3.3.5.
m. Raise flow through the assembly to 100% of rated flow. Repeat steps c through i 1250 times.
n. After 5000 total cycles the assembly shall be tested to determine if the assembly complies with 10.1.2.3.3.4 and 10.1.2.3.3.5. Failure to comply with these tests shall be cause for rejection.
o. Perform steps k1 through k5.
p. The assembly shall then be disassembled and inspected for any damage or permanent deformation to any component(s), other than the elastomer discs. Evidence of such shall be cause for rejection.

10.1.2.3.3.9

Purpose: To determine the overall compliance of the assembly with the Standard as set forth in Chapter 10 of this manual.

Requirement: All sections of the Standard of this manual that are not specifically covered by the above tests shall be satisfactorily met.

Steps:

a. For each pertinent section of the Standard inspect, test or otherwise be assured that the assembly meets the minimum conditions of this Standard.

Table 10-7
*Rated Flow and Maximum Allowable Pressure Loss for
Various Sizes of Atmospheric (AVB) and
Pressure (PVB & SVB) Vacuum Breakers*

Maximum Allowable Pressure Loss

Size of the Assembly (1)		Rated Flow (2)		Atmospheric AVB (3)		Pressure PVB & SVB (4)	
(inches)	(mm)	(gpm)	(L/s)	(psi)	(KPa)	(psi)	(KPa)
1/4	(6)	1	(0.31)	24	(165.45)	10	(68.9)
3/8	(10)	3	(0.50)	22	(151.66)	10	(68.9)
1/2 ***	(15)	5	(0.75)	22	(151.66)	10	(68.9)
1/2 ****	(15)	7.5	(0.47)	NA		10	(68.9)
3/4	(20)	30	(1.89)	20	(137.88)	10	(68.9)
1	(25)	50	(3.15)	18	(124.09)	10	(68.9)
1 1/4	(32)	75**	(4.73)	18	(124.09)	10	(68.9)
1 1/2	(40)	100	(6.31)	16	(110.30)	10	(68.9)
2	(50)	160	(10.10)	16	(110.30)	10	(68.9)
2 1/2	(65)	225	(14.20)	16	(110.30)	10	(68.9)
3	(80)	320	(20.19)	15	(103.41)	10	(68.9)
4	(100)	500	(31.55)	14	(96.51)	10	(68.9)
6	(150)	1000	(63.10)	14	(96.51)	10	(68.9)
8	(200)	1600	(100.96)	14	(96.51)	10	(68.9)
10	(250)	2300	(145.13)	14	(96.51)	10	(68.9)

* Extrapolated
** Interpolated
*** AVB and PVB only
**** SVB only

Note: The pressure losses as shown in Column (3) and Column (4) represent the maximum permissible pressure loss at any flow rate up to and including the rated flow shown in Column (2).

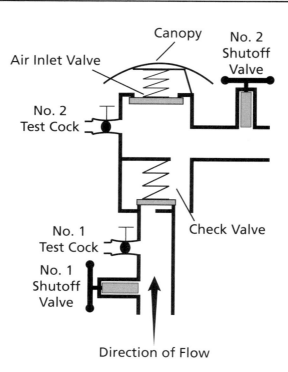

Figure 10.6 Pressure Vacuum Breaker Backsiphonage Prevention Assembly (PVB)

10.1.2.4 Design, Operational and Evaluation Standard for Pressure Vacuum Breaker Backsiphonage Prevention Assemblies (PVB)

10.1.2.4.1 Design and Operational Requirements

a. This assembly shall include an approved internally loaded check valve and a loaded air inlet valve opening to atmosphere on the discharge side of the check valve between two tightly closing resilient seated shutoff valves; it shall include two properly located resilient seated test cocks. (See figure 10.6).
b. The air inlet valve of the pressure vacuum breaker shall open when the internal pressure is a minimum of 1 psi (6.89 KPa). It shall be fully open when the water drains from the body.
c. The maximum allowable pressure drop across the assembly, from the upstream face of the No. 1 shutoff valve to the downstream face of the No. 2 shutoff valve, shall not exceed 10 psi (68.9 KPa) for any rate of flow up to and including the rated flow listed in Table 10-7.
d. The check valve shall be internally loaded and shall at all times be drip-tight in the normal direction of flow with the inlet pressure at 1 psi (6.89 KPa) and the outlet under atmospheric pressure.
e. The effective size of the air inlet port(s) of the assembly shall be governed by the vacuum dissipation test. If an air inlet port shield or canopy is used, it shall extend down around the body of the assembly to the lowest portion of the port(s). To reduce potential for fouling, the minimum clearance between the air inlet port(s) and the shield or canopy shall be 3/16 inch (4.76 mm).

10.1.2.4.2 Laboratory Evaluation

One assembly shall be inspected and evaluated in the following order. Unless otherwise noted, tests shall be performed at ambient temperature.

a. Conformance to general, design and material requirements outlined in 10.1.
b. Conformance to the operational requirements outlined in 10.1.2.

c. Conformance to the working drawings and materials specifications.
d. Hydrostatic tests: the test assembly shall be subjected to the conditions of 10.1.1.1.5.
e. Pressure loss characteristics for flow rates up to the rated conditions set forth in Table 10-7.
f. Test cock flow test
g. Conformance to the air inlet requirements.
h. Static closing point of the check valve.
i. Interdependence of Components
j. Thermal Test: the assembly shall be evaluated at the greater of 140°F (60°C) or the maximum working water temperature (MWWT), maximum working water pressure (MWWP) and specified flow rate.
k. Life cycle test.

10.1.2.4.3 Evaluation Procedure

10.1.2.4.3.1

Purpose: To determine the capability of the assembly to withstand the required hydrostatic test pressure.

Requirement: All components of the assembly shall withstand a hydrostatic pressure of twice the maximum working water pressure (MWWP), as stated by the manufacturer, for a period of ten minutes without any damage, permanent deformation or impairment of operation.

Steps:

a. With both shutoff valves closed and all of the air bled from the assembly, raise pressure from atmospheric to a hydrostatic pressure of twice the rated working water pressure supplied through the No. 1 test cock for a period of ten minutes. Close test cock No. 1 and open test cock No. 2 to relieve pressure. Any evidence of leakage up to and including hydrostatic pressure shall be cause for rejection.
b. Raise pressure from atmospheric to the required hydrostatic pressure supplied through test cock No. 2 with test cock No. 1 open to atmosphere for a period of ten minutes. Close test cock No. 2 and isolate assembly from pressure source. Open test cock No. 2 to relieve pressure. Any evidence of leakage shall be cause for rejection.
c. The assembly shall be disassembled and inspected for any damage, permanent deformation, or impairment of operation. Evidence of such shall be cause for rejection.

10.1.2.4.3.2

Purpose: To determine the overall pressure loss of the complete assembly as a function of the rate of flow.

Requirement: At any rate from static (0.0 gpm [0.0 L/s]) up to and including the rated flow, the overall pressure loss shall not exceed the values shown in Table 10-7. The assembly shall withstand 200% of the rated flow without damage, permanent deformation or impairment of operation.

Steps:

a. Install the assembly in a suitable hydraulic test line having a flow capacity of twice the rated flow conditions (see Table 10-7) and an available line pressure of at least 150 psi (1034 KPa).
b. Install, in the supply line at a suitable location upstream from the assembly, a piezometer ring; install a similar piezometer ring at a suitable location downstream from the assembly (located as indicated in ANSI/ISA Standard S75.02). Then install a mercury manometer or other suitable means of measuring pressure differentials between the two piezometer rings to measure the observed overall pressure loss at various rates of flow.

c. Record differential pressures between piezometer rings at static condition. Increase flow rate and record steady state differential pressure at a sufficient number of increments to fully define the flow curve characteristic. Data shall be taken for both increasing and decreasing flow conditions. The flow rate shall then be increased to 200% of the rated flow for five minutes and then returned to static.
d. Remove the assembly from the line and couple the downstream piping directly to the supply piping. Then observe data for a friction head correction curve for the pipe fittings required to adapt the test assembly to the supply line between the existing piezometer rings. The data of this correction curve are then subtracted from the observed overall pressure loss data of the assembly from the upstream face of the upstream shutoff valve to the downstream face of the downstream shutoff valve.
e. Exceeding the maximum allowable pressure loss listed in Table 10-7 for any rate of flow up to and including rated flow, shall be cause for rejection. The assembly shall be disassembled and inspected for any damage, permanent deformation or impairment of operation. Evidence of such shall be cause for rejection.

10.1.2.4.3.3

Purpose: To determine if there is a continuous flow of water from each of the test cocks.

Requirement: There shall be a continuous flow of water for each test cock when opened one at a time.

Steps:

a. Install the assembly in a suitable hydraulic test line, which is capable of maintaining an inlet pressure of at least 150 psi (1034 KPa) during the following steps.
b. Close the No. 2 shutoff valve while maintaining the No. 1 shutoff valve fully open.
c. Slowly open (4 seconds ± 1 second) test cock No. 1 until fully open. Then slowly close (4 seconds ± 1 second) test cock No. 1.
d. Repeat step c with test cock No. 2.
e. A non-continuous flow of water from a test cock shall be cause for rejection.

10.1.2.4.3.4

Purpose: To test the opening pressure differential of the air inlet valve.

Requirement: The air inlet valve shall open when the pressure in the body is a minimum of 1.0 psi (6.89 KPa) above atmospheric pressure. The air inlet valve shall be fully open when the water drains from the body.

Steps:

a. Install the assembly in a suitable hydraulic test line with a minimum pressure of 150 psi (1034 KPa). Remove air inlet valve canopy.
b. Install the high side hose of the differential pressure gage or other suitable means of measuring pressure differentials, to test cock No. 2 and bleed air from the hose and gage.
c. Close shutoff valve No. 2; and then close shutoff valve No. 1.
d. Slowly (0.2 psid (1.38 KPa)/second maximum) open the high side bleed needle valve on the gage. Record the pressure differential at which the air inlet valve opens. Failure to open at a value of 1.0 psi (6.89KPa) or greater shall be cause for rejection.
e. Allow water to drain out of the body through test cock No. 2 to determine if the air inlet valve opens fully. Failure to open fully shall be cause for rejection.

10.1.2.4.3.5

Purpose: To determine the static pressure drop across the check valve.

Requirement: The check valve shall be drip tight in the normal direction of flow when the inlet pressure is at least 1 psi (6.89 KPa) and the outlet pressure is atmospheric.

Steps:

a. Install a transparent tube approximately 6 feet (180 cm) long on test cock No. 1. Open test cock No. 1 to fill the tube with water, then close test cock No. 1.
b. Close shutoff valve No. 2, then close shutoff valve No. 1.
c. Open test cock No. 2, then open test cock No. 1. The air inlet valve will open; and, the water in the tube will fall until closure of the check valve is attained. When no further fall of the water in the tube is observed, record the height of water in the transparent tube above the centerline of test cock No. 2.
d. Failure to maintain a minimum level of 27 3/4 inches (704.85 mm) shall be cause for rejection.

10.1.2.4.3.6

Purpose: To determine that the check valve and air inlet valve shall be free to operate independently through their entire movement.

Requirement: With the assembly in its intended orientation(s), the movement of either the check valve or air inlet valve through their full limit of travel shall not affect the operation of the other.

Steps:

a. Visually inspect the movement of the check valve through its full limit of travel to determine if it contacts and affects (the proper operation of the air inlet valve shall be confirmed by performing test 10.1.2.4.3.4) the operation of the air inlet valve. If necessary, remove or disconnect the loading on the check valve to allow for free movement of the check valve.
b. Visually inspect the movement of the air inlet valve through its full limit of travel to determine if it contacts and affects (the proper operation of the check valve shall be confirmed by performing test 10.1.2.4.3.5 to determine static pressure drop across first check valve) the operation of the check valve.
c. Visually inspect the movement of each of the shutoff valves through their full limit of travel to determine if they contact and affect the operation of either the check valve or air inlet valve.
d. Failure in steps a, b or c shall be cause for rejection.

10.1.2.4.3.7

Purpose: To compare the air flow capacity of the air inlet valve to that of the effective water throughway of the assembly.

Requirement: The time required to dissipate a vacuum (25 to 5 inches of mercury [84.5 to 16.9 KPa]) through the air inlet valve shall be less than the time required to dissipate the same vacuum through the water throughway.

Steps:

a. Install the assembly in a normal operating position, with the outlet of the assembly attached to a vacuum source.

b. Close No. 1 shutoff valve to simulate the check valve being closed. (No. 2 shutoff valve remains fully open during entire test procedure.)
c. With the air inlet valve open, record the time required to dissipate the vacuum from 25 to 5 inches of mercury vacuum (84.5 to 16.9 KPa vacuum) when this flow is controlled by means of a quick opening valve. Repeat this step a minimum of three times, then average the recorded times.
d. Open No. 1 shutoff valve and attach a 12 inch (30.5 cm) long pipe nipple, the same nominal pipe diameter as the assembly, to the inlet of the No. 1 shutoff valve.
e. Hold the air inlet valve closed and hold the check valve in the fully open position; then record the time required to dissipate the vacuum from 25 to 5 inches of mercury vacuum (84.5 to 16.9 KPa vacuum) when this flow is controlled by means of a quick opening valve. Repeat this step a minimum of three times, then average the recorded times.
f. Based upon the average of at least three tests, the time required for step c should be less than the time required for step e. Failure to comply with this requirement shall be cause for rejection.

10.1.2.4.3.8

Purpose: To test that a backsiphonage condition together with a fouled check valve shall not allow a carry-over of downstream water back into the inlet.

Requirement: Water shall not be backsiphoned more than 6 inches (152 mm) above the free surface of water in the discharge pipe standing 12 inches (305 mm) below the bottom or critical level line of the assembly.

Steps:

a. Install the assembly in a normal operating position, with the inlet of the assembly connected to the vacuum source and the discharge side equipped with a transparent tube that extends down into a vessel of water where the free surface of the water is 12 inches (305 mm) below the bottom or critical level line of the assembly.
b. A wire of the size shown in Table 10-8 shall be placed across the seating surface in the lower quadrant of a hinged or horizontally moving poppet check valve or at a single point of a vertical moving check valve. The wire shall be placed so that the entire width of the seating contact surface is covered.
c. By the application of vacuum conditions ranging from 0 to 25 inches of mercury (84.5 KPa vacuum) under both steady flow and instantaneous conditions, the water in the discharge tube shall not rise more than 6 inches (152 mm) above the original free surface of the water in the vessel. Evidence of such shall be cause for rejection.

Table 10-8
Size of Fouling Wire for Backsiphonage Tests

Size of Assembly		Diameter of Wire	
(inches)	(mm)	(inches)	(mm)
1/2 and below	12	0.032	0.81
5/8×3/4	19	0.040	1.01
3/4	20	0.040	1.01
1	25	0.048	1.22
1 1/4	35	0.056	1.42
1 1/2	40	0.064	1.62
2	50	0.080	2.03
2 1/2	65	0.096	2.44
3	80	0.112	2.84
4	100	0.144	3.66
6	150	0.210	5.33
8	200	0.275	6.98
10	250	0.340	8.64
12	300	0.400	10.16
14	350	0.460	11.68
16	400	0.525	13.34

10.1.2.4.3.9

Purpose: To determine if all components of the assembly shall operate properly under rated temperature and pressure conditions and flow rates as specified in Table 10-6.

Requirements: All components of the assembly shall operate at and withstand a thermal test at the greater of 140°F (60°C) or at the maximum working water temperature (MWWT), maximum working water pressure (MWWP) and flow rate as specified in Table 10-6 for a period of 100 hours without any damage, permanent deformation or impairment of operation.

Steps:

a. Install assembly into suitable test line that can generate the following minimum parameters:
   Maximum working water pressure (minimum of 150 psi [1034 KPa])
   Maximum working water temperature (minimum of 140°F [60°C])
   Flow requirements per Table 10-6
b. The dimensions and durometer hardness of all elastomer components shall be inspected and recorded.
c. Set controls to the rated temperature and pressure conditions of the assembly, as well as rate of flow as specified in Table 10-6.
d. The temperature and pressure conditions shall be continuously monitored and recorded.
e. During and after 100 hours at rated temperature and pressure the assembly shall be tested to determine if the assembly operates satisfactorily per 10.1.2.4.3.4 and 10.1.2.4.3.5. Failure to comply with these tests shall be cause for rejection.
f. The assembly shall then be disassembled and inspected for any internal damage or permanent deformation. Evidence of such shall be cause for rejection.

g. Once the assembly returns to ambient temperature conditions, the assembly shall then be tested per 10.1.2.4.3.1. Failure to comply with this test shall be cause for rejection.

10.1.2.4.3.10

Purpose: To determine if the assembly sustains any damage, permanent deformation or impairment of operation following the specified cycles.

Requirement: The assembly shall withstand the specified cycle without leakage, damage, permanent deformation or impairment of operation.

Steps:

a. Install assembly into suitable test line, as shown in figure 10.4, that can generate the following parameters:
   Rated flow of assembly under test (per Table 10-7)
   Line pressure of 60 psi ± 10 psi (414 KPa ± 69 KPa)
   Temperature of 110°F +30/-0°F (43°C. +17/-0°C)
   Control valves capable of closing/opening within 5 seconds.*
   *(Resulting rise of pressure shall not exceed MWWP of the assembly.)
b. With Valve #3 (Valve #3A closed and Valve #3B open) and Valve #4 (figure 10.4) in the closed position and Valve #1 and Valve #2 in the open position, establish flow through the assembly equal to 25% of the maximum rated flow per Table 10-7.
c. Maintain the flow through the assembly for a minimum of 6 seconds. Close Valve #1.
d. After 5 seconds, close Valve #2.
e. After 3 seconds, open Valve #3.
f. After 10 seconds, close Valve #3.
g. After 3 seconds, open Valve #2.
h. After 3 seconds, open Valve #1.
i. Repeat steps c through h 1250 times. The assembly shall be tested to determine if the assembly complies with 10.1.2.4.3.4 and 10.1.2.4.3.5. Failure to comply with these tests shall be cause for rejection.
j. Raise flow through the assembly to 50% of rated flow. Repeat steps c through h 1250 times. The assembly shall be tested to determine if the assembly complies with 10.1.2.4.3.4 and 10.1.2.4.3.5

   1. With both shut off valves closed and all air bled from the assembly, open test cock No. 1 to atmosphere and maintain a pressure of 150 psi (1034 KPa) at test cock No. 2 for ten minutes.
   2. Lower pressure at test cock No. 2 to 10 psi (68.9 KPa) and maintain for ten minutes.
   3. Leakage during steps j1 or j2 shall be cause for rejection.

k. Raise flow through the assembly to 75% of rated flow. Repeat steps c through h 1250 times. The assembly shall be tested to determine if the assembly complies with 10.1.2.4.3.4 and 10.1.2.4.3.5.
l. Raise flow through the assembly to 100% of rated flow. Repeat steps c through h 1250 times.
m. After 5000 total cycles the assembly shall be tested to determine if the assembly complies with 10.1.2.4.3.4 and 10.1.2.4.3.5. Failure to comply with these tests shall be cause for rejection. Perform steps j1 through j3.
n. The assembly shall then be disassembled and inspected for any damage or permanent deformation to any component(s), other than the elastomer discs. Evidence of such shall be cause for rejection.

10.1.2.4.3.11

Purpose: To determine the overall compliance of the assembly with the Standard as set forth in Chapter 10 of this manual.

Requirement: That all sections of the Standard of this manual that are not specifically covered by the above tests are satisfactorily met.

Steps:

a. For each pertinent section of the Standard inspect, test or otherwise be assured that the assembly meets the minimum conditions of this Standard.

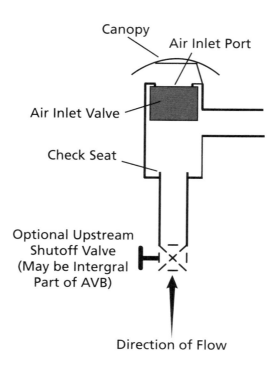

*Figure 10.7
Atmospheric Vacuum Breaker Backsiphonage Prevention Assembly (AVB)*

10.1.2.5 Design, Operational and Evaluation Standard for Atmospheric Vacuum Breaker Backsiphonage Prevention Assemblies (AVB)

10.1.2.5.1 Design and Operational Requirements

a. This assembly shall include an air inlet valve or float check, a check seat and air inlet port(s). A tightly closing integral shutoff valve may only be included on the upstream side of the air inlet valve. See figure 10.7.
b. The air inlet valve of the atmospheric vacuum breaker shall open when the internal pressure is atmospheric. It shall be fully open when the water drains from the body.
c. The maximum allowable pressure drop across the assembly, from the upstream face of the assembly to the downstream face of the assembly, for any rate of flow up to and including the rated flow shall not exceed the values given in Table 10-7.
d. The effective size of the air inlet port(s) of the assembly shall be governed by the vacuum dissipation test. If an air inlet port shield or canopy is used, it shall extend down around the body of the assembly to the lowest portion of the port(s). To reduce the potential for fouling, the minimum clearance between the air inlet port(s) and the shield or canopy shall be 3/16 inch (4.76 mm).

10.1.2.5.2 Laboratory Evaluation

One assembly shall be inspected and evaluated in the following order. Unless otherwise noted, tests shall be performed at ambient temperature.

a. Conformance to general, design and material requirements outlined in 10.1.
b. Conformance to the operational requirements outlined in 10.1.2.
c. Conformance to the working drawings and materials specifications.

d. Hydrostatic tests: the test assembly shall be isolated and subjected to the conditions of 10.1.1.1.5.
e. Pressure loss characteristics for flow rates up to the rated conditions.
f. Conformance to the air inlet requirements.
g. Thermal Test: the assembly shall be evaluated at the greater of 140°F (60°C) or the maximum working water temperature (MWWT), maximum working water pressure (MWWP) and specified flow rate.
h. Life cycle test.

10.1.2.5.3 Evaluation Procedure

10.1.2.5.3.1
Purpose: To determine the capability of the assembly to withstand the required hydrostatic test pressure.

Requirement: All components of the assembly shall withstand a hydrostatic pressure of twice the maximum working water pressure (MWWP), as stated by the manufacturer, for a period of 10 minutes without any damage, permanent deformation or impairment of operation.

Steps:

a. With the assembly isolated both upstream and downstream and all of the air bled from the assembly, a hydrostatic pressure of twice the maximum working water pressure shall be supplied through the inlet of the assembly for a period of ten minutes. Any evidence of leakage shall be cause for rejection.
b. The assembly shall be disassembled and inspected for any internal damage as well as damage to the external cover plates or other components. Evidence of such shall be cause for rejection.

10.1.2.5.3.2
Purpose: To determine the overall pressure loss of the complete assembly as a function of the rate of flow.

Requirement: At any rate from static (0.0 gpm [0.0 L/s]) up to and including the rated flow the overall pressure loss shall not exceed the values shown in Table 10-7. The assembly shall withstand 200% of the rated flow without any damage, permanent deformation or impairment of operation.

Steps:

a. Install the assembly in a suitable hydraulic test line having a flow capacity of twice the rated flow conditions (see Table 10-7) and an available line pressure of at least 150 psi (1034 KPa).
b. Install in the supply line at a suitable location upstream from the assembly a piezometer ring; and install a similar piezometer ring at a suitable location downstream from the assembly (located as indicated in ANSI/ISA Standard S75.02). Then install a mercury manometer or other suitable means of measuring pressure differentials between the two piezometer rings to measure the observed overall pressure loss at various rates of flow.
c. Record differential pressure between piezometer rings at static condition. Increase flow rate and record steady state differential at a sufficient number of increments to fully define flow curve characteristics. Data shall be taken for both increasing and decreasing flow conditions. The flow rate shall then be increased to 200% of the rated flow for 5 minutes and then returned to static.
d. Remove the assembly from the line and couple the downstream piping directly to the supply piping. Then observe data for a friction head correction curve for the pipe fittings required to adapt the test assembly to the supply line between the existing piezometer rings. The data

of this correction curve are then subtracted from the observed overall pressure loss data of the assembly from the upstream face of the assembly to the downstream face of the assembly.

e. Exceeding the maximum allowable pressure loss listed in Table 10-7 for any rate of flow up to and including rated flow, shall be cause for rejection. The assembly shall be disassembled and inspected for any damage, permanent deformation or impairment of operation. Evidence of such shall be cause for rejection.

10.1.2.5.3.3

Purpose: To test the opening of the air inlet valve.

Requirement: The air inlet valve shall open when the pressure in the body is atmospheric. And, the air inlet valve shall be fully open when the water drains from the body.

Steps:

a. Remove air inlet valve canopy.
b. Install the assembly in a suitable hydraulic test line with a minimum pressure of 150 psi (1034 KPa).
c. Isolate the assembly in the test line both upstream and downstream.
d. Lower the pressure in the body to atmospheric, the air inlet valve shall begin to open. Continue to drain the water from the body. Once all of the water has drained out, the air inlet valve shall be fully open.
e. Failure to fully open shall be cause for rejection.

10.1.2.5.3.4

Purpose: To compare the airflow capacity of the air inlet valve to that of the effective water throughway of the assembly.

Requirement: The time required to dissipate a vacuum 25 to 5 inches of mercury vacuum (84.5 to 16.9 KPa vacuum) through the air inlet valve shall be less than the time required to dissipate the same vacuum through the water throughway.

Steps:

a. Install the assembly in the normal operating position, with the discharge outlet of the assembly attached to a vacuum source.
b. Close the inlet by plugging or other suitable means. (If an integral upstream shutoff valve is present, close the shutoff valve) With the air inlet valve open, record the time required to dissipate the vacuum from 25 to 5 inches of mercury vacuum (84.5 to 16.9 KPa vacuum) when this flow is controlled by means of a quick opening valve. Repeat this step a minimum of three times, then average the recorded values.
c. Open the inlet (if integral upstream shutoff valve is present, open shutoff valve) and attach a 12 inch (30.5 cm) long pipe nipple, the same nominal pipe diameter as the inlet connection of the assembly.
d. Hold the air inlet valve closed; then record the time required to dissipate the vacuum from 25 to 5 inches of mercury vacuum (84.5 to 16.9 KPa vacuum) when this flow is controlled by means of a quick opening valve. Repeat this step a minimum of three times, then average the recorded times.
e. Based upon the average of at least three tests, the time required for step b should be less than the time required for step d. Failure to comply with this requirement shall be cause for failure.

10.1.2.5.3.5

Purpose: To test that a backsiphonage condition together with a fouled check valve shall not allow a carry-over of downstream water back into the inlet.

Requirement: Water shall not be backsiphoned more than 3 inches (76 mm) above the free surface of water in the discharge pipe standing 6 inches (152 mm) below the bottom or critical level line of the assembly.

Steps:

a. Install the assembly in a normal operating position, with the inlet of the assembly connected to the vacuum source and the discharge side equipped with a transparent tube that extends down into a vessel of water where the free surface of the water is 6 inches (152 mm) below the bottom or critical level line of the assembly.
b. A wire of the size shown in Table 10-8 shall be placed across the seating surface in the lower quadrant of a hinged or horizontally moving poppet check valve or at a single point of a vertical moving check valve. The wire shall be placed so that the entire width of the seating contact surface is covered.
c. By the application of vacuum ranging from 0 to 25 inches of mercury (84.5 KPa vacuum) under both steady flow and instantaneous conditions, the water in the discharge tube shall not rise more than 3 inches above the original free surface of the water in the vessel. Evidence of such shall be cause for rejection.

10.1.2.5.3.6

Purpose: To determine if all components of the assembly shall operate properly under rated temperature and pressure conditions and flow rates as specified in Table 10-6.

Requirements: All components of the assembly shall operate at and withstand a thermal test at 140°F (60°C) or at the maximum working water temperature (MWWT) whichever is greater, maximum working water pressure (MWWP) and flow rate as specified in Table 10-6, for a period of 100 hours without any damage, permanent deformation or impairment of operation.

Steps:

a. Install assembly into suitable test line that can generate the following minimum parameters:
   Maximum working water pressure (minimum of 150 psi [1034 KPa])
   Maximum working water temperature (minimum of 140°F [60°C])
   Flow requirements per Table 10-6.
b. The dimensions and durometer hardness of all elastomer components shall be inspected and recorded.
c. Set controls to the rated temperature and pressure conditions of the assembly as well as rate of flow as specified in Table 10-6.
d. The temperature and pressure conditions shall be continuously monitored and recorded.
e. During and after 100 hours at rated temperature and pressure the assembly shall be tested to determine if the assembly operates satisfactorily per 10.1.2.5.3.3. Failure to comply with this test shall be cause for rejection.
f. The assembly shall then be disassembled and inspected for any internal damage or permanent deformation. Evidence of such shall be cause for rejection.
g. Once the assembly returns to ambient temperature conditions, the assembly shall then be tested per 10.1.2.5.3.1. Failure to comply with this test shall be cause for rejection.

10.1.2.5.3.7

Purpose: To determine if the assembly sustains any damage, permanent deformation or impairment of operation following the specified cycles.

Requirement: The assembly shall withstand the specified cycle without leakage, damage, permanent deformation or impairment of operation.

Steps:

a. Install assembly into suitable test line, as shown in figure 10.4, that can generate the following parameters:
   Rated flow of assembly under test (per Table 10-7)
   Line pressure of 60 psi ± 10 psi (414 KPa ± 69 KPa)
   Temperature of 110°F +30/-0°F (43°C +17/-0°C)
   Control valves capable of closing/opening within 5 seconds.*
   *(Resulting rise of pressure shall not exceed MWWP of the assembly.)
b. With Valve #3 (Valve 3A closed and Valve 3B open) and Valve #4 (Fig. 10.4) in the closed position and Valve #1 and Valve #2 in the open position, establish flow through the assembly equal to 25% of the maximum rated flow per Table 10-7.
c. Maintain the flow through the assembly for a minimum of 6 seconds. Close Valve #1.
d. After 5 seconds, close Valve #2
e. After 3 seconds, open Valve #3.
f. After 10 seconds, close Valve #3.
g. After 3 seconds, open Valve #2.
h. After 3 seconds, open Valve #1.
i. Repeat steps c through h 5000 times.
j. After 5000 total cycles the assembly shall be tested to determine if the assembly complies with the following:

   1. With the assembly isolated downstream, raise inlet pressure from atmospheric up to 10 psi (68.9 KPa) and hold for 10 minutes.
   2. Raise inlet pressure to 150 psi (1034 KPa) and hold for 10 minutes.
   3. Leakage during steps j1 or j2 shall be cause for rejection.

k. The assembly shall be tested to determine if the assembly complies with 10.1.2.5.3.3. Failure to comply with this test shall be cause for rejection.
l. The assembly shall then be disassembled and inspected for any damage or permanent deformation to any component(s) other than the elastomer disc. Evidence of such shall be cause for rejection.

10.1.2.5.3.8

Purpose: To determine the overall compliance of the assembly with the Standard as set forth in Chapter 10 of this manual.

Requirement: All sections of the Standard of this manual that are not specifically covered by the above tests shall be satisfactorily met.

Steps:

a. For each pertinent section of the Standard inspect, test or otherwise be assured that the assembly meets the minimum conditions of this Standard.

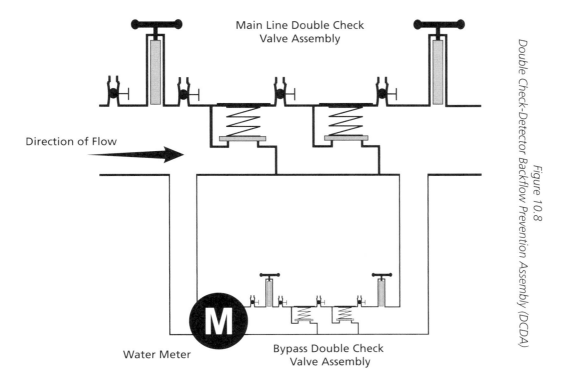

Figure 10.8
Double Check-Detector Backflow Prevention Assembly (DCDA)

10.1.2.6 Design, Operational and Evaluation Standard for Double Check Detector Backflow Prevention Assemblies (DCDA)

10.1.2.6.1 Design and Operational Requirements

a. This assembly shall include a line-size approved double check valve assembly with a parallel water meter and parallel approved double check valve assembly (see figure 10.8). The purpose of this assembly is to provide double check valve protection for the distribution system and at the same time provide a telltale of the fire sprinkler system showing any system leakage or unauthorized use of water.
b. All low flow demands up to a minimum of 2 gpm (0.126 L/s) shall pass only through the bypass water meter and bypass double check valve assembly and be accurately recorded.
c. All flows above that of paragraph b) above, shall be permitted to pass through both the line-size double check valve assembly and bypass without accurate registration by or damage to the water meter.
d. The double check valve assemblies used for this complete assembly shall meet all of the specifications for double check valve assemblies as set forth in 10.1.2.3 of this manual.
e. The bypass piping must attach to the line-size assembly between the No. 1 shutoff valve and the No. 1 check valve and between the No. 2 check valve and the No. 2 shutoff valve. This is to allow for testing and maintenance of the bypass assembly.
f. The test cock No. 4 of the line-size assembly shall not be located on the bypass piping and test cock No. 4 shall not be attached to the main-line body at the same location as the bypass piping.

10.1.2.6.2 Laboratory Evaluation
One assembly shall be inspected and evaluated in the following order. Unless otherwise noted, tests shall be performed at ambient temperature.

a. Conformance to the general design and material requirements outlined in 10.1.

b. Conformance of each individual double check valve assembly to the operational requirements outlined in 10.1.2.3.
c. Conformance to the working drawings and materials specifications.
d. Hydrostatic tests: the test assembly shall be subjected to the conditions of 10.1.1.1.5.
e. Pressure loss characteristics for flow rates up to the rated conditions.
f. Cross-over point from bypass to line-size assembly.
g. Thermal Test: the assembly shall be evaluated at the greater of 140°F (60°C) or the maximum working water temperature (MWWT), maximum working water pressure (MWWP) and specified flow rate.
h. Body strength test

10.1.2.6.3 Evaluation Procedure

10.1.2.6.3.1

Purpose: To determine the capability of the assembly to withstand the required hydrostatic test pressure.

Requirement: All components of the assembly shall withstand a hydrostatic pressure of twice the maximum working water pressure (MWWP), as stated by the manufacturer, for a period of 10 minutes without any damage, permanent deformation or impairment of operation.

Steps:

a. With both shutoff valves closed on the line-size assembly and all of the air bled from the assembly, raise pressure from atmospheric to a hydrostatic pressure of twice the maximum working water pressure shall be supplied through the No. 2 test cock of the line-size assembly for a period of 10 minutes. Any evidence of leakage up to and including the hydrostatic pressure shall be cause for rejection.
b. The assemblies shall be disassembled and inspected for any damage, permanent deformation or impairment of operation. Evidence of such shall be cause for rejection.

10.1.2.6.3.2

Purpose: To determine the overall pressure loss of the complete assembly as a function of the rate of flow.

Requirement: At any rate of flow from zero (static) up to and including the rated flow the overall pressure loss shall not exceed the values shown in Table 10-1 for double check valve assemblies. The assembly shall withstand 200% of the rated flow without damage, permanent deformation or impairment of operation.

Steps:

a. Install the assembly in a suitable hydraulic test line having a flow capacity of twice the rated flow conditions (see Table 10-1) and an available line pressure of at least 150 psi (1034 KPa).
b. Install in the supply line at a suitable location upstream from the assembly a piezometer ring; and, install a similar piezometer ring at a suitable location downstream from the assembly (located as indicated in ANSI/ISA Standard S75.02). Then install a mercury manometer or other suitable means of measuring pressure differentials between the two piezometer rings to measure the observed overall pressure loss at various rates of flow.
c. Record differential pressure between piezometer rings at static condition. Increase flow rate and record steady state differential pressure at a sufficient number of increments to fully define the flow curve characteristic. Data shall be taken for both increasing and decreasing flow conditions. The flow rate shall then be increased to 200% of the rated flow for 5 minutes and then returned to static.
Manifold Assembly (10.1.1.2.18) Only: Close all shutoff valves so that only one of the

individual line size assemblies of the manifold assembly is open. Gradually increase the flow of water through the individual assembly until the rated flow of the manifold assembly is reached, then return to static condition. Repeat flow test for each individual line size assembly contained in the manifold assembly.

d. Remove the assembly from the line and couple the downstream piping directly to the supply piping. Then observe data for a friction head correction curve for the pipe fittings required to adapt the test assembly to the supply line between the existing piezometer rings. The data of this correction curve are then subtracted from the observed overall pressure loss data of the assembly from the upstream face of the upstream shutoff valve to the downstream face of the downstream shutoff valve.

e. Exceeding the maximum allowable pressure loss listed in Table 10-1 for double check valve assemblies for any rate of flow up to and including rated flow, shall be cause for rejection. The assembly shall be disassembled and inspected for any damage, permanent deformation or impairment of operation. Evidence of such shall be cause for rejection.

10.1.2.6.3.3

Purpose: To determine the flow rate at which the line-size double check valve assembly begins to open.

Requirement: All low flow demands up to a minimum of 2 gpm (0.126 L/s) shall pass only through the bypass meter and bypass double check valve assembly.

Steps:

a. Install the assembly in a suitable hydraulic test line with an available line pressure of at least 150 psi (1034 KPa).
b. Attach a control valve to the discharge end of the double check-detector assembly, so that it discharges into a weigh tank or other suitable means of measuring flow rate.
c. Open the control valve until approximately 1 gpm (0.06 L/s) is discharging into the weigh tank or other means of measuring flow rate. Determine the flow rate through the bypass meter and the flow rate through the entire assembly.
d. Increase the flow rate through the control valve at small increments and at each increase determines what the flow rates are through the bypass meter and the entire assembly. When the total flow rate into the weigh tank or other means of measuring flow rate, exceeds the flow rate through the bypass meter, then the line-size assembly has begun to open. This crossover point must be equal to or greater than 2 gpm (0.126 L/s).
e. Failure to crossover at 2 gpm (0.126 L/s) or above shall be cause for rejection.

10.1.2.6.3.4

Purpose: To determine if all components of the assembly shall operate properly under rated temperature and pressure conditions and flow rates as specified in Table 10-6.

Requirements: All components of the assembly shall operate at and withstand a thermal test at the greater of 140°F (60°C) or at the maximum working water temperature (MWWT), maximum working water pressure MWWP and the flow rate as specified in Table 10-6, for a period of 100 hours without any damage, permanent deformation or impairment of operation.

Steps:

a. Install assembly into suitable test line that can generate the following minimum parameters:
   Maximum working water pressure (minimum of 150 psi [1034 KPa])
   Maximum working water temperature (minimum of 140°F [60°C])
   Flow requirements per Table 10-6
b. The dimensions and durometer hardness of all elastomer components shall be inspected and recorded.

c. Set controls to the rated temperature and pressure conditions of the assembly as well as rate of flow as specified in Table 10-6.
d. The temperature and pressure conditions shall be continuously monitored and recorded.
e. During and after 100 hours at rated temperature and pressure the assembly shall be tested to determine if both the line-size assembly and bypass assembly operates satisfactorily per 10.1.2.3.3.4 and 10.1.2.3.3.5. Failure to comply with these tests shall be cause for rejection
f. The assembly shall then be disassembled and inspected for any damage or permanent deformation. Evidence of such shall be cause for rejection.
g. After reaching ambient temperature, the assembly shall then be tested per 10.1.2.3.3.1. Failure to comply with this test shall be cause for rejection.

10.1.2.6.3.5

Purpose: To determine the capability of the assembly to withstand the body strength test pressure.

Requirement: All components of the assembly shall withstand a hydrostatic pressure of:

For line sizes up to 6 inch (150 mm), the greater of 875 psi (6,032 KPa) or five times the maximum working water pressure (MWWP).

For line sizes 8 inch (200 mm) and larger, the greater of 700 psi (4,825 KPa) or four times the MWWP for a period of one minute without any structural damage which allows leakage.

Steps:

a. Plug the inlet and outlet of line-size assembly. With both shutoff valves open on the line-size assembly and all of the air bled from the assembly, a hydrostatic pressure of the following shall be supplied though the No. 2 test cock of the line-size assembly for a period of one minute:

- For line sizes up to 6 inch (150 mm), the greater of 875 psi (6,032 KPa) or five times the maximum working water pressure (MWWP).

- For line sizes 8 inch (200 mm) and larger, the greater of 700 psi (4,825 KPa) or four times the MWWP.

The bypass piping shall be open and the water meter shall be permitted to be removed and replaced with a section of equally sized piping.

b. Any evidence of leakage from structural damage shall be cause for rejection. Note: Leakage of seals or gaskets at flanges or threaded joints shall not be cause for rejection.

10.1.2.6.3.6

Purpose: To determine the overall compliance of the assembly with the Standard as set forth in Chapter 10 of this manual.

Requirement: All sections of the Standard of this manual that are not specifically covered by the above tests shall be satisfactorily met.

Steps:

a. For each pertinent section of the Standard inspect, test or otherwise be assured that the assembly meets the minimum conditions of this Standard.

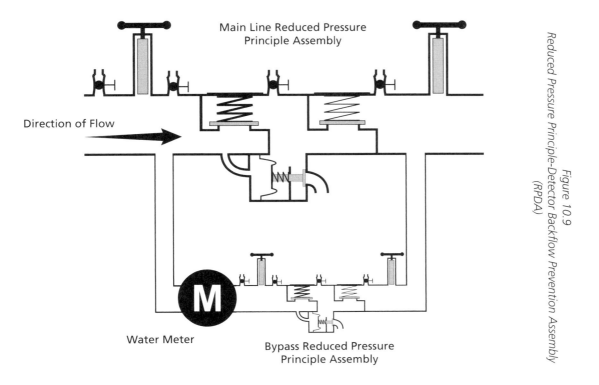

Figure 10.9
Reduced Pressure Principle-Detector Backflow Prevention Assembly (RPDA)

10.1.2.7 Design, Operational and Evaluation Standard for Reduced Pressure Principle Detector Backflow Prevention Assemblies (RPDA)

10.1.2.7.1 Design and Operational Requirements

a. This assembly shall include a line-size approved reduced pressure principle backflow prevention assembly with a parallel water meter and parallel approved reduced pressure principle backflow prevention assembly (see figure 10.9). The purpose of this assembly is to provide reduced pressure principle protection for the distribution system and at the same time provide a telltale of the fire sprinkler system showing any system leakage or unauthorized use of water.
b. All low flow demands up to a minimum of 2 gpm (0.126 L/s) shall pass only through the bypass water meter and bypass reduced pressure principle backflow prevention assembly and is accurately recorded.
c. All flows above that of paragraph b) above, shall be permitted to pass through both the line-size reduced pressure principle backflow prevention assembly and bypass without accurate registration by or damage to the water meter.
d. The reduced pressure principle backflow prevention assemblies used for this complete assembly shall meet all of the specifications for reduced pressure principle backflow prevention assemblies as set forth in 10.1.2.2 of this manual. The maximum allowable pressure loss through the RPDA shall be 2 psi (13.78 KPa) above the values stated for the reduced pressure principle backflow prevention assemblies in Table 10-1 for each specific size.
e. The bypass piping shall attach to the line-size assembly between the No. 1 shutoff valve and the No. 1 check valve and between the No. 2 check valve and the No. 2 shutoff valve. This is to allow for testing and maintenance of the bypass assembly.
f. Test cock No. 4 of the line-size assembly shall not be located on the bypass piping and test cock No. 4 shall not be attached to the main-line body at the same location as the bypass piping.

10.1.2.7.2 Laboratory Evaluation

One assembly shall be inspected and evaluated in the following order. Unless otherwise noted, tests shall be performed at ambient temperature.

a. Conformance to the general design and material requirements outlined in 10.1.
b. Conformance of each individual reduced pressure principle assembly to the operational requirements outlined in 10.1.2.2.
c. Conformance to the working drawings and materials specifications.
d. Hydrostatic tests: the test assembly shall be subjected to the conditions of 10.1.1.1.5.
e. Pressure loss characteristics for flow rates up to the rated conditions.
f. Cross-over point from bypass to line-size assembly.
g. Thermal test: the assembly shall be evaluated at the greater of 140°F (60°C) or the maximum working water temperature (MWWT), maximum working water pressure (MWWP) and specified flow rate.
h. Body strength test

10.1.2.7.3 Evaluation Procedure

10.1.2.7.3.1

Purpose: To determine the capability of the assembly to withstand the required hydrostatic test pressure.

Requirement: All components of the assembly shall withstand a hydrostatic pressure of twice the maximum working water pressure (MWWP), as stated by the manufacturer, for a period of 10 minutes without any damage, permanent deformation or impairment of operation.

Steps:

1. With both shutoff valves closed on the line-size assembly and all of the air bled from the assembly, raise pressure from atmospheric to a hydrostatic pressure of twice the maximum working water pressure shall be supplied through the No. 2 test cock for a period of 10 minutes. Any evidence of leakage up to and including the hydrostatic pressure shall be cause for rejection.
2. The assembly shall be disassembled and inspected for any damage, permanent deformation or impairment of operation. Evidence of such shall be cause for rejection.

10.1.2.7.3.2

Purpose: To determine the overall pressure loss of the complete assembly as a function of the rate of flow.

Requirement: At any rate of flow from zero (static) up to and including the rated flow the overall pressure loss shall not exceed the values detailed in 10.1.2.7.1d. The assembly shall withstand 200% of the rated flow without damage, permanent deformation or impairment of operation.

Steps:

a. Install the assembly in a suitable hydraulic test line having a flow capacity of twice the rated flow conditions (see Table 10-1) and an available line pressure of at least 150 psi (1034 KPa).
b. Install in the supply line at a suitable location upstream from the assembly a piezometer ring; and, install a similar piezometer ring at a suitable location downstream from the assembly (located as indicated in ANSI/ISA Standard S75.02). Then install a mercury manometer or other suitable means of measuring pressure differentials between the two piezometer rings to measure the observed overall pressure loss at various rates of flow.

c. Record differential pressure between piezometer rings at static condition. Increase flow rate and record steady state differential pressure at a sufficient number of increments to fully define flow curve characteristic. Data shall be taken for both increasing and decreasing flow conditions. The flow rate shall then be increased to 200% of the rated flow for 5 minutes and then returned to static.
Manifold Assembly (10.1.1.2.19) Only: Close all shutoff valves so that only one of the individual line-size assemblies of the manifold assembly is open. Gradually increase the flow of water through the individual assembly until the rated flow of the manifold assembly is reached, then return to static condition. Repeat flow test for each individual line-size assembly contained in the manifold assembly.
d. Remove the assembly from the line and couple the downstream piping directly to the supply piping. Then observe data for a friction head correction curve for the pipe fittings required to adapt the test assembly to the supply line between the existing piezometer rings. The data of this correction curve are then subtracted from the observed overall pressure loss data of the assembly from the upstream face of the upstream shutoff valve to the downstream face of the downstream shutoff valve.
e. Exceeding the maximum allowable pressure loss detailed in 10.1.2.7.1d for any rate of flow up to and including rated flow or discharge from either differential pressure relief valve, shall be cause for rejection. The assembly shall be disassembled and inspected for any damage, permanent deformation or impairment of operation. Evidence of such shall be cause for rejection.

10.1.2.7.3.3

Purpose: To determine the flow rate at which the line-size reduced pressure principle backflow prevention assembly begins to open.

Requirement: All low flow demands up to a minimum of 2 gpm (0.126 L/s) shall pass only through the bypass meter and bypass reduced pressure principle backflow prevention assembly.

Steps:

a. Install the assembly in a suitable hydraulic test line with an available line pressure of at least 150 psi (1034 KPa).
b. Attach a control valve to the discharge end of the reduced pressure principle-detector assembly, so that it discharges into a weigh tank or other suitable means of measuring flow rate.
c. Open the control valve until approximately 1 gpm (0.06 L/s) is discharging into the weigh tank or other means of measuring flow rate. Determine the flow rate through the bypass meter and the flow rate through the entire assembly.
d. Increase the flow rate through the control valve at small increments and at each increase determines what the flow rates are through the bypass meter and the entire assembly. When the total flow rate into the weigh tank or other suitable means of measuring flow rate, exceeds the flow rate through the bypass meter, then the line-size assembly has begun to open. This crossover point must be equal to or greater than 2 gpm (0.126 L/s).
e. Failure to crossover above 2 gpm (0.126 L/s) shall be cause for rejection.

10.1.2.7.3.4

Purpose: To determine if all components of the assembly shall operate properly under rated temperature and pressure conditions and flow rates as specified in Table 10-6.

Requirements: All components of the assembly shall operate at and withstand a thermal test at the greater of 140°F (60°C) or at the maximum working water temperature (MWWT), maximum working water pressure (MWWP) and the flow rate as specified in Table 10-6, for a period of 100 hours without any leakage, damage, permanent deformation or impairment of operation.

Steps:

a. Install assembly into suitable test line that can generate the following minimum parameters:
   Maximum working water pressure (minimum of 150 psi [1034 KPa])
   Maximum working water temperature (minimum of 140°F (60°C])
   Flow requirements per Table 10-6

b. The dimensions and durometer hardness of all elastomer components shall be inspected and recorded.

c. Set controls to the rated temperature and pressure conditions of the assembly as well as rate of flow as specified in Table 10-6.

d. The temperature and pressure conditions shall be continuously monitored and recorded.

e. During and after 100 hours at rated temperature and pressure the assembly shall be tested to determine if both the line-size assembly and bypass assembly operates satisfactorily per 10.1.2.2.3.3, 10.1.2.2.3.5 and 10.1.2.2.3.6. Failure to comply with these tests shall be cause for rejection.

f. The assembly shall then be disassembled and inspected for any damage or permanent deformation. Evidence of such shall be cause for rejection.

g. Once the assembly returns to ambient temperature conditions, the assembly shall then be tested per 10.1.2.2.3.1. Failure to comply with this test shall be cause for rejection.

10.1.2.7.3.5

Purpose: To determine the capability of the assembly to withstand the body strength test pressure.

Requirement: All components of the assembly shall withstand a hydrostatic pressure of:

For line sizes up to 6 inch (150 mm), the greater of 875 psi (6,032 KPa) or five times the maximum working water pressure (MWWP).

For line sizes 8 inch (200 mm) and larger, the greater of 700 psi (4,825 KPa) or four times the MWWP

for a period of one minute without any structural damage which allows leakage.

Steps:

a. Plug inlet and outlet of line-size assembly. With both shutoff valves open on the line-size assembly and all of the air bled from the assembly, a hydrostatic pressure of the following shall be supplied though the No. 2 test cock of the line-size assembly for a period of one minute:

   - For line sizes up to 6 inch (150 mm), the greater of 875 psi (6,032 KPa) or five times the maximum working water pressure (MWWP).

   - For line sizes 8 inch (200 mm) and larger, the greater of 700 psi (4,825 KPa) or four times the MWWP.

   The bypass piping shall be open and the water meter shall be permitted to be removed and replaced with a section of equally sized piping.

b. Any evidence of leakage from structural damage shall be cause for rejection.
   Note: Leakage of seals or gaskets at flanges or threaded joints shall not be cause for rejection. If the differential pressure relief valve discharges during the test, it shall be permitted to block the relief valve closed.

10.1.2.7.3.6

Purpose: To determine the overall compliance of the assembly with the Standard as set forth in Chapter 10 of this manual.

Requirement: All sections of the Standard of this manual that are not specifically covered by the above tests shall be satisfactorily met.

Steps:

a. For each pertinent section of the Standard inspect, test or otherwise be assured that the assembly meets the minimum conditions of this Standard.

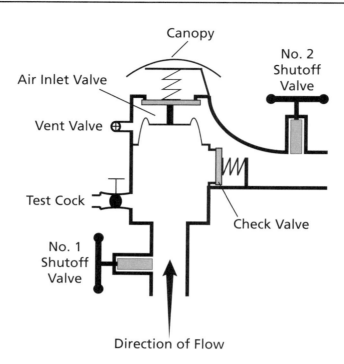

Figure 10.10 Spill-Resistant Pressure Vacuum Breaker Backsiphonage Prevention Assembly (SVB)

10.1.2.8 Design, Operational and Evaluation Standard for Spill-Resistant Pressure Vacuum Breaker Backsiphonage Prevention Assemblies (SVB)

10.1.2.8.1 Design and Operational Requirements

a. This assembly shall include an internally loaded check valve and a loaded air inlet valve opening to atmosphere on the discharge side of the check valve between two tightly closing resilient seated shutoff valves; and, will include one properly located resilient seated test cock and one vent valve. (Fig. 10.10.)
b. The air inlet valve of the spill-resistant vacuum breaker shall open when the inlet pressure is 1 psi (6.89 KPa) or greater and the outlet pressure is atmospheric. The air inlet valve shall be fully open when the inlet pressure is atmospheric and the outlet pressure is atmospheric.
c. The maximum allowable pressure drop across the assembly, from the upstream face of the No. 1 shutoff valve to the downstream face of the No. 2 shutoff valve, shall not exceed 10 psi (68.94 KPa) for any rate of flow up to and including the rated flow listed in Table 10-7.
d. The check valve shall be internally loaded and shall at all times be drip-tight in the normal direction of flow with the inlet pressure at 1 psi (6.89 KPa) and the outlet under atmospheric pressure.
e. The effective size of the air inlet port(s) of the assembly shall be governed by the vacuum dissipation test. If an air inlet port shield or canopy is used, it shall extend down around the body of the assembly to the lowest portion of the port(s). To reduce potential for fouling, the minimum clearance between the air inlet port(s) and the shield or canopy shall be 3/16 inch (4.76 mm).
f. There shall be no water leakage from the air inlet port(s) when the assembly is pressurized from atmospheric to working pressure.

10.1.2.8.2 Laboratory Evaluation

One assembly shall be inspected and evaluated in the following order. Unless otherwise noted, tests shall be performed at ambient temperatures.

a. Conformance to general, design and material requirements outlined in 10.1.
b. Conformance to the operational requirements outlined in 10.1.2.

c. Conformance to the working drawings and materials specifications.
d. Hydrostatic tests: the test assembly shall be subjected to the conditions of 10.1.1.1.5.
e. Pressure loss characteristics for flow rates up to the rated conditions set forth in Table 10-7.
f. Test cock flow test
g. Conformance to the air inlet requirements.
h. Static closing point of the check valve.
i. Interdependence of Components
j. Thermal Test: the assembly shall be evaluated at the greater of 140°F (60°C) or the maximum working water temperature (MWWT), maximum working water pressure (MWWP) and specified flow rate.
k. Life cycle test.

10.1.2.8.3 Evaluation Procedure

10.1.2.8.3.1
Purpose: To determine the capability of the assembly to withstand the required hydrostatic test pressure.

Requirement: All components of the assembly shall withstand a hydrostatic pressure of twice the maximum working water pressure (MWWP), as stated by the manufacturer, for a period of 10 minutes without any damage, permanent deformation or impairment of operation.

Steps:

a. With both shutoff valves closed, slowly raise pressure from atmospheric up to a hydrostatic pressure of twice the maximum working water pressure shall be supplied through the test cock for a period of ten minutes. Close test cock and open shutoff valve No. 2 to relieve pressure. Any evidence of leakage up to and including the hydrostatic pressure shall be cause for rejection.
b. With all of the air bled from the assembly, raise pressure from atmospheric up to the required hydrostatic pressure supplied through shutoff valve No. 2 with the test cock and shutoff valve No. 1 open to atmosphere for a period of ten minutes. Isolate pressure from No. 2 shutoff valve and depressurize assembly. Any evidence of leakage up to and including the hydrostatic pressure shall be cause for rejection.
c. The assembly shall be disassembled and inspected for any damage, permanent deformation or impairment of operation. Evidence of such shall be cause for rejection.

10.1.2.8.3.2
Purpose: To determine the overall pressure loss of the complete assembly as a function of the rate of flow.

Requirement: At any rate from static (0.0 gpm [0.0 L/s]) up to and including the rated flow the overall pressure loss shall not exceed the values shown in Table 10-7. The assembly shall withstand 200% of the rated flow without damage, permanent deformation or impairment of operation.

Steps:

a. Install the assembly in a suitable hydraulic test line having a flow capacity of twice the rated flow conditions (see Table 10-7) and an available line pressure of at least 150 psi (1034 KPa).
b. Install, in the supply line at a suitable location upstream from the assembly, a piezometer ring; and, install a similar piezometer ring at a suitable location downstream from the assembly (located as indicated in ANSI/ISA Standard S75.02). Then install a mercury manometer or other suitable means of measuring pressure differentials between the two piezometer rings to measure the observed overall pressure loss at various rates of flow.

c. Record differential pressures between piezometer rings at static condition. Increase flow rate and record steady state differential pressure at a sufficient number of increments to fully define flow curve characteristic. Data shall be taken for both increasing and decreasing flow conditions. The flow rate shall then be increased to 200% of the rated flow for 5 minutes and then returned to static.

d. Remove the assembly from the line and couple the downstream piping directly to the supply piping. Then observe data for a friction head correction curve for the pipe fittings required to adapt the test assembly to the supply line between the existing piezometer rings. The data of this correction curve are then subtracted from the observed overall pressure loss data of the assembly from the upstream face of the upstream shutoff valve to the downstream face of the downstream shutoff valve.

e. Exceeding the maximum allowable pressure loss listed in Table 10-7 for any rate of flow up to and including rated flow, shall be cause for rejection. The assembly shall be disassembled and inspected for any damage, permanent deformation or impairment of operation. Evidence of such shall be cause for rejection.

10.1.2.8.3.3

Purpose: To determine if there is a continuous flow of water from the test cock.

Requirement: There shall be a continuous flow of water from the test cock when fully opened.

Steps:

a. Install the assembly in a suitable hydraulic test line, which is capable of maintaining an inlet pressure of at least 150 psi (1034 KPa) during the following steps.
b. Close the No. 2 shutoff valve while maintaining the No. 1 shutoff valve fully open.
c. Slowly open (4 seconds ± 1 second) test cock until fully open. Then slowly close (4 seconds ± 1 second) test cock.
d. A non-continuous flow of water from the test cock shall be cause for rejection.

10.1.2.8.3.4

Purpose: To test the opening pressure differential of the air inlet valve and to determine if there is leakage from the air inlet port(s) when the assembly is pressurized from atmospheric pressure.

Requirement: The air inlet valve shall open when the inlet pressure is a minimum of 1.0 psi (6.89 KPa) above atmospheric pressure and the outlet is atmospheric. The air inlet valve shall be fully open when inlet pressure is at atmospheric and the outlet is atmospheric. There shall be no leakage from the air inlet port(s) when the assembly is pressurized from atmospheric pressure to 10 psi (68.9 KPa).

Steps:

a. Install the assembly in a suitable hydraulic test line with a minimum pressure of 150 psi (1034 KPa). Remove the air inlet valve canopy.
b. Install the high side hose of the differential pressure gage or other suitable means of measuring pressure differentials, to the test cock, open the test cock and bleed air from the hose and gage.
c. Close shutoff valve No. 2 and then close shutoff valve No. 1.
d. Open the vent valve to reduce the outlet pressure to atmospheric.
e. Slowly (0.2 psid/second maximum) open the high side bleed needle valve on the gage. Record the pressure differential at which the air inlet valve opens. Failure to open at a value of 1.0 psi (6.89 KPa) or greater shall be cause for rejection.

f. Allow water to drain out of the test cock until the inlet pressure is atmospheric, verify that the air inlet valve opens fully. Failure to open fully shall be cause for rejection.
g. Close the test cock and vent valve. Slowly (0.2 psid/second maximum) raise pressure through shutoff valve No. 1 from atmospheric pressure to 10 psi (68.9 KPa). Should there be a discharge of water from the air inlet port(s) during the increase of pressure, stop the increase of pressure and observe if the leakage continues. Continuous leakage from the air inlet valve shall be cause for rejection.

10.1.2.8.3.5

Purpose: To determine the static pressure drop across the check valve.

Requirement: The check valve shall be drip-tight in the normal direction of flow when the inlet pressure is at least 1 psi (6.89 KPa) and the outlet pressure is atmospheric.

Steps:

a. Install a transparent tube approximately 6 feet (180 cm) long on the test cock. Open the test cock to fill the tube with water and then close the test cock.
b. Close shutoff valve No. 2, then close shutoff valve No. 1.
c. Open vent valve, then open the test cock. The water in the tube will fall until closure of the check valve is attained. When no further fall of the water in the tube is observed, record the height of water in the transparent tube above the centerline of the vent valve.
d. Failure to maintain a minimum level of 27 3/4 inches (704.85 mm) shall be cause for rejection.

10.1.2.8.3.6

Purpose: To determine that the check valve and air inlet valve shall be free to operate independently through their entire movement.

Requirement: With the assembly in its intended orientation(s), the movement of either the check valve or air inlet valve through their full limit of travel shall not affect the operation of the other.

Steps:

a. Visually inspect the movement of the check valve through its full limit of travel to determine if it contacts and affects (the proper operation of the air inlet valve shall be confirmed by performing test 10.1.2.8.3.4) the operation of the air inlet valve. If necessary, remove or disconnect the loading on the check valve to allow for free movement of the check valve.
b. Visually inspect the movement of the air inlet valve through its full limit of travel to determine if it contacts and affects (the proper operation of the check valve shall be confirmed by performing test 10.1.2.8.3.5 to determine static pressure drop across first check valve) the operation of the check valve.
c. Visually inspect the movement of each of the shutoff valves through their full limit of travel to determine if they contact and affect the operation of either the check valve or air inlet valve.
d. Failure in step a, b or c shall be cause for rejection.

10.1.2.8.3.7

Purpose: To compare the air flow capacity of the air inlet valve to that of the effective water throughway of the assembly.

Requirement: The time required to dissipate a vacuum (25 to 5 inches of mercury [84.5 to 16.9 KPa]) through the air inlet valve shall be less than the time required to dissipate the same vacuum through the water throughway.

Steps:

a. Install the assembly in a normal operating position, with the outlet of the assembly attached to a vacuum source.
b. Close No. 1 shutoff valve to simulate the check valve being closed. (No. 2 shutoff valve remains fully open during entire test procedure.)
c. With the air inlet valve open, record the time required to dissipate the vacuum from 25 to 5 inches of mercury vacuum (84.5 to 16.9 KPa vacuum) when this flow is controlled by means of a quick opening valve. Repeat this step a minimum of three times, then average the recorded times.
d. Open No. 1 shutoff valve and attach a 12 inch (30.5 cm) long pipe nipple, the same nominal pipe diameter as the assembly, to the inlet of the No. 1 shutoff valve.
e. Hold the air inlet valve closed and the check valve in the fully open position, then record the time required to dissipate the vacuum from 25 to 5 inches of mercury vacuum (84.5 to 16.9 KPa vacuum) when this flow is controlled by means of a quick opening valve. Repeat this step a minimum of three times, then average the recorded times.
f. Based upon the average of at least three tests, the time required for step c shall be less than the time required for step e. Failure to comply with this requirement shall be cause for rejection.

10.1.2.8.3.8

Purpose: To test that a backsiphonage condition together with a fouled check valve shall not allow a carry-over of downstream water back into the inlet.

Requirement: Water shall not be backsiphoned more than 6 inches (152 mm) above the free surface of water in the discharge pipe standing 12 inches (305 mm) below the bottom or critical level line of the assembly.

Steps:

a. Install the assembly in a normal operating position, with the inlet of the assembly connected to the vacuum source and the discharge side equipped with a transparent tube that extends down into a vessel of water where the free surface of the water is 12 inches (305 mm) below the bottom or critical level line of the assembly.
b. A wire of the size shown in Table 10-8 shall be placed across the seating surface in the lower quadrant of a hinged or horizontally moving poppet check valve or at a single point of a vertical moving check valve. The wire shall be placed so that the entire width of the seating contact surface is covered.
c. By the application of vacuum conditions ranging from 0 to 25 inches of mercury (84.5 KPa vacuum) under both steady flow and instantaneous conditions, the water in the discharge tube shall not rise more than 6 inches (152 mm) above the original free surface of the water in the vessel. Evidence of such shall be cause for rejection.

10.1.2.8.3.9

Purpose: To determine if all components of the assembly shall operate properly under rated temperature and pressure conditions and flow rates as specified in Table 10-6.

Requirements: All components of the assembly shall operate at and withstand a thermal test at the greater of 140°F (60°C) or at the maximum working water temperature (MWWT), maximum working water pressure (MWWP) and flow rate as specified in Table 10-6 for a period of 100 hours without any damage, permanent deformation or impairment of operation.

Steps:

a. Install assembly into suitable test line that can generate the following minimum parameters:
   Maximum working water pressure (minimum of 150 psi [1034 KPa])
   Maximum working water temperature (minimum of 140°F [60°C])
   Flow requirements per Table 10-6
b. The dimensions and durometer hardness of all elastomer components shall be inspected and recorded.
c. Set controls to the rated temperature and pressure conditions of the assembly, as well as rate of flow as specified in Table 10-6.
d. The temperature and pressure conditions shall be continuously monitored and recorded.
e. During and after 100 hours at rated temperature and pressure the assembly shall be tested to determine if the assembly operates satisfactorily per 10.1.2.8.3.4 and 10.1.2.8.3.5. Failure to comply with these tests shall be cause for rejection.
f. The assembly shall then be disassembled and inspected for any internal damage or permanent deformation. Evidence of such shall be cause for rejection.
g. Once the assembly returns to ambient temperature conditions. The assembly shall then be tested per 10.1.2.8.3.1. Failure to comply with this test shall be cause for rejection.

10.1.2.8.3.10

Purpose: To determine if the assembly sustains any damage, permanent deformation or impairment of operation following the specified cycles.

Requirement: The assembly shall withstand the specified cycle without leakage, damage, permanent deformation or impairment of operation.

Steps:

a. Install assembly into suitable test line, as shown in figure 10.4, that can generate the following parameters:
   Rated flow of assembly under test (per Table 10-7)
   Line pressure of 60 psi ± 10 psi (414 KPa ± 69 KPa)
   Temperature of 110°F +30/-0°F (43°C. +17/-0°C)
   Control valves capable of closing/opening within 5 seconds. *
   *(Resulting rise of pressure shall not exceed MWWP of the assembly.)
b. With Valve #3 (Valve #3A closed and Valve #3B open) and Valve #4 (see figure 10.4) in the closed position and Valve #1 and Valve #2 in the open position, establish flow through the assembly equal to 25% of the maximum rated flow per Table 10-7.
c. Maintain the flow through the assembly for a minimum of 6 seconds. Close Valve #1.
d. After 5 seconds, close Valve #2.
e. After 3 seconds, open Valve #3.
f. After 3 seconds, open Valve #4. Maintain Valve #4 open for 6 seconds. Close Valve #4.
g. After 2 seconds, close Valve #3.
h. After 3 seconds, open Valve #2.
i. After 2 seconds open Valve #1.

j. Repeat steps c through i 1250 times. The assembly shall be tested to determine if the assembly complies with 10.1.2.8.3.4 and 10.1.2.8.3.5. Failure to comply with these tests shall be cause for rejection.

k. Raise flow through the assembly to 50% of rated flow. Repeat steps c through i 1250 times. The assembly shall be tested to determine if the assembly complies with 10.1.2.8.3.4 and 10.1.2.8.3.5.

1. Slowly raise pressure through shutoff valve No. 1 from atmospheric up to a pressure of 10 psi. Close shutoff valve No. 1 and open shutoff valve No. 2 to maintain 10 psi in the body. Open test cock and maintain 10 psi in body for ten minutes.
2. Raise pressure through shutoff valve No. 2 to 150 psi and maintain for ten minutes.
3. Leakage during steps k1 or k2 shall be cause for rejection.

l. Raise flow through the assembly to 75% of rated flow. Repeat steps c through i 1250 times. The assembly shall be tested to determine if the assembly complies with 10.1.2.8.3.4 and 10.1.2.8.3.5. Failure to comply with these tests shall be cause for rejection.

m. Raise flow through the assembly to 100% of rated flow. Repeat steps c through i 1250 times.

n. After 5000 total cycles the assembly shall be tested to determine if the assembly complies with 10.1.2.8.3.4 and 10.1.2.8.3.5. Failure to comply with these tests shall be cause for rejection. Perform steps k1 through k3.

o. The assembly shall then be disassembled and inspected for any damage or permanent deformation to any component(s), other than the elastomer disc. Evidence of such shall be cause for rejection.

10.1.2.8.3.11

Purpose: To determine the overall compliance of the assembly with the Standard as set forth in Chapter 10 of this manual.

Requirement: That all sections of the Standard of this manual that are not specifically covered by the above tests are satisfactorily met.

Steps:

a. For each pertinent section of the Standard inspect, test or otherwise be assured that the assembly meets the minimum conditions of this Standard.

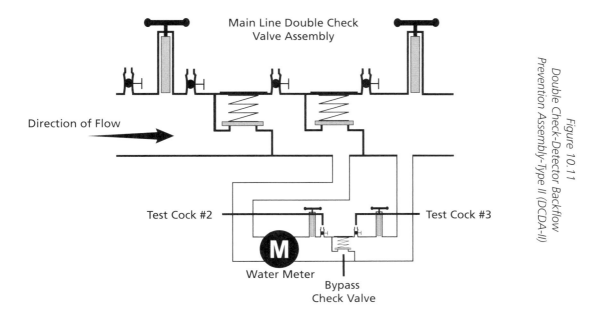

Figure 10.11
Double Check-Detector Backflow Prevention Assembly-Type II (DCDA-II)

10.1.2.9 Design, Operational and Evaluation Standard for Double Check-Detector Backflow Prevention Assemblies-Type II (DCDA-II)

10.1.2.9.1 Design and Operational Requirements

a. This assembly shall include a line-size approved double check valve assembly with a parallel bypass assembly (see figure 10.11). The purpose of this assembly is to provide double check valve protection for the distribution system and at the same time provide a telltale of the fire sprinkler system showing any system leakage or unauthorized use of water.
b. The bypass assembly shall consist of a water meter and a single check valve assembly.
c. All low flow demands up to a minimum of 2 gpm (0.126 L/s) shall pass only though the bypass water meter and bypass single check valve assembly and be accurately recorded.
d. All flows above that of paragraph c above shall be permitted to pass though both the line-size double check valve assembly and bypass assembly without accurate registration by or damage to the water meter.
e. The line-size double check valve assembly used for this complete assembly shall meet all of the Standard for double check valve assemblies as set forth in 10.1.2.3 of this manual.
f. The single check valve assembly used for this complete assembly shall meet all of the Standard for double check valve backflow prevention assemblies as set forth in 10.1.2.3 excluding the requirement for a second check valve and a test cock located between the two check valves.
g. The bypass piping must attach to the line-size assembly between the No. 1 check valve and the No. 2 check valve and between the No. 2 check valve and the No. 2 shutoff valve.
h. Test cock No. 4 of the line-size assembly shall not be located on the bypass piping and test cock No. 4 shall not be attached to the main-line body at the same location as the bypass piping.

10.1.2.9.2 Laboratory Evaluation
One assembly shall be inspected and evaluated in the following order. Unless otherwise noted, tests shall be performed at ambient temperature.

a. Conformance to the general design and material requirements outlined in 10.1.
b. Conformance of the line-size double check valve assembly and the single check valve assembly to the operational requirements outlined in 10.1.2.3.

c. Conformance to the working drawings and materials specifications.
d. Hydrostatic tests: the test assembly shall be subjected to the conditions of 10.1.1.1.5.

e. Pressure loss characteristics for flow rates up to the rated conditions.
f. Cross-over point from bypass to line-size assembly.
g. Thermal Test: the assembly shall be evaluated at the greater of 140°F (60°C) or the maximum working water temperature (MWWT), maximum working water pressure (MWWP) and specified flow rate.
h. Body strength test

10.1.2.9.3 Evaluation Procedure

10.1.2.9.3.1

Purpose: To determine the capability of the assembly to withstand the required hydrostatic test pressure.

Requirement: All components of the assembly shall withstand a hydrostatic pressure of twice the maximum working water pressure (MWWP), as stated by the manufacturer, for a period of 10 minutes without any damage, permanent deformation or impairment of operation.

Steps:

a. With both shutoff valves closed on the line-size assembly and all of the air bled from the assembly, raise pressure from atmospheric up to a hydrostatic pressure of twice the maximum working water pressure shall be supplied though the No. 2 test cock of the line-size assembly for a period of ten minutes. Close No. 2 test cock and isolate assembly from pressure and open No. 4 test cock on the line size assembly to relieve pressure, then close No. 4 test cock. Any evidence of leakage shall be cause for rejection.
b. The assemblies shall be disassembled and inspected for any damage, permanent deformation or impairment of operation. Evidence of such shall be cause for rejection.

10.1.2.9.3.2

Purpose: To determine the overall pressure loss of the complete assembly as a function of the rate of flow.

Requirement: At any rate of flow from static (0.0 gpm [0.0 L/s]) up to and including the rated flow the overall pressure loss shall not exceed the values shown in Table 10-1 for double check valve assemblies. The assembly shall withstand 200% of the rated flow without damage, permanent deformation or impairment of operation.

Steps:

a. Install the assembly in a suitable hydraulic test line having a flow capacity of twice the rated flow conditions (see Table 10-1) and an available line pressure of at least 150 psi (1034 KPa).
b. Install in the supply line at a suitable location upstream from the assembly a piezometer ring; and install a similar piezometer ring at a suitable location downstream from the assembly (located as indicated in ANSI/ISA Standard S75.02). Then install a mercury manometer or other suitable means of measuring pressure differentials between the two-piezometer rings to measure the observed overall pressure loss at various rates of flow.
c. Record differential pressure between piezometer rings at static condition. Increase flow rate and record steady state differential pressure at a sufficient number of increments to fully define the flow curve characteristic. Data shall be taken for both increasing and decreasing flow conditions. The flow rate shall then be increased to 200% of the rated flow for five minutes and then returned to static.

Manifold Assembly (10.1.1.2.18) Only. Close all shutoff valves so that only one of the individual line-size assemblies of the manifold assembly is open. Gradually increase the flow of water through the individual assembly until the rated flow of the manifold assembly is reached, then return to static condition. Repeat flow test for each individual line size assembly contained in the manifold assembly.

d. Remove the assembly from the line and couple the downstream piping directly to the supply piping. Then observe data for a friction head correction curve for the pipe fittings required to adapt the test assembly to the supply line between the existing piezometer rings. The data of this correction curve are then subtracted from the observed overall pressure loss data of the assembly from the upstream face of the upstream shutoff valve to the downstream face of the downstream shutoff valve.

e. Exceeding the maximum allowable pressure loss listed in Table 10-1 for double check valve assemblies for any rate of flow up to and including rated flow, shall be cause for rejection. The assembly shall be disassembled and inspected for any damage, permanent deformation or impairment of operation. Evidence of such shall be cause for rejection.

10.1.2.9.3.3

Purpose: To determine the flow rate at which the line-size double check valve assembly begins to open.

Requirement: All low flow demands up to a minimum of 2 gpm (0.126 L/s) shall pass only though the bypass assembly.

Steps:

a. Install the assembly in a suitable hydraulic test line with an available line pressure of at least 150 psi (1034 KPa).
b. Attached a control valve to the discharge end of the double check-detector assembly, so that it discharges into a weigh tank or other suitable means of measuring flow rate.
c. Open the control valve until approximately 1 gpm (0.06 L/s) is discharging into the weigh tank or other means of measuring flow rate. Determine the flow rate through the bypass meter and the flow rate though the entire assembly.
d. Increase the flow rate though the control valve at small increments and at each increase determine what the flow rates are through the bypass meter and the entire assembly. When the total flow rate into the weigh tank or other means of measuring flow rate, exceeds the flow rate through the bypass meter, then the line-size assembly has begun to open. This crossover point must be equal to or greater than 2 gpm (0.126 L/s).
e. Failure to crossover at 2 gpm (0.126 L/s) or above shall be cause for rejection.

10.1.2.9.3.4

Purpose: To determine if all components of the assembly shall operate properly under rated temperature and pressure conditions and flow rates as specified in Table 10-6.

Requirements: All components of the assembly shall operate at and withstand a thermal test at the greater of 140°F (60°C) or at the maximum working water temperature (MWWT), maximum working water pressure (MWWP) and the flow rate as specified in Table 10-6, for a period of 100 hours without any damage, permanent deformation or impairment of operation.

Steps:

a. Install assembly into suitable test line that can generate the following minimum parameters:
   Maximum working water pressure (minimum of 150 psi [1034 KPa])
   Maximum working water temperature (minimum of 140°F [60°C])
   Flow requirements per Table 10-6
b. The dimensions and durometer hardness of all elastomer components shall be inspected and recorded.
c. Set controls to the rated temperature and pressure conditions of the assembly, as well as rate of flow as specified in Table 10-6.
d. The temperature and pressure conditions shall be continuously monitored and recorded.
e. During and after 100 hours at rated temperature and pressure, the assembly shall be tested to determine if the line-size assembly operates satisfactorily per 10.1.2.3.3.4 and 10.1.2.3.3.5 and the bypass assembly operates satisfactorily per 10.1.2.3.3.4. Failure to comply with these tests shall be cause for rejection.
f. The assembly shall then be disassembled and inspected for any damage or permanent deformation. Evidence of such shall be cause for rejection.
g. After reaching ambient temperature, both the line-size assembly and bypass assembly shall then be tested per 10.1.2.3.3.1. Failure to comply with this test shall be cause for rejection.

10.1.2.9.3.5

Purpose: To determine the capability of the assembly to withstand the body strength test pressure.

Requirement: All components of the assembly shall withstand a hydrostatic pressure of:

For line sizes up to 6 inch (150mm), the greater of 875 psi (6,032 KPa) or five times the maximum working water pressure (MWWP).

For line sizes 8 inch (200 mm) and larger, the greater of 700 psi (4,825 KPa) or four times the MWWP.

for a period of one minute without any structural damage which allows leakage.

Steps:

a. Plug inlet and outlet of line-size assembly. With both shutoff valves open on the line-size assembly and all of the air bled from the assembly, a hydrostatic pressure of the following shall be supplied though the No. 2 test cock of the line-size assembly for a period of one minute:

   - For line sizes up to 6 inch (150 mm), the greater of 875 psi (6,032 KPa) or five times the maximum working water pressure (MWWP).

   - For line sizes 8 inch (200 mm) and larger, the greater of 700 psi (4,825 KPa) or four times the MWWP. The bypass piping shall be open and the water meter shall be permitted to be removed and replaced with a section of equally sized piping.

b. Any evidence of leakage from structural damage shall be cause for rejection. Note: Leakage of seals or gaskets at flanges or threaded joints shall not be cause for rejection.

10.1.2.9.3.6

Purpose: To determine the overall compliance of the assembly with the Standard as set forth in Chapter 10 of this manual.

Requirement: All sections of the Standard of this manual that are not specifically covered by the above tests shall be satisfactorily met.

Steps:

    a.   For each pertinent section of the Standard inspect, test or otherwise be assured that the assembly meets the minimum conditions of this Standard.

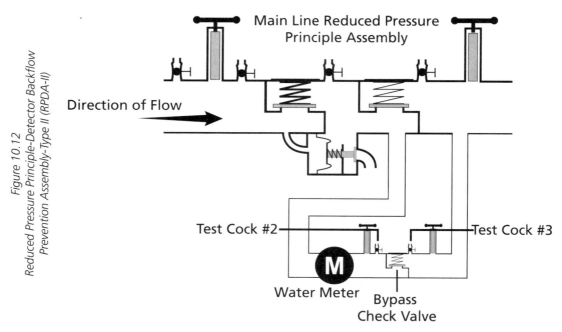

*Figure 10.12
Reduced Pressure Principle-Detector Backflow Prevention Assembly-Type II (RPDA-II)*

10.1.2.10 Design, Operational and Evaluation Standard for Reduced Pressure Principle-Detector Backflow Prevention Assemblies-Type II (RPDA-II)

10.1.2.10.1 Design and Operational Requirements

a. This assembly shall include a line-size approved reduced pressure principle backflow prevention assembly with a parallel bypass assembly (see Fig 10.12). The purpose of this assembly is to provide reduced pressure principle backflow prevention assembly protection for the distribution system and at the same time provide a telltale of the fire sprinkler system showing any system leakage or unauthorized use of water.
b. The bypass assembly shall consist of a water meter and a single check valve assembly.
c. All low flow demands up to a minimum of 2 gpm (0.126 L/s) shall pass only though the bypass water meter and bypass assembly and be accurately recorded.
d. All flows above that of paragraph c above shall be permitted to pass though both the line-size reduced pressure principle assembly and bypass assembly without accurate registration by or damage to the water meter.
e. The line-size reduced pressure principle assembly used for this complete assembly shall meet all of the standard for reduced pressure principle backflow prevention assemblies as set forth in 10.1.2.2 of this manual.
f. The single check valve assembly used for this complete assembly shall meet all of the standard for double check valve backflow prevention assemblies as set forth in 10.1.2.3 excluding the requirement for a second check valve and a test cock located between the two check valves.
g. The bypass piping must attach to the line-size assembly between the No. 1 check valve and the No. 2 check valve and between the No. 2 check valve and the No. 2 shutoff valve.
h. Test cock No. 4 of the line-size assembly shall not be located on the bypass piping and test cock No. 4 shall not be attached to the main-line body at the same location as the bypass piping.

10.1.2.10.2 Laboratory Evaluation

One assembly shall be inspected and evaluated in the following order. Unless otherwise noted, tests shall be performed at ambient temperature.

a. Conformance to the general design and material requirements outlined in 10.1.

b. Conformance of the line-size reduced pressure principle backflow prevention assembly and the single check valve assembly to the operational requirements outlined in 10.1.2.2 and respectively 10.1.2.3.
c. Conformance to the working drawings and materials specifications.
d. Hydrostatic tests: the test assembly shall be subjected to the conditions of 10.1.1.1.5.
e. Pressure loss characteristics for flow rates up to the rated conditions.
f. Cross-over point from bypass to line-size assembly.
g. Thermal Test: the assembly shall be evaluated at the greater of 140°F (60°C) or the maximum working water temperature (MWWT), maximum working water pressure (MWWP) and specified flow rate.
h. Body strength test

10.1.2.10.3 Evaluation Procedure

10.1.2.10.3.1

Purpose: To determine the capability of the assembly to withstand the required hydrostatic test pressure.

Requirement: All components of the assembly shall withstand a hydrostatic pressure of twice the maximum working water pressure (MWWP), as stated by the manufacturer, for a period of 10 minutes without any damage, permanent deformation or impairment of operation.

Steps:

a. With both shutoff valves closed on the line-size assembly and all of the air bled from the assembly, raise pressure from atmospheric up to a hydrostatic pressure of twice the maximum working water pressure shall be supplied though the No. 2 test cock of the line-size assembly for a period of ten minutes. Close No. 2 test cock and isolate assembly from pressure and open No. 4 test cock on the line size assembly to relieve pressure, then close No. 4 test cock. Any evidence of leakage shall be cause for rejection.
b. The assemblies shall be disassembled and inspected for any damage, permanent deformation or impairment of operation. Evidence of such shall be cause for rejection.

10.1.2.10.3.2

Purpose: To determine the overall pressure loss of the complete assembly as a function of the rate of flow.

Requirement: At any rate of flow from static (0.0 gpm [0.0 L/s]) up to and including the rated flow the overall pressure loss shall not exceed the values shown in Table 10-1 for reduced pressure principle assemblies. The assembly shall withstand 200% of the rated flow without damage, permanent deformation or impairment of operation.

Steps:

a. Install the assembly in a suitable hydraulic test line having a flow capacity of twice the rated flow conditions (see Table 10-1) and an available line pressure of at least 150 psi (1034 KPa).
b. Install in the supply line at a suitable location upstream from the assembly a piezometer ring; and install a similar piezometer ring at a suitable location downstream from the assembly (located as indicated in ANSI/ISA Standard S75.02). Then install a mercury manometer or other suitable means of measuring pressure differentials between the two-piezometer rings to measure the observed overall pressure loss at various rates of flow.
c. Record differential pressure between piezometer rings at static condition. Increase flow rate and record steady state differential pressure at a sufficient number of increments to fully define the flow curve characteristic. Data shall be taken for both increasing and decreasing flow

conditions. The flow rate shall then be increased to 200% of the rated flow for 5 minutes and then returned to static.

Only for the manifold assembly (10.1.1.2.18) close all shutoff valves so that only one of the individual line-size assemblies of the manifold assembly is open. Gradually increase the flow of water through the individual assembly until the rated flow of the manifold assembly is reached, then return to static condition. Repeat flow test for each individual line size assembly contained in the manifold assembly.

d. Remove the assembly from the line and couple the downstream piping directly to the supply piping. Then observe data for a friction head correction curve for the pipe fittings required to adapt the test assembly to the supply line between the existing piezometer rings. The data of this correction curve are then subtracted from the observed overall pressure loss data of the assembly from the upstream face of the upstream shutoff valve to the downstream face of the downstream shutoff valve.

e. Exceeding the maximum allowable pressure loss listed in Table 10-1 for reduced pressure principle assemblies for any rate of flow up to and including rated flow or discharge from the differential pressure relief valve, shall be cause for rejection. The assembly shall be disassembled and inspected for any damage, permanent deformation or impairment of operation. Evidence of such shall be cause for rejection.

10.1.2.10.3.3

Purpose: To determine the flow rate at which the line-size reduced pressure principle backflow prevention assembly begins to open.

Requirement: All low flow demands up to a minimum of 2 gpm (0.126 L/s) shall pass only though the bypass assembly.

Steps:

a. Install the assembly in a suitable hydraulic test line with an available line pressure of at least 150 psi (1034 KPa).
b. Attached a control valve to the discharge end of the reduced pressure principle-detector assembly, so that it discharges into a weigh tank or other suitable means of measuring flow rate.
c. Open the control valve until approximately 1 gpm (0.06 L/s) is discharging into the weigh tank or other means of measuring flow rate. Determine the flow rate through the bypass meter and the flow rate though the entire assembly.
d. Increase the flow rate though the control valve at small increments and at each increase determine what the flow rates are through the bypass meter and the entire assembly. When the total flow rate into the weigh tank or other means of measuring flow rate, exceeds the flow rate through the bypass meter, then the line-size assembly has begun to open. This crossover point must be equal to or greater than 2 gpm (0.126 L/s).
e. Failure to crossover at 2 gpm (0.126 L/s) or above shall be cause for rejection.

10.1.2.10.3.4

Purpose: To determine if all components of the assembly shall operate properly under rated temperature and pressure conditions and flow rates as specified in Table 10-6.

Requirements: All components of the assembly shall operate at and withstand a thermal test at the greater of 140°F (60°C) or at the maximum working water temperature (MWWT), maximum working water pressure (MWWP) and the flow rate as specified in Table 10-6, for a period of 100 hours without any damage, permanent deformation or impairment of operation.

Steps:

a. Install assembly into suitable test line that can generate the following minimum parameters:
   Maximum working water pressure (minimum of 150 psi [1034 KPa])
   Maximum working water temperature (minimum of 140°F [60°C])
   Flow requirements per Table 10-6.
b. The dimensions and durometer hardness of all elastomer components shall be inspected and recorded.
c. Set controls to the rated temperature and pressure conditions of the assembly, as well as rate of flow as specified in Table 10-6.
d. The temperature and pressure conditions shall be continuously monitored and recorded.
e. During and after the 100 hours at rated temperature and pressure, the assembly shall be tested to determine if the line size assembly operates satisfactorily per 10.1.2.2.3.3, 10.1.2.2.3.5 and 10.1.2.2.3.6 in addition to the bypass assembly operating satisfactorily per 10.1.2.3.3.4. Failure to comply with these tests shall be cause for rejection.
f. The assembly shall then be disassembled and inspected for any damage or permanent deformation. Evidence of such shall be cause for rejection.
g. After reaching ambient temperature, the line size assembly shall then be tested per 10.1.2.2.3.1 and the bypass assembly shall be tested per 10.1.2.3.3.1. Failure to comply with this test shall be cause for rejection.

10.1.2.10.3.5

Purpose: To determine the capability of the assembly to withstand the body strength test pressure.

Requirement: All components of the assembly shall withstand a hydrostatic pressure of:

> For line sizes up to 6 inch (150 mm), the greater of 875 psi (6,032 KPa) or five times the maximum working water pressure (MWWP).
>
> For line sizes 8 inch (200 mm) and larger, the greater of 700 psi (4,825 KPa) or four times the MWWP.

for a period of one minute without any structural damage which allows leakage.

Steps:

a. Plug inlet and outlet of line-size assembly. With both shutoff valves open on the line-size assembly and all of the air bled from the assembly, a hydrostatic pressure of the following shall be supplied though the No. 2 test cock of the line-size assembly for a period of one minute:

   - For line sizes up to 6 inch (150 mm), the greater of 875 psi (6,032 KPa) or five times the maximum working water pressure (MWWP).
   - For line sizes 8 inch (200 mm) and larger, the greater of 700 psi (4,825 KPa) or four times the MWWP

   shall be supplied though the No. 2 test cock of the line-size assembly for a period of one minute. The bypass piping shall be open and the water meter shall be permitted to be removed and replaced with a section of equally sized piping.

b. Any evidence of leakage from structural damage shall be cause for rejection.
   Note: Leakage of seals or gaskets at flanges or threaded joints shall not be cause for rejection. If the differential pressure relief valve discharges during the test, it shall be permitted to block the relief valve closed.

10.1.2.10.3.6

Purpose: To determine the overall compliance of the assembly with the Standard as set forth in Chapter 10 of this manual.

Requirement: All sections of the Standard of this manual that are not specifically covered by the above tests shall be satisfactorily met.

Steps:

a. For each pertinent section of the Standard inspect, test or otherwise be assured that the assembly meets the minimum conditions of this Standard.

10.2 Differential Pressure Gage Field Test Kits Used for Field Testing of Backflow Prevention Assemblies (Field Test Kit)

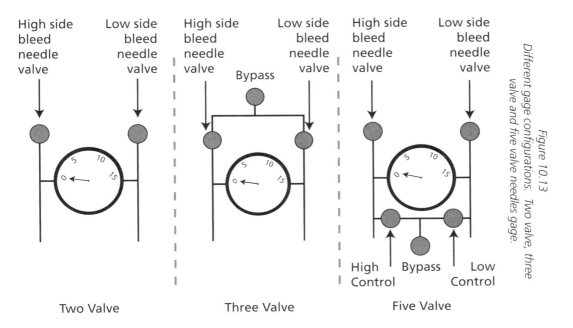

Figure 10.13 Different gage configurations. Two valve, three valve and five valve needles gage.

10.2.1 Scope
This Standard pertains to a portable differential pressure gage field test kit utilized for the in situ field testing of backflow prevention assemblies. The portable differential pressure gage field test kit shall be comprised of all needle valves, connecting hoses and differential pressure gage body(s). A line pressure gage shall be permitted to be an integral component of the field test kit.

10.2.2 Definitions
Accuracy: The conformity of a gage indication to an accepted standard or true value. Accuracy is the difference (error) between the true value and the gage indication. It is the combined effects of method, observer, apparatus and environment. Accuracy error includes hysteresis and repeatability errors, but not friction error.

Accuracy Verification: The checking of a gage by comparison with a given standard to determine the indication error at specified points on the scale.

Differential Pressure Gage: A gage having two pressure connections and an indication of the difference between the two applied pressures.

Hysteresis Error: The difference between increasing and decreasing differential pressure readings at any point on the scale obtained during a pressure cycle after lightly tapping the field test kit to eliminate friction error.

Line pressure Gage: Pressure gage used for the purpose of indicating line pressure on a backflow prevention assembly.

Normal Operating Conditions: the environmental conditions in which the stated accuracy applies.

10.2.3 General Design Requirements

10.2.3.1 Case
To minimize damage in transport, a carrying case shall be permitted.

10.2.3.2 Display
This Standard pertains to digital and analog field test kits.

10.2.3.3 Range/Resolution
The gage field test kit shall have a minimum range of 0-15 psid (0 KPa - 103.4 KPa) and be capable of resolving 0.1 psid (0.689 KPa). Dual scales such as 0-15 PSID/0-100 KPa shall be permitted.

10.2.3.4 Operating Parameters
Pressure - The gage field test kit shall be so designed to function through the maximum working water pressure (MWWP), but not less than 175 psi (1206 KPa).

Temperature - The field test kit shall be so designed to function through the maximum working water temperature (MWWT), but not less than 140°F (60°C).

Field test kit shall have an minimum accuracy of ± 0.2 psid (1.378 KPa).

10.2.3.5 External Calibration Adjustment
External calibration adjustment(s) shall be tamper resistant or designed to show evidence of tampering.

10.2.3.6 Plumbing of the Field Test Kit
High side and low side of field test kit must be hydraulically separated by a minimum of two needle valves.

Typical needle valve arrangements (see figure 10.13)

- Two Needle valve
- Three Needle Valve
- Five Needle Valve

The needle valves shall withstand all hydraulic testing of the field test kit, in both the open and closed positions, without leakage, damage or permanent deformation. No needle valve(s) shall shut off pressure to the high-pressure side or the low-pressure side of the gage.

10.2.3.7 Marking
The markings on a field test kit shall include:

a. Make: Name or Trademark
b. Model
c. Serial Number
d. Maximum working water pressure (MWWP)
e. Maximum working water temperature (MWWT)

All markings shall be easily read and shall be either cast, molded, stamped, engraved or etched on the body of the gage or on a nameplate permanently affixed to the body of the gage or printed on the dial. The nameplate shall be either brass or stainless steel. Letters shall be at least 10 pt font.

High side and Low side pressure connections shall be identified. Field test kit shall contain Critical Level(s) markings for both Horizontal/Vertical orientation(s).

All markings shall be in English units (i.e., PSI, °F). Metric equivalents (i.e., KPa, °C) shall be permitted in addition to the English units at the manufacturer's discretion.

10.2.4 Design Requirements

10.2.4.1 Valving: Minimum C_V flow rates

Rate of flow through gage from high side hose to low side hose shall be a minimum of 0.02 gpm (0.001L/s) at 1.0 psig (6.89 KPa) inlet pressure with the high side control needle valve fully open and the low side control needle valve open 1/4 turn (see figure 10.14).

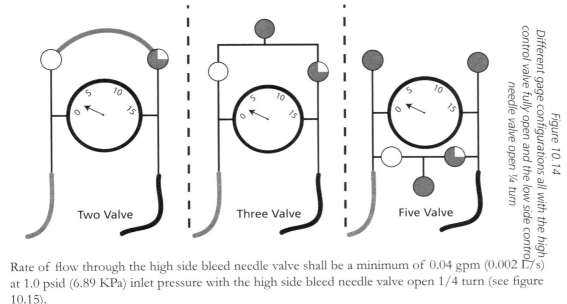

Figure 10.14
Different gage configurations all with the high control valve fully open and the low side control needle valve open ¼ turn

Rate of flow through the high side bleed needle valve shall be a minimum of 0.04 gpm (0.002 L/s) at 1.0 psid (6.89 KPa) inlet pressure with the high side bleed needle valve open 1/4 turn (see figure 10.15).

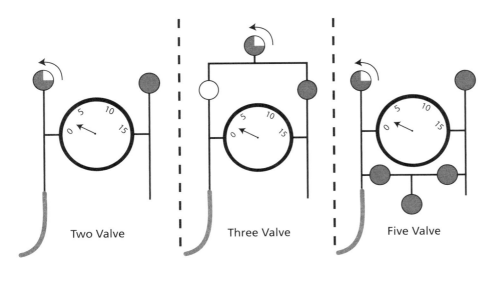

Figure 10.15
Different gage configurations all with the high side bleed needle valve open ¼ turn

10.2.4.2 Hoses
Minimum 3 foot length with female 1/4 inch flared type connectors per SAE J513. Hoses shall have a minimum burst strength of four times the field test kit MWWP.

10.2.5 Material Requirements

10.2.5.1 Wetted Components
Wetted components shall be constructed of valve bronze which conforms to ASTM Designation B61 or B62 or B584 UNS number C84400 or other bronze alloy containing a minimum 79% copper; ASTM B197 Beryllium-Copper UNS number C17200, UNS C36000, or stainless steel that conforms to ASTM Designation A276 UNS No. S30400, S30500 or S31600; or engineered plastic.

10.2.5.2 Non Wetted Components
Shall be corrosion resistant equivalent to non-ferrous materials.

10.2.6 Evaluation of Design and Performance

10.2.6.1 General

10.2.6.1.1 Drawings and Specifications
A full set of working drawing and specifications of all materials shall be furnished with each model of field test kit that is submitted for evaluation.

10.2.6.1.2 Gages Required for Evaluation
Three field test kits of each model shall be submitted for evaluation and one field test kit of each model selected at random shall undergo a complete laboratory evaluation of its design and operating characteristics, with the dial/display in the vertical, under the supervision of the Foundation for Cross-Connection Control and Hydraulic Research.

10.2.6.1.3 Approval
In addition to the satisfactory completion of the laboratory evaluation contained in this Standard, the manufacturer shall supply to the Foundation for Cross-Connection Control and Hydraulic Research the sales literature and service/maintenance literature for the model being evaluated under this Standard.

Upon the satisfactory completion of the laboratory evaluation and acceptable review of the manufacturer's literature, the manufacturer shall supply to the approving agency the serial number of the first field test kit manufactured in the approved configuration. The Foundation for Cross-Connection Control and Hydraulic Research shall then grant an approval for the specific model evaluated under this Standard. This approval shall be valid for a period of no more than three years and may be rescinded for cause before that time. The approving agency shall issue a list of all field test kits that are currently approved by said agency.

10.2.6.1.4 Renewal of Approval
Continuing verification of compliance with this standard and field performance shall be accomplished at a maximum of once every three years to the satisfaction of the Foundation for Cross-Connection Control and Hydraulic Research. The Foundation for Cross-Connection Control and Hydraulic Research, at its own discretion retains the right of determining the extent of re-evaluation required before renewal is granted. The manufacturer's current sales, service and maintenance literature shall be submitted for review at this time. Acceptable re-evaluation, literature, past performance of the field test kit under field operating conditions, shall be considered before the renewing of an approval. Failure to meet these requirements shall result in the automatic withdrawal of the approval

for that model of field test kit. The latest listing of approved assemblies shall reflect the current status of a particular model field test kit.

10.2.6.1.5 Change of Design, Materials or Operation

The Foundation for Cross-Connection Control and Hydraulic Research shall be notified by the manufacturer in writing of any proposed change in design, materials or operation of an approved field test kit. The Foundation for Cross-Connection Control and Hydraulic Research at its own discretion shall be permitted to require another laboratory evaluation. Failure to notify the Foundation for Cross-Connection Control and Hydraulic Research of any changes of design, materials or operation prior to implementation shall be cause for the rescinding of the approval for the model of field test kit involved.

10.2.6.2 Design, Operational and Evaluation Standard

10.2.6.2.1 Laboratory Evaluation

All tests shall be performed on one field test kit in the following order. Unless otherwise noted, tests shall be performed at ambient temperature. All gage hoses shall be attached in their respective normal operating position. All digital field test kits must be turned to the on position before each of the following tests. All needle valves shall be in the closed position unless otherwise noted.

a. Conformance to general design and material requirements outlined in 10.2.3, 10.2.4 and 10.2.5.
b. Accuracy Test
c. Hydrostatic tests
d. Accuracy rated temperature and pressure test
e. Pressurizing Fatigue Test
f. Valving flow test
g. Pressure Dissipation Test

10.2.6.2.2 Evaluation Procedure

10.2.6.2.2.1 Accuracy Test

Purpose: To test the accuracy of the field test kit at various line pressures.

Requirement: The field test kit shall maintain an accuracy of ± 0.2 psid (1.378 KPa) for increasing and decreasing differential pressure readings over the scale of 1.0 psid to 15 psid (6.9kPa - 103.4 KPa). The field test kit shall be tested for accuracy with water in the vertical position and different inlet pressures.

Verification reference sources shall have a maximum permissible error of ± 0.05 PSIG (0.344 KPa). Verification reference sources shall have their calibration traceable to National Institute of Standards and Technology (NIST).

Steps:

1. Attach high side of verification reference source (VRS) and high side hose of field test kit under test to a common high side pressure source capable of MWWP.
2. Attach low side of VRS and low side hose of field test kit to a common low side pressure source capable of MWWP.
3. Open high side pressure source and vent the air through the field test kit by using the field test kit's high side bleed needle valve.
4. Open low side pressure source and vent the air through the field test kit by using the field test kit's low side bleed needle valve.

5. With the field test kit in the vertical orientation and equal pressure on the high and low side of the VRS and field test kit, verify that VRS and field test kit are reading 0.0 psid (0.0 KPa).
6. Reduce low-pressure source until the VRS reads 1.0 psid (6.89 KPa) and record the reading on the field test kit. Repeat at values of 2.0 (13.78 KPa), 5.0 (34.47 KPa, 8.0 (55.15 KPa) and 15.0 (103.4 KPa) psid.
7. Increase low-pressure source until the VRS reads 8.0 (55.15 KPa), then record the reading on the field test kit. Repeat at values of 5.0 (34.47 KPa), 2.0 (13.78 KPa), 1.0 (6.89 KPa), 0.0 (0.0 KPa) psid.
8. Lower the high side and low side pressure sources to atmospheric and verify that the VRS and field test kit are reading 0.0 psid (0.0 KPa).
9. With the low side pressure source at atmospheric, increase the high side pressure source until the VRS reads 1.0 psid (6.89 KPa). Record the field test kit reading. Repeat at values of 2.0 (13.78 KPa), 5.0 (34.47 KPa), 8.0 (55.15 KPa) and 15.0 (103.4 KPa) psid.
10. Decrease the high side pressure source until the VRS reads 7.5, 8.0 (55.15 KPa). Record the field test kit reading. Repeat at 5.0 (34.47 KPa), 2.0 (13.78 KPa), 1.0 (6.89 KPa) and 0.0 (0.0 KPa) psid.
11. Exceeding the accuracy of ± 0.2 psid (1.378 KPa) shall be cause for rejection.

10.2.6.2.2.2 Hydrostatic Test

Purpose: To determine the capability of the field test kit to withstand the required hydrostatic test pressure.

Requirement: All components of the field test kit shall withstand a hydrostatic pressure of twice the rated water working pressure (MWWP) for a period of ten minutes without any leakage, damage, permanent deformation or impairment of operation. The line pressure gage, if utilized, does not have to maintain accuracy but shall remain leak free.

Steps:

a. With both the high and low side hoses connected to a water supply, bleed all air from the field test kit. Close all needle valves.
b. Raise pressure to a hydrostatic pressure of twice the maximum working water pressure for a period of ten minutes.
c. Evidence of leakage, damage, permanent deformation or impairment of operation shall be cause for rejection. The line pressure gage, if utilized, does not have to maintain accuracy but shall remain leak free.

10.2.6.2.2.3 Accuracy at Rated Temperature and Pressure Test

Purpose: To test the accuracy of the field test kit at rated temperature and rated pressure.

Requirement: The field test kit shall maintain an accuracy of ± 0.2 psid (1.378 KPa) while at the greater of 140°F. (60°C) or rated temperature (MWWT) and rated pressure (MWWP), without any leakage, damage, permanent deformation or impairment of operation.

Steps:

a. Attach high and low side hoses of the gage field test kit to source of water at MWWP and the greater of 140°F. (60°C) or MWWT. Bleed water through field test kit by fully opening high and low side needle valves for 30 seconds.
b. While at MWWP and MWWT check accuracy per 10.2.6.2.2.1, steps 5 thru 7.
c. Exceeding the accuracy of ± 0.2 psid (1.378 KPa) shall be cause for rejection. Evidence of leakage, damage, permanent deformation or impairment of operation shall be cause for rejection too.

10.2.6.2.2.4 Pressurizing Fatigue Test

Purpose: To determine the capability of the field test kit to withstand pressurizing fatigue test.

Requirement: The field test kit shall maintain required accuracy after pressurizing fatigue test and shall not leak.

Steps:

a. Attach high side hose of gage to source of water pressure and bleed all air from the field test kit. Slowly pressurize high side of gage from 0 to MWWP or minimum of 175 psi (1206 KPa), at a rate of 25psi/sec (172 KPa/sec) ± 10%. Release the pressure to 0 psi (0 KPa) at a rate of 50 psi/sec (344 KPa) ± 10%. The application and release of the pressure shall be as smooth as practicable, so as not to subject the instrument mechanism to excessive upscale or down scale accelerations or high amplitude impulses (pressure spikes).
b. Repeat step a 1000 times.
c. Attach low side hose of gage to source of water pressure and bleed all air from the field test kit. Slowly pressurize low side of gage from 0 psi (0 KPa) to MWWP or minimum of 175 psi (1206 KPa), at a rate of 25 psi/sec (172 KPa/sec) ± 10%. Release the pressure to 0 psi (0 KPa) at a rate of 50 psi/sec (344 KPa/sec) ± 10%. The application and release of the pressure shall be as smooth as practicable, so as not to subject the instrument mechanism to excessive upscale or down scale accelerations or high amplitude impulses (pressure spikes).
d. Repeat step c 1000 times.
e. The gage shall be tested for accuracy, 10.2.6.2.2.1 steps 8 to 11, following the set number of cycles.
f. Exceeding the accuracy of ± 0.2 psid (1.37 KPa) shall be cause for rejection. Evidence of leakage, damage, permanent deformation or impairment of operation shall be cause for rejection too.

10.2.6.2.2.5 Field Test Kit Flow Test

Purpose: To determine the capability of the field test kit to handle the specified rate of flow.

Requirement: With an inlet pressure of 1.0 psig (6.89 KPa) the field test kit must discharge a minimum rate of flow of:

a. 0.02 gpm (0.001L/s) from high side hose to low side hose with the high side control needle valve fully open and the low side control needle valve open 1/4 turn (Fig. 10.16).

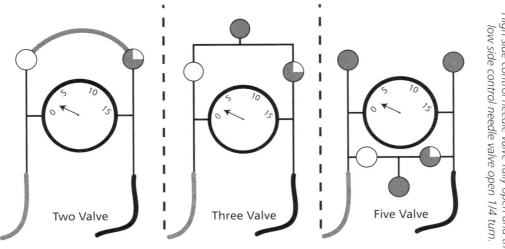

Figure 10.16 High side control needle valve fully open and the low side control needle valve open 1/4 turn.

b. 0.04 gpm (0.002 L/s) from the high side bleed needle valve with the high side bleed needle valve open 1/4 turn (see figure. 10.17)

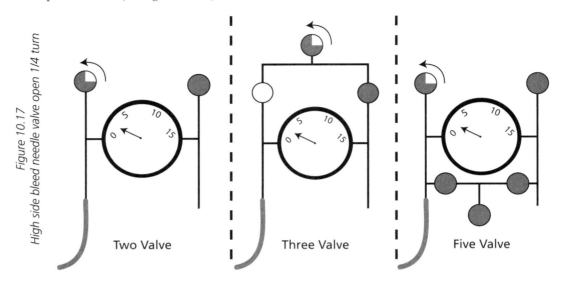

Figure 10.17 High side bleed needle valve open 1/4 turn

Steps:

a. Attach high side hose of field test kit to 1.0 psig (6.89 KPa) water source and flush all air from field test kit by opening all needle valves. Close all needle valves
b. Open high side control needle valve to the fully open position. Slowly open low side control needle valve until water begins to flow from low side hose. Close low side control needle until water just stops flowing from low side hose. This establishes the closed position of the low side control needle valve. Open low side control needle valve 1/4 turn (90 degrees) and measure the rate of flow discharging from the low side hose. Close all needle valves
c. Open high side bleed needle valve until water begins to flow. Close high side bleed needle valve until water just stops flowing. This establishes the closed position of the high side bleed needle valve. Open side bleed needle valve 1/4 turn (90 degrees) and measure the rate of flow discharging from the high side bleed needle valve.
d. Failure of the field test kit to discharge the minimum the specified rate of flow of 0.02 gpm (0.001L/s) in step a and 0.04 gpm (0.002 L/s) in step b, shall be cause for rejection.

10.2.6.2.2.6 Pressure Dissipation Test

Purpose: To determine the time required to dissipate pressure from the field test kit gage.

Requirement: The field test kit gage shall dissipate a pressure source of the greater of 175 psi (1206 KPa) or MWWP to 1 psi (6.89 KPa) in 5 seconds or less.

Steps:

a. The high side hose of field test kit shall be connected to the pressure source by means of an accumulator with an internal volume of 80 in^3 (1310 ml). Bleed all air from the field test kit and accumulator. Raise pressure in accumulator to the greater of 175 psi (1206 KPa) or MWWP. Isolate accumulator from its pressure source.
b. Quickly open 1/4 inch ⌀ valve on accumulator to dissipate pressure.
c. Record time from opening of 1/4 inch ⌀ valve until the field test kit reading reaches 1.0 psid (6.894KPa).
d. Repeat steps a to c two more times.
e. Exceeding 5.0 seconds for the average of the three readings shall be cause for failure.

10.2.6.2.2.7

Purpose: To determine the overall compliance of the field test kit with the Standard as set forth in 10.2 of this manual and capability of performing the field test procedures as set forth in Chapter 9 of this manual.

Requirement: All sections of the Standard of this manual that are not specifically covered by the above tests shall be satisfactorily met.

Steps:

    a. For each pertinent section of the Standard inspect, test or otherwise be assured that the assembly meets the minimum conditions of this Standard. Verify capability of field test kit to perform the field test procedures as set forth in Chapter 9 of this manual.

Reference Standards
ASME B40.1 - Gauges - Pressure Indicating Dial Type - Elastic Element
ASME/ANSI PTC 19.2 - 1987 Pressure Measurement

Standards

Chapter 11

Backflow Incidents

Backflow Incidents

The following pages contain backflow incidents as reported to the Foundation for Cross-Connection Control and Hydraulic Research at the University of Southern California. The date and location of the incident is included with as much detail as possible. Some incidents have a great deal of detail, while others have very limited information. The reference number will refer to the reference source that was used to report the specific backflow incident. This is shown at the end of the list of incidents.

Additionally, this table is available in a sortable MS Excel version on the CD that accompanied this manual.

11 Backflow Incidents

Source	Month - Year	City	State	Country	Cause (What Happened?)	Type of Establishment
*1	1908	Auburn	NY	USA	Cross-connections.	Factory
*2	September-20	Neenah	WI	USA	Cross-connection in Industrial Plant.	County
*3	October-20	Amsterdam	NY	USA	Cross-connection to polluted creek.	Town
*5	1922	Franklin	NJ	USA	Single check valve failed.	
*2	1923	Wausau	WI	USA	Cross-connection with contaminated on-site well.	Factory
*5	May-23	Rockaway	NJ	USA	Cross-connection with raw river water.	City
*6	July-23	Everett	WA	USA	Cross-connection with mill fire system.	Factory
*7	1924	Oakland	CA	USA	Cross-connection in plant pumped molasses into city mains.	Factory
*9	January-24	Bloomington	IL	USA	Cross-connection with contaminated Industrial on-site supply.	Town
*2	June-24	Wausau	WI	USA	Cross-connection with contaminated industrial on-site supply.	Factory
*10	Sep-24	Newport	VT	USA	Cross-connection.	Town
*11	January-25	Yonkers	NY	USA	Cross-connection.	Town
*9	February-25	Sterling	IL	USA	Cross-connections with contaminated factory on-site supply.	Factory
*8	July-25	Winona Lake	IN	USA	Cross-connection between town mains and contaminated lake.	Town
*12	October-25	Eco City	MI	USA	Temporary cross-connections with contaminated stream.	Town
*3	January-26	Woodbury		USA	Cross-connection with contaminated surface waters.	City
*13	March-26	Ada	OK	USA	Cross-connections between city and industrial line from contaminated lake.	Plant
*14	May-26	Tiverton	NJ	USA	Cross-connections.	Town
*7	August-26	Richmond	CA	USA	High pressure salt water forced into city mains.	Oil Company
*15	December-26	Rochester	NY	USA	Cross-connection in Department store with fire sprinkler supply.	Store
*16	February-27	Western	PA	USA	Cross-connection.	Town

Backflow Incidents

Source	Month - Year	City	State	Country	Cause (What Happened?)	Type of Establishment
*13	February-27	Beaverton	OR	USA	Cross-connection.	Town
*17	August-27	Oswego	NY	USA	Cross-connection to auxiliary river supply.	Factory
*15	September-27	Cohoes	NY	USA	Cross-connection to auxiliary river supply.	Factory
*15	October-27	Albany	NY	USA	Cross-connection to auxiliary river supply.	Factory
*7	1929	Oakland	CA	USA	Cross-connection with high pressure triplex pumps forced soap mix into mains.	Garage
*20	1929	Ft. Wayne	PA	USA	Cross-connection to contaminated railroad.	City
*2	January-29	Two Rivers	WI	USA	Cross-connection in plant.Industrial Plant	
*18	March-29	Ft. Wayne	IN	USA	Cross-connection with contaminated on-site reservoir.	
*19	June-29	So. St. Paul	MN	USA	Cross-connection.	Plant
*7	January-30	Oakland	CA	USA	Cross-connection with on-site high pressure well supply.	Dry Dock
*7	February-30	Oakland	CA	USA	Cross-connection with liquefied pigment under high pressure.	Chemical
*21	August-30	Milford	DE	USA	Cross-connection to contaminated stream.	Town
*12	December-30	Lawrence	MA	USA	Cross-connection to contaminated private fire sprinkler supply.	City
*12	June-32	Lapeer	MI	USA	Cross-connection with contaminated creek.	Town
*22	June-33	Chicago	IL	USA	Cross-connection with contaminated fire sprinkler system.	Plant
*41	June-33	Chicago	IL	USA	Cross-connection with sewer system.	Hotel
*22	July-33	Marseilles	IL	USA	Cross-connection within camp.	Camp
*22	December-33	Chicago	IL	USA	Cross-connection with sewer in hotel.	Hotel
*22	December-33	Chicago	IL	USA	Cross-connection with contaminated fire sprinkler system.	Factory
*10	January-34	Winooski	VT	USA	8" cross-connection within woolen mill.	Factory
*20	May-34	Brentwood	CA	USA	Backsiphonage.	Home
*20	December-34	Burbank	CA	USA	Cross-connection to boiler fed water.	Winery
*12	May-35	Worcester	MA	USA	Cross-Connection.	City
*12	June-35	North Adams	MA	USA	Cross-connection with contaminated private supply.	Town

11 Backflow Incidents

Source	Month - Year	City	State	Country	Cause (What Happened?)	Type of Establishment
*23	July-35	Indianapolis	IN	USA	Check valve on pump discharge failed.	Town
*20	1936	Pomona	CA	USA	Several Cross-connections on campus.	College
*7	January-36	Oakland	CA	USA	Ship pumped salt water into potable water system.	Railroad yard
*17	July-36	India Lake	NY	USA	Cross-connection with contaminated lake.	Town
*24	August-36	Schroon	NY	USA	Cross-connection within Hotel.	Hotel
*20	September-36	Los Angeles	CA	USA	Backsiphonage.	Apartment House
*20	December-36	Cleveland	MS	USA	Cross-connection at construction project.	Road Building
*20	June-38	Camarillo	CA	USA	Backsiphonage from irrigation ditch.	Building Project
*25	August-38	Mamakating	NY	USA	Backsiphonage.	Summer Hotel
*25/pg 46	November-38	Middleboro	MA	USA	Cross-connection.	Industrial Plant
*26	January-39	Lansing	MI	USA	Backsiphonage.	College Lab
*25/pg 51	January-39	Palmer	MA	USA	Cross-connection with contaminated river.	Town
*27	March-39	Los Angeles	CA	USA	Backsiphonage at nurses quarters.	Hospital
*28	June-39	Fullerton	CA	USA	High pressure from oil well forced into water system.	Oil Well
*25/pg 53	June-39	Hopkinsville	KY	USA	Backsiphonage	Town
*25/pg 51	August-39	Boston	MA	USA	Faulty Plumbing.	Coop work rooms
*29	September-39	Cleveland	OH	USA	Liquid paraffin pumped into city water mains.	Oil Refinery
*7	October-39	Oakland	CA	USA	Cross-connection in car wash pumped soap into city mains.	Service Station
*30	January-40	Joliet	IL	USA	Backpressure from private well.	Soda Factory
*28	January-40	Joliet	IL	USA	Backpressure forced beer into city mains.	Brewery
*25/pg 56	April-40	So. Hadley	MA	USA	Cross-connection with contaminated fire sprinkler system.	College
*165	May-40	Albany	NY	USA	Numerous internal cross-connections.	Hotel
*25/pg 56	May-40	Canastota	NY	USA	Backflow of sewage through faulty valve.	Village
*25/pg 56	July-40	Coxsackie	NY	USA	Cross-connection with contaminated well.	Children's Camp
*31	August-40	Vallejo	CA	USA	Salt water backpressured into potable water mains on base.	Navy Base
*25/pg 56	September-40	Louisville	KY	USA	Cross-Connection.	Apartment House
*25/pg 56	November-40	Clymer	NY	USA	Backsiphonage from toilets.	School

Backflow Incidents

Source	Month - Year	City	State	Country	Cause (What Happened?)	Type of Establishment
*25/pg 56	December-40	Rochester	NY	USA	Cross-connection to sewage.	City
*25/pg 63	June-41	Fall River	MA	USA	Cross-connection with contaminated river.	Mill
*25/pg 62	August-41	Oro Grande	NM	USA	Cross-connection with contaminated source.	Town
*7	September-41	Berkeley	CA	USA	Lake water pumped into public supply.	Park
*25/pg 64	October-41	High Point	NC	USA	Cross-connection to sewage.	City
*33	1942	Los Angeles	CA	USA	Harbor water backpressured into domestic and city mains.	Shipyard
*32	1942	Los Angeles	CA	USA	Cross-connections.	Aircraft Factory
*20	February-42	Los Angeles	CA	USA	Several Cross-connections.	Labor Camp
*25/pg 71	March-42	Baltimore	MD	USA	Cross-Connection.	Dept. Store
*25/pg 72	March-42	St. Mary's	OH	USA	Cross-connection with contaminated canal.	City
*33	May-42	San Francisco	CA	USA	Ship pumped harbor water ashore.	Pier 16
*25/pg 70	July-42	Lee County	IL	USA	Open valve between contaminated on-site supply & public supply.	Ordnance Plant
*25/pg 70	July-42	Pittsburgh	PA	USA	Cross-connection to raw and treated waters.	Industrial Plant
*25/pg 71	July-42	Fallsburg	NY	USA	Cross-connection to contaminated supply.	Summer Hotel
*25/pg 72	August-42	Laurel Hills	NY	USA	Cross-connection to contaminated creek.	
*25/pg 70	August-42	Superior	WI	USA	Cross-connection, broken main at yard.	Shipyard
*36	August-42	Los Angeles	CA	USA	Partial vacuum in drinking water backsiphoned toilet contents into water line.	
*36	August-42	Los Angeles	CA	USA	Heavy use caused venturi effect resulting in backflow from toilets.	School
*33	August-42	Los Angeles	CA	USA	Ship pumped harbor water into domestic system.	Army Post Outpost
*25/pg 70	September-42	Newton	KS	USA	Cross-connection to sewer system.	Shipyard Building
*25/pg 71	October-42	Camden	NJ	USA	Cross-connection to contaminated river.	City
*33	October-42	Los Angeles	CA	USA	Local contamination, source not determined.	Shipyard
*33	November-42	Los Angeles	CA	USA	Ship pumped harbor water into both ship & base domestic system.	Port of Embarkation
*33	November-42	San Francisco	CA	USA	Ship pumped harbor water from fire system into city system.	Navy Base
*33	December-42	Los Angeles	CA	USA	Ship pumped harbor water through its fresh water tanks into city system.	Pier 25
						Shipyard

11 Backflow Incidents

Source	Month - Year	City	State	Country	Cause (What Happened?)	Type of Establishment
*36	December-42	Los Angeles	CA	USA	Cross-connection to high pressure process water in plant.	Plant Mach. Shop
*33	December-42	San Francisco	CA	USA	Ship fire system backpressured salt water into city system	Pier 20
*7	1943	Alameda	CA	USA	Pump test, backpressured salt water into plant's system.	Engr. Co.
*35	February-43	San Pedro	CA	USA	985 cross-connections found during inspection.	Army Post
*36	February-43	Van Nuys	CA	USA	Cross-connection to vacuum pump in plant.	Indust. Plant
*25/pg 77	April-43	Seattle	WA	USA	Cross-connection with untreated supply.	Shipyard
*28	August-43	Fall River	MA	USA	Cross-connection with raw river water for fire sprinkler system test.	Factory
*33	August-43	San Francisco	CA	USA	Cross-connection with ship fire system.	Pier 23, 25, 29
*36	September-43	Los Angeles	CA	USA	Cross-connection backsiphoned toilet contents into domestic system.	Army Field HDQ
*7	September-43	Orinda	CA	USA	Cross-connection between on-site spring and public supply.	Estate
*33	September-43	Long Beach	CA	USA	Ship pumped harbor water into city mains.	Naval Base
*33	October-43	San Pedro	CA	USA	Ship pumped harbor water through fire line into city mains.	S.S. Avalon
*33	October-43	San Francisco	CA	USA	Ship pumped harbor water into both on-board tanks and city mains.	Pier 46
*25/pg 77	October-43	Portland	OR	USA	Cross-connection with contaminated river.	Shipyard
*20	1944	Los Angeles	CA	USA	618 cross-connections found during campus inspection.	College
*25/pg 82	January-44	Forest City	NC	USA	Backsiphonage of waste water into city storage tank.	City
*33	February-44	Los Angeles	CA	USA	Ship pumped harbor water into both on-board tanks and city mains.	Pier 232
*33	February-44	San Diego	CA	USA	Ship pumped harbor water into city mains.	Destroyer Base
*33	February-44	Terminal Is.	CA	USA	Ship pumped harbor water through fire lines to city mains.	Berth 232

Backflow Incidents

Source	Month - Year	City	State	Country	Cause (What Happened?)	Type of Establishment
*33	March-44	Oakland	CA	USA	Ship pumped harbor water through fire lines into city mains.	Grove St. pier
*25/pg 81	March-44	Elizabethtown	KY	USA	Backsiphonage through frost-proof toilets.	Town
*25/pg 82	March-44	Bay City	TX	USA	Possible backsiphonage during fire or contamination during flood.	
*25/pg 82	April-44	Oklahoma	OK	USA	Cross-Connections.	City
*33	May-44	Alameda	CA	USA		School
*33	June-44	San Francisco	CA	USA	Ship pumped harbor water through fire system into city mains.	Pier
*33	July-44	Oakland	CA	USA	Ship pumped harbor water through fire system into city mains.	Pier 23
*33	July-44	San Pedro	CA	USA	Ship pumped harbor water through fire system into city mains.	Pier 17
*25/pg 81	August-44	Southern	MI	USA	Cross-connection to fire sprinkler system.	Port of Embarkation
*25/pg 81	August-44	Blooming Grive	NY	USA	Cross-connection permitted backsiphonage.	College
*37	December-44	Los Angeles	CA	USA	Cross-connection with unapproved spring.	Resort Lodge
*33	February-45	Los Angeles	CA	USA	Pumps on fire line forced harbor water into domestic lines. Berth 195-197.	Terminal Co.
*25/pg 86	February-45	Green Bayou	TX	USA	Many cross-connection found during inspection.	Hospital
*35	March-45	Los Angeles	CA	USA	Cross-connection with flush water.	Navy Freighter
*28	June-45	Compton	CA	USA	Ship pumped harbor water thru fire lines into city water mains.	Pier Shipbuilders
*20	October-45	Terminal Is.	CA	USA	Cross-connection between gas and water mains allowed gas to enter water mains.	Flushing Conn.
*25/pg 86	October-45	Chicago	IL	USA	Ship fire pumps forced harbor water into city mains.	Pier
*38	January-46	North Central	IN	USA	Backflow from toilet connection.	School
*33	October-47	Sawtelle	CA	USA	Cross-connection with only one gate valve.	Industrial Plant
*28	August-48	New York	NY	USA	Cross-connection with unapproved well.	Vet. Adm.
*35	December-48	Van Nuys	CA	USA	Backsiphonage through low inlets into city mains.	Office Bldg.
*166	June-49	Clinton	IA	USA	Backsiphonage of Sodium Arsenate through unprotected hose.	Paving Co.
					Cross-connection, single check removed.	Factory

11 Backflow Incidents

Source	Month - Year	City	State	Country	Cause (What Happened?)	Type of Establishment
*39	August-49	El Segundo	CA	USA	Zinc Sulfate pumped into city mains.	Chemical Plant
*28	December-50	San Diego	CA	USA	7200 ppm's of Chloride backsiphoned into city mains.	Residential
*385	1952	Huskerville	NE	USA	Backsiphonage from toilets into potable looped system.	65 Unit Housing
*20	July-52	Venice	CA	USA	Tret-o-lite solution pumped back into city mains.	Oil Well
*20	July-55	San Pedro	CA	USA	Ship pumped harbor water through five obsolete check valves into city mains.	Shipyard
*41	August-56	Shreveport	LA	USA	Cross-connections between contaminated bayou and city mains through lawn system.	Private Res.
*42	October-56	Renton	WA	USA	Cross-connection between petroleum product transmission line @130 psi and city mains @70 psi.	Pumping Station
*20	October-58	Sunland	CA	USA	Weedicide pumped into city main.	Orchard
*41	1960	Center Point	AL	USA	Sludge from septic tank pumped into drinking water system through pump priming line. Existing check valve failed.	School
*41	1960	Crestview	FL	USA	Gas from 250 gallon butane tank backpressured into city mains.	Cafe
*200	1960	Atlanta	GA	USA	Blood backsiphoned from water operated aspirator into drinking fountain.	Mortuary
*76	April-63	Bay City	MI	USA	Shipboard fire pumps backpressured harbor water into city mains.	Shipyard
*44	December-64	Michigan	MI	USA	Backsiphonage from unprotected autopsy table contaminated hospital water.	Hospital
*76	1966	Wyoming	MI	USA	Plating tanks backsiphoned into city mains.	Industrial Plant
*107	October-67	Renton	WA	USA	Cross-connection of gasoline pipeline with city water system caused 2000 gallons of gasoline to enter the water system.	Pipeline Company
*167	October-67	Renton	WA	USA	Gasoline flows from hydrant due to direct connection between flammable liquids transmission line and the water main line.	Commercial

Backflow Incidents

Source	Month - Year	City	State	Country	Cause (What Happened?)	Type of Establishment
*216	July-68	Spokane	WA	USA	Boiler treatment chemicals backflowed into the potable water system when the water supply was shut down.	Institution
*45	April-69	Berkeley	CA	USA	Backsiphonage from treated hot water heating system into potable system.	8-Story Building
*169	April-69	Manhasset	NY	USA	A vending machine dispensing drinks was improperly connected to the recirculating hot-water system.	Commercial
*46	August-69	Elmhurst	IL	USA	Fire hose used to flush out gasoline pipeline to airport.	Pipeline
*382	August-69	Worcester	MA	USA	83 football team members/coaching staff stricken with infectious hepatitis via sunken hose-bibs backsiphonage due to local fire.	Football Field
*158	September-69	Peoria	AZ	USA	Sodium Arsenate backsiphoned from tank of weed killer.	Farm
*170	September-69	Philadelphia	PA	USA	Chromium seeped from a defective part within an air conditioning system into its water pipes.	Institution
*171	September-69	Worcester	MA	USA	Contaminated water back-siphoned into the drinking water system due to pressure drop in the water system.	Institution
*48	November-69	Atlanta	GA	USA	Chromate treated cooling water backflowed to potable system.	Commercial Bldg.
*48	November-69	Atlanta	GA	USA	Backsiphonage from air conditioning system treated with bichromate of soda.	Hotel
*49	December-69	Silverton	OH	USA	Wine flushed into city mains.	Wine Cellar
*387	December-69	Cincinnati	OH	USA	Wine enters the potable water lines.	Wine Cellar
*50	1970	New York	NY	USA	Cross-connection of chromate treated cooling tower system with domestic system.	College
*43	January-70	Atlanta	GA	USA	Cross-connection of chromate treated cooling system with potable water system.	Bank
*51	January-70	Vancouver	BC	Canada	Cross-connection on board ship.	Cruise Ship

11 Backflow Incidents

Source	Month - Year	City	State	Country	Cause (What Happened?)	Type of Establishment
*417	June-70	Mattoon	IL	USA	Hot wash water from asphalt plant backpressured into city mains coupled with flow testing of fire hydrants in the area.	Asphalt Plant
*52	June-70			USA	Radioactive water backpressured into plant potable water system.	Nuclear Plant
*99	Jun-70	Boston	MA	USA	Chromates backpressured from boiler.	High School
*54	July-70	San Bruno	CA	USA	Strong detergent water backpressured through a faulty check valve into city water mains.	Car Wash
*53	September-70	New Jersey	NJ	USA	Soft drink vending machine cross-connected to a building heating system which contained Hexavalent Chromium.	Caddy House
*172	February-71	Mattoon	IL	USA	Hot, dirty water backflowed through a wheel-operated gate valve into the city main.	Commercial
*99	1972	Boston	MA	USA	Sodium Chromate/caustic soda backpressured from cooling a tower.	Office Building
*57	1972	Lyndon	OR	USA	Liquid aluminum backsiphoned into city mains.	Factory
*157	March-72	Huachuca City	AZ	USA	Backsiphonage of pesticide through garden hose submerged in mixing tank.	Exterminating Co
*58	April-72	Bakersfield	CA	USA	Aldrite (termite pesticide) back-siphoned through cross-connection.	200 Home Area
*173	May-72	Riverside	CA	USA	Contents of a mixing tank backflowed into the public water mains.	Residences
*396	July-72	San Antonio	TX	USA	Private well pumping into city water system. Meter was found running backwards.	Packing Company
*159	September-72	Salt Lake City	UT	USA	Boiler compound backpressured into potable system through leaking check valve.	Apartment Bldg,
*59	September-72	Bydgoszcz, Poland			Single, faulty valve permitted beer to backflow into city mains.	Brewery
*57	1973	Klamath Falls	OR	USA	Contaminated irrigation water backpressured into city mains.	Farm

Backflow Incidents

Source	Month - Year	City	State	Country	Cause (What Happened?)	Type of Establishment
*60	February-73	Roebling	NJ	USA	Cross-connection between fire line using contaminated river and service connection to a building.	U.S. Post Office
*108	March-73	Seattle	WA	USA	Cross-connection between hydrant drain and the sewer.	City
*438	April-73	Greenville	SC	USA	Suspected cross-connection in contaminated well water.	Residence
*77	May-73	Morganville	NJ	USA	Submerged hose in Chlordane mix backsiphoned during a water main break.	Residence
*159	May-73	Salt Lake City	UT	USA	Backsiphonage of slag from submerged inlets in a plating process.	Plating Company
*99	1974	Woburn	MA	USA	Fungicide backpressured from chemical application system through single check valve.	Greenhouse
*99	1974	Boston	MA	USA	Harbor water backpressured from ship into city main.	Harbor
*397	February-74	San Antonio	TX	USA	Salt water from water softener equipment backsiphoned into internal water system.	Pharmacy
*109	June-74	Soldiers Grove	IL	USA	Backsiphonage of herbicide Balan to water system.	Village
*174	June-74	Boston	MA	USA	A faulty valve allowed air-conditioning water to backflow into the drinking water supply.	Commercial
*110	July-74	Seattle	WA	USA	Chemical De-Germ and other pollutants backsiphoned into water system.	Sea-Tac Airport
*77	September-74	New York	NY	USA	Backpressure from a Chromate treated air conditioning system into a building potable water system.	Office Bldg.
*432	December-74	South Hills	PA	USA	Water tank was contaminated with industrial chemicals and fed to residences.	Residence
*396	December-74	San Antonio	TX	USA	Chemically treated steam backpressured into potable system through faulty check valve.	Food Manufacturer
*383	December-74	No. Carolina	NC	USA	Chemically treated boiler feed backpressured into city mains.	Fertilizer Plant

11 Backflow Incidents

Source	Month - Year	City	State	Country	Cause (What Happened?)	Type of Establishment
*99	1975	Chelmsford	MA	USA	Backflow of Ethylene Glycol from solar heating system.	Commercial Bldg.
*159	1975	Salt Lake City	UT	USA	Scale from water heater backsiphoned, causing the users water systems to get plugged.	Beverage Company
*63	March-75	Redwood City	CA	USA	Detergent laden wash water backpressured into city mains.	Car Wash
*175	April-75	Riverside	CA	USA	Chromate backsiphoned from a cooling tower make-up line into a building's water supply.	Business
*64	October-75	Langley	SC	USA	Ethylacroglate backsiphoned from an industrial plant into city mains.	Residences
*176	October-75	Langley City	SC	USA	Pressure drop from a city's water system causing a contaminated water well to backpressure into the city's system.	Industrial
*41	November-75	East Monitro	WA	USA	Endrin treated water (pesticide) backsiphoned into 21 homes.	Fire Hydrant
*111	March-76	Chattanooga	TN	USA	City water main break caused backsiphonage of Chlordane.	Residence
*390	March-76	San Antonio	TX	USA	Contaminated private lake back-pressured through lawn sprinkler system into city mains.	Private Grounds
*396	April-76	San Antonio	TX	USA	Cross-connection between private lake used for irrigation and city water system, caused backpressure through an untested double check valve assembly.	Residential
*63	May-76	San Jose	CA	USA	Backsiphonage to a camp site's water system occurred through a direct cross-connection between an automatic water softener and a sewer line.	Private Camp
*395	June-76	San Antonio	TX	USA	6" and 4" services looped through a hospital causing backflow.	Hospital
*66	October-76	Washington	PA	USA	Pesticide pump discharge backpressured through hose connected to house hose bib.	140 Homes

Backflow Incidents

Source	Month - Year	City	State	Country	Cause (What Happened?)	Type of Establishment
*431	1977	Washington	MD	USA	Washington County. Cross-connection between drinking water and fertilizer mixing tank.	Factory
*80	1977	North Dakota	ND	USA	DDT backsiphoned from two homes into city main.	Residences
*61	January-77	San Antonio	TX	USA	Resin from water softener backsiphoned into house system due to a nearby main break.	Private Home
*112	February-77	Seattle	WA	USA	Cross-connection between heating system and city water system resulted in backflow of Borate-Nitrite from the heating boiler.	Hospital
*75	March-77	Burlington	VT	USA	CO_2 and carbonated water backflowed into copper piping at a hospital causing copper poisoning.	Hospital
*178	March-77	Oak Hill	SC	USA	Air pressure in a cleaning apparatus overcame the water pressure and backpressured solvents into the water system.	Commercial
*274	June-77	Keota	IA	USA	Backflow from poultry processing plant contaminates water system with Salmonella.	Poultry Processing Plant
*394	July-77	San Antonio	TX	USA	Failure of a control switch pumped scale inhibitor into domestic water lines.	Industrial Bldg.
*67	July-77	Kulm	ND	USA	Backsiphonage from garden sprayers contaminated city mains with DDT.	
*179	July-77	US Naval Vessel		USA	Hydroquinone backflowed into a ship's potable water system from a 40-gallon tank.	Institution
*396	July-77	San Antonio	TX	USA	Malfunction of cooling tower caused 50 gallons of Algaecide-Slimcide to backpressure into potable water supply.	Office Bldg.
*68	August-77	Chico	CA	USA	Detergent ladened water from a car wash backpressured into city water main.	Car Wash
*69	October-77	Chicago	IL	USA	Cross-connection at an apartment caused Shigella in one case and acute symptoms in others.	Apartment House

11 Backflow Incidents

Source	Month - Year	City	State	Country	Cause (What Happened?)	Type of Establishment
*393	October-77	San Antonio	TX	USA	Black water from fire sprinkler system polluted major portions of a hospital's internal water system.	Hospital
*422	October-77	San Bernardino	CA	USA	Contaminated water backflowed into the distribution system due to plumbing changes.	Residential
*159	1978	Salt Lake City	UT	USA	Backflow of Carbon Dioxide from a beverage dispenser results in copper contamination.	Restaurant
*70	January-78	Hinsdale	IL	USA	High pressure air from hospital escaped through a stuck valve into water main.	Hospital
*114	January-78	Vancouver	BC	Canada	Backflow of Carbon Dioxide from a beverage dispenser results in copper contamination.	Theater
*396	February-78	Kelly AFB	TX	USA	Chrome residue backsiphoned into potable water supply.	Air Force Base
*181	February-78	San Antonio	TX	USA	Backsiphonage of chrome residue from plating process into potable water supply. Several people hospitalized.	Institution
*115	March-78	Vancouver	BC	Canada	Backflow of corrosion inhibitor (BRAMCO 750) from apartment's heating boiler into water system.	High-rise Apartment
*113	May-78	Pierce County	WA	USA	Backsiphonage from a septic tank sewage into a school's water system resulted in 33% student absenteeism.	School
*392	July-78	San Antonio	TX	USA	Backsiphonage of 'blue water' from toilet without a vacuum breaker.	Residence
*182	August-78	Warrenville	SC	USA	Chlordane was backsiphoned into water system through a garden hose.	Institution
*41	1976	Suffolk Co.	NY	USA	Suspected backsiphonage in cafeteria from a slop sink.	Public Bldg.
*427	February-79	San Antonio	TX	USA	Steam from chemically treated boiler backpressured into city mains.	Plastics Plant

Backflow Incidents

Source	Month - Year	City	State	Country	Cause (What Happened?)	Type of Establishment
*79	February-79	Seattle	WA	USA	Cross-connection allowed backpressure of industrial detergent into 100 city blocks.	Car Wash
*116	March-79	Kulm	ND	USA	Backsiphonage of DDT from an aspirator sprayer caused contamination of municipal water system.	City
*41	April-79	Marshalltown	IA	USA	Cross-connection between waste water line and new on-site well caused 1.1 million lb. of pork contamination.	Packing Plant
*184	May-79	Spartanburg	SC	USA	Raw sewage was backpressured into the potable water line.	Institution
*185	May-79	Marshalltown	IA	USA	Cross-connection between waste water line and new on-site well caused 100,000 lb. of pork contamination.	Commercial
*118	June-79	Meridian	ID	USA	Stagnant water containing high bacterial count backsiphoned through a leaking check valve.	City
*106	June-79	Meridian	ID	USA	Leaky alarm check on fire line system.	Supermarket
*73	June-79	Adair Co.	MO	USA	Backsiphonage through submerged hose in herbicide tank.	Farm
*73	June-79	Callaway Co.	MO	USA	Backsiphonage through submerged hose in herbicide tank.	Farm
*105	June-79	Lake Havasu	AZ	USA	Backpressure from effluent tree bubbler system into potable water supply.	Marina
*377	June-79	Meridian	ID	USA	Higher than normal concentrations of bacteria flowed in water supply.	Residential
*74	August-79	Los Alamos	NM	USA	Backsiphonage of treated boiler water into school potable water system.	Pueblo Jr. High School
*140	August-79	San Antonio	TX	USA	Backsiphonage of "blue water" from several toilet tanks entered the potable water supply.	Apartment Bldg.
*117	September-79	Portland	OR	USA	Detergent water backflows through faulty RP device.	Dairy

11 Backflow Incidents

Source	Month - Year	City	State	Country	Cause (What Happened?)	Type of Establishment
*186	October-79	Roanoke	VA	USA	Chlordane mixture backsiphoned into a home plumbing network due to pressure drop when water supply was shut off.	Residential
*391	October-79	San Antonio	TX	USA	Blue water from commode tank backsiphoned into potable water supply.	Residence
*119	October-79	Seattle	WA	USA	Backflow of cooling solution into building's water system due to faulty plant control valve.	Industrial
*396	November-79	San Antonio	TX	USA	Backsiphonage of "blue water" from unprotected toilet.	Residence
*81	March-80	Manchester	NH	USA	Chemical backsiphoned from hot water system into potable supply during heavy draft for fire.	Office Bldg.
*396	April-80	San Antonio	TX	USA	Carbonated water backpressured through faulty check valve from soda dispensing equipment.	Restaurant
*396	June-80	San Antonio	TX	USA	Swimming pool water backsiphoned into potable water supply.	Residence
*187	July-80	Cheraw	SC	USA	Copper leached from copper piping by carbon dioxide gas from a soft drink machine, causing copper poisoning.	Commercial
*159	August-80	Salt Lake City	UT	USA	Backpressure through unprotected make-up line from heating/cooling system on 18th floor of a building.	High-rise
*121	September-80	Unalaska	AL	USA	Sewage contaminates water system due to cross-connection aboard a crab processing ship.	Ship
*120	October-80	Tacoma	WA	USA	Oil is pumped into city water system due to leaks in a heat exchanger.	Pumping Station
*396	November-80	San Antonio	TX	USA	Main break caused backsiphonage of blue water from toilets into water system.	Residence
*159	November-80	Salt Lake City	UT	USA	Detergent backsiphoned from a hose submerged in a large tank.	Warehouse Complex
*188	November-80	Robinson	PA	USA	Chemical mixed in a tank truck backflowed into the potable water system when the water supply was shut off.	Residential

Backflow Incidents

Source	Month - Year	City	State	Country	Cause (What Happened?)	Type of Establishment
*82	December-80	Beechview	PA	USA	Cross-connection caused backsiphonage of Chlordane from exterminator truck into water mains. 25,000 residences out of water for weeks due to contamination.	Residential
*159	December-80	Salt Lake City	UT	USA	Backpressure of chemicals from a solar heating system through a hose.	Office Building
*189	December-80	Pittsburgh	PA	USA	Gasoline compound with small amounts of pesticide were found in the drinking water system.	Residential
*122	1981	South Bend	IN	USA	Cross-connection between gas line and water line, causes water to enter the gas lines.	
*190	January-81	Norfolk	VA	USA	Cross-connection backpressured harbor water from fire system into city mains.	Shipyard & Restaurant
*191	January-81	Norfolk	VA	USA	Sea water pumped from shipyard's fire protection system into the public water distribution system through a cross-connection.	Commercial
*418	January-81	Port Arthur	TX	USA	Rice tainted with Cholera was contaminated by water due to an open valve.	Floating Oil Rig
*396	February-81	San Antonio	TX	USA	Backsiphonage of filter material into hospital's internal system.	Hospital
*159	March-81	Salt Lake City	UT	USA	Backpressure of corrosion inhibitor through bypass around RP on make up line to chiller.	Office Building
*159	March-81	Salt Lake City	UT	USA	Contractor broke water main which then washed out sewer line. Sewage backsiphoned into water main when it was shut down.	Construction Area
*396	April-81	San Antonio	TX	USA	Backsiphonage of chrome plating solution into customer system.	Plating Plant
*83	June-81	Chester	VA	USA	Polyvinyl garden hose with pistol type nozzle left in sun permitted gas from hose to enter resident's water system, producing taste & odor problem.	Residence
*159	July-81	Salt Lake City	UT	USA	Backpressure of onsite recycled water.	Garage

11 Backflow Incidents

Source	Month - Year	City	State	Country	Cause (What Happened?)	Type of Establishment
*192	July-81	Robinson	PA	USA	Chemical mixed in a tank truck was backpressured into the potable water system when the water supply was shut off.	Residential
*396	September-81	San Antonio	TX	USA	Chemically treated water from a heat exchanger backpressured through faulty gage valve.	Hospital
*84	November-81	Chicago	IL	USA	Cross-connection allowed contaminated fluids from industrial plant into city mains.	Industrial Plant
*85	November-81	Bangor	ME	USA	Cross-connection backsiphoned a home's hot water heating system into water mains during main repairs.	Residence
*86	December-81	Bailey	CO	USA	Cross-connection allowed antifreeze in solar heating system to contaminate school potable water system.	School
*396	January-82	San Antonio	TX	USA	Carbonated water backpressured into water supply through leaking check valve on soda dispenser.	Convenience Store
*125	March-82	Tacoma	WA	USA	A leaking check valve permitted soapy water from a car wash to be pumped into city water system.	Car Wash Facility
*396	March-82	San Antonio	TX	USA	Carbonated water from soda dispenser backpressured into water system. Several illnesses reported.	Restaurant
*193	April-82	Tampa	FL	USA	Contaminant siphoned from toilet tanks into the drinking water system when water was turned off for main repair.	Residential
*194	April-82	Shelby County	TN	USA	Chiller water backflowed into the potable water system.	Industrial
*87	May-82	Bancroft	MI	USA	Cross-connection backsiphoned Marathon from garden hose applicator during water main shutdown.	Residence

435

Backflow Incidents

Source	Month - Year	City	State	Country	Cause (What Happened?)	Type of Establishment
*195	May-82	Bailey	CO	USA	Antifreeze seeped into the water pipes form a school's solar heating system through a malfunctioning valve.	Institution
*196	May-82	Bancroft	MI	USA	Insecticide backsiphoned into the water system through a garden hose while the system was shut down for repairs.	Residential
*396	June-82	San Antonio	TX	USA	Carbonated water backpressured into water supply through leaking check valve on soda dispenser.	Restaurant
*88	June-82	Buffalo	NY	USA	Cross-connection backpressured from industrial plant into 10-block area producing bad taste & odor in the drinking water.	Industrial
*123	June-82	Springfield	OR	USA	Backflow from irrigation hose resulted in an insect larvae entering a food processing vat.	Food Processing Plant
*162	July-82	Laguna Hills	CA	USA	Hydro seed mulch backpressured into fire hydrant.	Fire Hydrant
*89	July-82	North Andover	MA	USA	Cross-connection allowed backpressure of Hexavalent Chromium into potable water system.	Mfg. Plant
*91	July-82	Jersey City	NJ	USA	Rupture of 72" main with inoperable isolation valves forced 300,000 customers to be without water.	City
*197	July-82	North Andover	MA	USA	A backpressure incident resulted in a toxic substance from a refrigeration chiller contaminating the drinking water.	Commercial
*198	August-82	Trumbull	CT	USA	Backflow of gas into water mains due to high pressure caused fires and blasts.	Commercial
*199	August-82	New England		USA	Propane backflowed into the city main, resulting in home two fires.	Industrial
*200	September-82	Sangamon	IL	USA	Kidney dialysis machines were contaminated with air-conditioning antifreeze due to a cracked valve causing patients deaths.	Institution

11 Backflow Incidents

Source	Month - Year	City	State	Country	Cause (What Happened?)	Type of Establishment
*90	September-82	Alice	TX	USA	Cross-connection backpressured petroleum products contaminating city mains for several days.	Petro-Chem Allegra
*201	September-82	Alice	TX	USA	Five barrels of black chemical contaminant were pumped in to the city water supply due to an unauthorized connection by a local business.	Commercial
*202	September-82	Bath	ME	USA	River water entered the shipyard drinking water system through a faulty cross-connection.	Commercial
*124	October-82	Springfield	OR	USA	Cross-connection between potable water system and processed water & fire system.	Lumbermill
*159	October-82	Salt Lake City	UT	USA	Backpressure of CO_2 from beverage dispenser.	Restaurant
*203	October-82	Palm Harbor	FL	USA	Water from chiller system backflow into the water supply through a jammed backflow preventer.	Institution
*204	October-82	Gwinnett County	GA	USA	Industrial contaminant backflowed into the drinking water system through a cross-connection.	Commercial
*205	October-82	Fayetteville	NC	USA	Liquid caustic soda was pumped by a malfunctioned pump into drinking water system.	Residential
*206	October-82	Monterey Park	CA	USA	Nearly 200 people got sick at a high school football game due to copper and CO_2 contamination from a soda dispenser.	Institution
*207	December-82	Roanoke	VA	USA	Water from chilled water compressor flowed into the potable water system through cross-connections.	Institution
*159	December-82	Salt Lake City	UT	USA	Backpressure of CO_2 from beverage dispenser.	Hospital Cafeteria
*208	January-83	St. Petersburg	FL	USA	Horticultural well water backflowed into the supply water system through an illegal connection.	Institution
*209	January-83	New Braunfels	TX	USA	Chemical backsiphoned into the potable water line at the Coleman Mfg. plant.	Industrial

Backflow Incidents

Source	Month - Year	City	State	Country	Cause (What Happened?)	Type of Establishment
*92	February-83	McKeesport	PA	USA	Giardia Lambia contamination from open reservoir coupled with cross-connection in filtration plant. 45,000 residents required to boil water.	Inlet
*210	February-83	Smyrna	GA	USA	Chemical contaminant was backsiphoned into the drinking water system when water main broke.	Residential
*211	April-83	St. Petersberg	FL	USA	Chlordane was backsiphoned into the drinking water system through a garden hose when water line broke.	Residential
*396	May-83	San Antonio	TX	USA	Blue water from commode tank backsiphoned into potable water supply.	Residential
*212	June-83	Woodsboro	MD	USA	Pesticide form a herbicide holding tank was sucked back into the water system when the system malfunctioned.	Residential
*94	August-83	Akron	OH	USA	Numerous cross-connections in major industrial plants caused contamination of several blocks of city water lines.	Industrial & Residential
*213	August-83	Indianapolis	IN	USA	Cooling water was cross-connected to the drinking water system.	Industrial
*95	September-83	Belleview	FL	USA	Cross-connection of gasoline from undetermined source.	Residential
*126	October-83	Vancouver	WA	USA	Cross-connection in dental office pumped air into the building's water system.	Dental Building
*194	October-83	Shelby County	TN	USA	Lead slurry solution backflowed into the potable water supply through a cross-connection.	Industrial
*154	December-83	Philadelphia	PA	USA	Direct connection to tank containing car wash product backsiphoned into adjacent buildings.	Laboratory
*398	December-83	San Antonio	TX	USA	Sewer from Riling Road Treatment Plant backpressured into plant lines during plant maintenance. Incident isolated to in-plant system by RP at service connection.	Sewage Treatment

11 Backflow Incidents

Source	Month - Year	City	State	Country	Cause (What Happened?)	Type of Establishment
*154	January-84	Philadelphia	PA	USA	Backsiphonage of antifreeze from heating system.	Clothing company
*215	January-84	Manhattan	KS	USA	Boiler water backflowed into the potable water line when the main line was shut off to repair a break.	Institution
*127	February-84	Riverbend	OR	USA	Cross-connection between a solar hot water system and domestic hot water system.	Mobile Home Park
*128	February-84	Seattle	WA	USA	Air conditioning water containing Nitrate-Borate was pumped into a high rise office building's & city's water system.	High-rise Office Building
*399	April-84	San Antonio	TX	USA	Chromates backpressured into internal system through faulty check valves and gate valve on a heat exchanger. City system protected by RP.	High School
*97	April-84	Willets	CA	USA	Internal cross-connection of private well with domestic water backpressured into company's distribution system.	Residential
*434	June-84	Melrose Park	IL	USA	Salmonella outbreak at a dairy plant.	Dairy Plant
*216	June-84	Farmington	NM	USA	Chromium used in the heating system backflowed into the water supply system through a leaky check valves.	Institution
*217	June-84	Perry	GA	USA	Herbicide was backsiphoned into the water system through a fire hydrant.	
*129	August-84	Bellevue	WA	USA	Backsiphonage of Sodium Silicate into the building's potable water system.nursing.	Home
*433	September-84	Haverhill	MA	USA	Green water runs from a faucet.	Building
*154	October-84	Philadelphia	PA	USA	Backpressure from air compressors into 12" service main.	Office Buildings
*218	October-84	Jefferson City	MO	USA	Suspected cause of the gastrointestinal illness outbreak was sewage-contaminated water supplied to an airport complex.	Institution

Backflow Incidents

Source	Month - Year	City	State	Country	Cause (What Happened?)	Type of Establishment
*98	November-84	Camarillo	CA	USA	CO2 backpressured from soft drink vending machine into copper piping contaminated water supply.	High School
*99	1985	Boston	MA	USA	Hexavalent Chromium backflow from cooling tower into potable system.	Condominium Complex
*99	1985	Boston	MA	USA	500-1000 gallons of water treated with ethylene glycol and hydrazine backflowed through temporary hose on chilled water system.	Hospital
*99	1985	Chelmsford	MA	USA	Wash water backpressures from car wash into city water mains.	Car Wash
*159	February-85	Salt Lake City	UT	USA	Backsiphonage from circuit board washer occurred when system was shut down for repairs.	Manufacturing Plant
*102	March-85	New York	NY	USA	Hospitalized woman died after being exposed to Ethylene Glycol while undergoing Hemodialysis. Backpressure came from an air conditioning system.	Hospital
*396	March-85	San Antonio	TX	USA	Backsiphonage of blue water from toilet.	Residence
*159	March-85	Salt Lake City	UT	USA	Backpressure of CO2 from beverage dispenser.	Post Office
*219	March-85	Denver	CO	USA	A solar hot-water heater leaked poisonous antifreeze into the drinking water at a fire station.	Institution
*220	March-85	Harleyville	SC	USA	LASSO herbicide backflowed into the water supply when the water main was shut down.	Residential
*396	April-85	San Antonio	TX	USA	Carbonated water backpressured into water supply through leaking check valve on soda dispenser.	Restaurant
*159	April-85	Salt Lake City	UT	USA	Air backpressured through dental equipment on three separate dates.	Dental Office
*130	June-85	Yakima	WA	USA	Cross-connection of an irrigation system resulted in pesticide contamination from a well supplying residential areas.	Private Well

11 Backflow Incidents

Source	Month - Year	City	State	Country	Cause (What Happened?)	Type of Establishment
*131	July-85	Arpelan	OK	USA	Break in waterline caused backsiphonage of Chlorine, Malathion, Sevin, and Diazanon into the public water system.	Residential
*221	July-85	McAlester	OK	USA	Pesticides were backsiphoned into the water lines during the water line repair.	Residential
*222	July-85	Arpelar	OK	USA	Pesticide backsiphoned into the water line due to a water break.	Residential
*223	July-85	Juneau	AK	USA	Toxic glycol in a generator's cooling system contaminated the first floor cold water pipes.	Commercial
*224	July-85	Fife	WA	USA	Sewer gases and sewage backsiphoned into a single family residence through a direct connection between a hose bib and a sanitary sewer.	Residential
*225	August-85	Marlboro	NJ	USA	800 gallons of liquid fertilizer were pumped into potable water system.	Residential
*226	September-85	W. Sacramento	CA	USA	Malathion backpressured through faulty check valves into potable system.	Port of Sacramento
*223	October-85	Juneau	AK	USA	An incorrect toilet tank assembly allowed unsafe blue disinfectant chemicals to siphon from the tank into all the cold water piping in a home.	Residence
*228	November-85	Atlantic City	NJ	USA	Antifreeze was pumped into the water system due to a partly opened valve in the bottom three floors of a Casino and Restaurant.	Commercial
*229	December-85	Marion County	IN	USA	Blue water from a toilet backflowed into the potable water line.	Institution
*132	January-86	Bonney Springs		USA	Pesticide Malathion was fed into the water system by an aspirator causing an employee to become ill.	Grain Mill
*230	January-86	Macon	GA	USA	Creosote normally flowed to the recycling tank but entered the water pipe during a fire hydrant installing.	Institution

Backflow Incidents

Source	Month - Year	City	State	Country	Cause (What Happened?)	Type of Establishment
*231	January-86	Bonner Springs	KS	USA	Insecticide flowed back to the drinking water system when a leak caused irregular pressure within the plant's water piping.	Industrial
*194	January-86	Shelby County	TN	USA	A direct cross connection involving a mold-a-pack machine discolored the water coming out of drinking fountain and restroom faucets.	Industrial
*396	April-86	San Antonio	TX	USA	Backsiphonage of blue water from toilet into potable water supply.	Residential
*194	March-86	Shelby County	TN	USA	Carbonated water was forced into the filtered water system through a cross-connection.	Industrial
*133	April-86	Withrow	WA	USA	Herbicide 2, 4-D was backsiphoned into the community's water system causing residents to go without water for four days.	Residential
*234	April-86	Hope Mills	NC	USA	Pesticide backflowed into the potable water system.	Residential
*159	April-86	Salt Lake City	UT	USA	Backsiphonage from ornamental ponds occurred when repairs were performed. More than 40 units affected with blue dyed water.	Condominium
*104	April-86	San Rafael	CA	USA	Air from a single jacketed air compressor cooler backpressure into potable system.	Auto Painting Co.
*154	April-86	Philadelphia	PA	USA	Backpressure through unprotected makeup line to chilled water system.	Office Bldg
*159	April-86	Salt Lake City	UT	USA	Carbonated water backpressured into water supply through leaking check valve on soda dispenser.	Golf Course Café
*235	April-86	Tempe	AZ	USA	Contamination backflowed into the potable water through cross-connections.	Commercial

11 Backflow Incidents

Source	Month - Year	City	State	Country	Cause (What Happened?)	Type of Establishment
*134	June-86	San Luis Obispo	CA	USA	Surface water was siphoned into the city water due to defective operating valves on a lawn sprinkler system.	Lawn Sprinkler System
*400	June-86	San Antonio	TX	USA	Oil slurry from asphalt plant backpressured into internal water system. City supply protected by RP and not affected.	Manufacturing Plant
*396	July-86	San Antonio	TX	USA	Carbonated water backpressured into water supply through leaking check valve on soda dispenser.	Convenience Store
*396	August-86	San Antonio	TX	USA	Recirculated wash water backpressured into internal water system. City supply protected by RP and not affected.	Car Wash
*236	September-86	Greenport	NY	USA	Detergent entered the water supply's distribution system when the pressure dropped in the water main.	Commercial
*237	October-86	Birmingham	AL	USA	Sodium Hydroxide backflowed to the water system due to a water main brake.	Industrial
*159	November-86	Salt Lake City	UT	USA	Blue water from a toilet on a 2nd floor backsiphoned when building water was shut off.	Condominium
*238	November-86	Nevada	IA	USA	Nitrogen fertilizer diluted with water in a tanker truck was accidentally flushed into the city's water system.	Commercial
*239	January-87	Northglenn	CO	USA	Cross-connections at a high school tests positive for coliform from inside and outside of school.	Institution
*396	February-87	San Antonio	TX	USA	Carbonated water backpressured into water supply through leaking check valve on soda dispenser.	Restaurant
*240	February-87	Sault Ste. Marie	MI	USA	Carbonated water backflowed into the copper waterline, resulting in several copper poisonings.	Commercial

Backflow Incidents

Source	Month - Year	City	State	Country	Cause (What Happened?)	Type of Establishment
*239	February-87	Northglenn	CO	USA	Positive for coliform during routine sampling from an inside faucet at an elementary school resulted from a slope sink with a hose cross-connections.	Institution
*206	February-87	British Columbia	BC	Canada	CO2 backflowed into the potable water system from the post-mix system and leached copper from the copper piping.	Commercial
*153	March-87	Santa Ana	CA	USA	Domestic water interconnected to industrial water line.	Modular Office
*153	March-87	Fountain Valley	CA	USA	Secondary effluent backpressured into domestic water within the plant.	Sewage Treatment Plant
*102	April-87	North Dakota	ND	USA	Heating system backpressure ethylene glycol into potable water which was used to mix a powdered beverage at a picnic.	Firehall
*194	April-87	Shelby County	TN	USA	With no vacuum breakers installed on deep sinks on each floor allowed pesticides to backflow into the potable water system.	Industrial
*243	April-87	San Diego	CA	USA	Backflow of carbon dioxide in a beverage dispensing system corroded the copper waterline.	Commercial
*200	May-87	Evanston	IL	USA	Water cooled air conditioner backflowed through leaky valve.	Hospital
*103	June-87	Gridley	KS	USA	Herbicide backsiphoned from tanker truck when main was broken during excavation.	Grain and Feed Store
*244	June-87	Kitchener	ON	Canada	Nickel solution leaked into the water system during repairs to a water pipe.	Industrial
*245	June-87	Hawthorne	NJ	USA	Four chemicals, heptachlor, chlordane, dursban, and lindane entered water pipes due to backsiphonage when the pipe broke or when the main was shutdown during bridge construction.	Institution
*239	June-87	Northglenn	CO	USA	Contamination backflowed into the potable water from the residential boiler system.	Residential

444

11 Backflow Incidents

Source	Month - Year	City	State	Country	Cause (What Happened?)	Type of Establishment
*435	July-87	Bentonia	MS	USA	Suspected cross-connection due to garden hose submerged in a tank of chemicals.	Residence
*396	July-87	San Antonio	TX	USA	Swimming pool water backpressured into potable water through make up connection to pool circulation system.	Recreation Center
*159	July-87	Salt Lake City	UT	USA	Liquid fertilizer backsiphoned from irrigation system with no protection.	Residence
*159	August-87	Salt Lake City	UT	USA	Backpressure of sodium silica through water flush line.	Chemical Co.
*246	August-87	Plantation	FL	USA	Backflow of CO_2 in the water line from carbonators created acid conditions that dissolved the copper.	Commercial
*396	September-87	San Antonio	TX	USA	Non potable water from heat exchanger backpressured through faulty check valve in ten story building. Numerous illnesses reported.	Office Bldg.
*100	October-87	Gainesville	FL	USA	Improperly connected pipes allowed backflow of chemically treated water from an air conditioning unit.	University of Florida
*247	October-87	Clifton Park	NY	USA	Pesticide in a tank was back siphoned into the drinking water system through a hose when a water main broke.	Residential
*396	October-87	San Antonio	TX	USA	Backsiphonage of blue water from toilet.	Apartment
*168	October-87	San Antonio	TX	USA	Carbonated water backpressured into water supply through leaking check valve on soda dispenser.	Restaurant
*396	November-87	San Antonio	TX	USA	Carbonated water backpressured into water supply through faulty check valve on soda dispenser.	Commercial
*163	December-87	Costa Mesa	CA	USA	Chromates from chilled water system backpressured into cafeteria.	Newspaper

Backflow Incidents

Source	Month - Year	City	State	Country	Cause (What Happened?)	Type of Establishment
*206	November-87	Crystal	MN	USA	Carbon Dioxide backflows from soft drink dispenser causing high levels of copper to contaminate restaurants drinking fountain.	Restaurant
*249	December-87	LaPoint	UT	USA	The hot water supply of the school was contaminated with high levels of boron, copper, zinc, and lead by the recirculating hot water system through a cross-connection.	Institution
*206	January-87	East Lansing	MI	USA	CO_2 entered the water supply through a open double check valve assembly in the soft drink dispenser carbonate unit and created a potential copper poisoning.	Institution
*396	March-88	San Antonio	TX	USA	Backsiphonage of blue water from a toilet into an apartment's water supply.	Apartment Complex
*396	March-88	San Antonio	TX	USA	Backsiphonage of green water from toilet into apartment's water supply.	Apartment Complex
*251	March-88	Cleveland	OH	USA	Diluted water-soluble oil containing potential toxic chemical additives backflowed into the city waterlines.	Industrial
*150	April-88	Edgewater	FL	USA	Propylene glycol backflowed due to valve malfunction.	Paint Co.
*159	April-88	Salt Lake City	UT	USA	Boiler compound backpressured through malfunctioning unapproved RP.	Office Bldg.
*252	April-88	Bushnell	IL	USA	Agricultural chemicals were back-siphoned and backflowed to the water distribution system when a water main broke.	
*396	April-88	San Antonio	TX	USA	Soapy water from a toilet backsiphonage into apartment's water supply.	Apartment Complex
*253	May-88	Belle Glade	FL	USA	A city worker killed by pesticides in his water bottle from a faucet often used to dilute pesticides.	Residential

11 Backflow Incidents

Source	Month - Year	City	State	Country	Cause (What Happened?)	Type of Establishment
*254	May-88	San Bernardino	CA	USA	Pond water backpressured into the distribution system when system pressure was low and booster pumps were used.	Residential
*396	June-88	San Antonio	TX	USA	Acid backpressured throughout internal system from air injected degreaser.	Maintenance
*151	June-88	Glendo	WY	USA	Synthetic detergent backsiphoned from truck tank.	Railroad Yard
*396	August-88	San Antonio	TX	USA	Carbonated water backpressured into water supply through leaking check valve on soda dispenser.	Restaurant
*155	August-88	Gilbert	AZ	USA	Interconnection between irrigation system and decorative pond.	Industrial Park
*152	August-88	Fresno	CA	USA	Sodium Nitrite backflowed into the city water system from an air-conditioning system due to city water system pressure loss.	Institution
*141	September-88	Edmonton		Canada	A faulty water check valve permitted backflow of water from a fire sprinkler system into a buildings potable supply.	Office Building
*256	September-88	Fresno	CA	USA	Make-up water to a cooling tower backflowed into the drinking water system through a cross-connection.	Institution
*257	September-88	Anchorage	AL	USA	Water with a slight glycol residue in the pressurized fire sprinkler system backflowed into the potable water system through a failed single check valve.	Residential
*258	November-88	Kansas City	MO	USA	Anti-rust chemicals from a boiler backflowed to the water supply system through two failed backflow valves during a new water pipe installation.	Institution
*149	December-88	Newport Beach	CA	USA	Backwash from sand filters backpressured into restaurant's beverage dispensers.	Restaurant

Backflow Incidents

Source	Month - Year	City	State	Country	Cause (What Happened?)	Type of Establishment
*259	January-89	Garland	TX	USA	Carbonated water from a soft drink mixer backflowed into the water line causing copper to leach due to a malfunctioning backflow preventer.	Commercial
*216	January-89	Riley County	KS	USA	Boiler water backflowed past the pressure regulator into the potable water system when the water main was shut off to repair a break.	Institution
*261	January-89	Edmonton	AB	Canada	Recycled cooling water containing a green dye backflowed into the potable water due to by-pass valve which was inadvertently left open.	Industrial
*262	February-89	Seattle	WA	USA	Air went into the water system due to an erroneously connected air compressor.	Institution
*156	March-89	Tucson	AZ	USA	Diazinon backflowed through garden hose from one home to another.	Residence
*401	March-89	San Antonio	TX	USA	Diesel fuel backpressured into internal system. Service connection protected by RP.	Asphalt Mfg. Plant
*261	March-89	Edmonton	AB	Canada	Treated boiler water from the heating system backflowed into the potable water due to a faulty single check valve during a water main shut off.	Residential
*194	April-89	Shelby County	TN	USA	Drinking fountain was connected to non-potable water.	Institution
*141	May-89	Edmonton			Backflow occurred through an alarm check valve on a fire sprinkler system into a department stores water system.	Department Store
*419	May-89	Lehigh	PA	USA	Detergent entered the hot water line through a broken valve.	Institution
*402	June-89	San Antonio	TX	USA	Carbonated water backpressured through faulty check valve in soda dispenser.	Restaurant

11 Backflow Incidents

Source	Month - Year	City	State	Country	Cause (What Happened?)	Type of Establishment
*159	June-89	Salt Lake City	UT		Propylene glycol backpressured through leaking check valve on a fire sprinkler system during shutdowns to replace sprinkler heads.	Office Bldg.
*396	June-89	San Antonio	TX	USA	Condensate from cooling system backpressured into internal water system. City supply protected by RP.	Retirement Center
*263	June-89	Redmond	OR	USA	Ethylene Glycol in an air conditioning system back lowed into the drinking water supply.	Institution
*261	June-89	Maryville	MO	USA	Water in from a water closet tank backsiphoned into the potable water system.	Residential
*144	July-89	Edmonton,		Canada	Leakage through three water check valves on a fire system allowed backflow of oily water into station's washrooms.	Rail Transit Station
*161	July-89	Cincinnati	OH	USA	Algae retardant backflowed into potable water from an under repair air conditioner. About 50 people reported nausea.	Office Bldg.
*396	July-89	San Antonio	TX	USA	Backsiphonage of blue water from toilet.	Residence
*144	August-89	Edmonton,		Canada	Leakage through water check on fire sprinkler system allowed backflow of brown water into printing shop restroom.	Printing Shop
*239	August-89	Northglenn	CO	USA	Contamination of potable water was traced to a cross-connection at a swamp cooler.	Residential
*436	September-89	Norwich	CT	USA	Suspected backsiphonage due to a faulty pump valve.	Residence
*396	September-89	San Antonio	TX	USA	Backsiphonage of blue water from toilet.	Residence
*396	September-89	San Antonio	TX	USA	Backsiphonage of blue water from toilet into apartment's water supply.	Apartment Complex
*141	October-89	Edmonton,		Canada	Water quality complaint lead to discovery of drinking fountain piped from a fire hose cabinet.	Building

Backflow Incidents

Source	Month - Year	City	State	Country	Cause (What Happened?)	Type of Establishment
*264	October-89	Riverside	CA	USA	Cross-connection between caustic soda chemical feed line and the soft water line.	Institution
*194	October-89	Shelby County	TN	USA	Toilet tank backsiphoned into internal water supply due to non-approved ballcock.	Residential
*403	December-89	San Antonio	TX	USA	Backsiphonage of blue water from toilet into apartment's water supply. Four illnesses report by health department.	Apartment Complex
*396	December-89	San Antonio	TX	USA	Backsiphonage of blue water from toilet into apartment's water supply.	Apartment Complex
*265	December-89	Cabool	MO	USA	Suspected backsiphonage due to meter settings and two water line breaks.	Residential
*266	December-89	San Antonio	TX	USA	Back-siphonage of blue-water from commode tank affected several apartments.	Residential
*426	January-90	Preston	CA	USA	Cross-connection between dual-plumbed recycled water system and dental office.	Building
*148	January-90	Brighton	CO	USA	Antifreeze in heating system backpressured into school water system. Nine students treated.	School
*267	January-90	Denver	CO	USA	Antifreeze from a building's boiler mixed with drinking water due to human error and faulty system.	Institution
*268	January-90	Brighton	CO	USA	Antifreeze in the heating system backflowed into the drinking water through an incomplete valve assembly.	Institution
*396	February-90	San Antonio	TX	USA	Back-siphonage of blue water from commode tank.	Residential
*409	February-90	Seattle	WA	USA	Pond water pumped into water system at a golf course.	Golf Course
*160	May-90	Tucson	AZ	USA	Backflow of cooling system water into potable water system.	Police Station

11 Backflow Incidents

Source	Month - Year	City	State	Country	Cause (What Happened?)	Type of Establishment
*147	May-90	Webster	NY	USA	Solvent backpressured through a hole in heat exchanger. Bypass around backflow preventer on water makeup line.	Manufacturing Plant
*270	May-90	Tucson	AZ	USA	Water from the building's cooling system contaminated the drinking water at a police station.	Institution
*239	June-90	Northglenn	CO	USA	Coliform contamination inside a residence was traced to a cross-connected swamp cooler.	Residential
*271	June-90	Tucumcari	NM	USA	Plant cooling water backflowed into the city water.	Industrial
*159	July-90	Salt Lake City	UT	USA	Boiler treatment compound backpressures into potable water system.	Elderly Housing
*145	August-90	Bellevue	WA	USA	Carbonated water backpressured into water supply through leaking check valve on a soda dispenser.	Restaurant
*272	August-90	Brentwood	TN	USA	A cross-connection between a potable water service and a non-potable source (contaminated with raw sewage) caused 1100 people sick at a country club.	Institution
*273	September-90	Johnson County	KS	USA	Backflow of CO_2 from a post-mix machine into the potable water supply made employees sick.	Commercial
*274	October-90	Sioux City	IA	USA	Water was contaminated due to backflow from a blood centrifuge that got plugged.	Industrial
*143	November-90	Memphis	TN	USA	Backsiphonage of blue water from toilet into apartment's water supply.	Apartment Complex
*275	November-90	Wheatfield	IN	USA	Antifreeze in the cooling system accidentally backflowed into the drinking water.	Institution
*276	December-90	Johnson County	KS	USA	Backflow of CO_2 from a post-mix machine.	Industrial
*426	January-91	Las Virgenes	CA	USA	Backpressure of recycled water into private residence.	Residence

Backflow Incidents

Source	Month - Year	City	State	Country	Cause (What Happened?)	Type of Establishment
*396	January-91	San Antonio	TX	USA	Water from heat exchanger backpressured through unprotected make up connection. City system protected by RP.	Office Bldg.
*146	February-91	Fountain Valley	CA	USA	Compressed air used for dental equipment backpressured into building water supply.	Medical Office
*142	February-91	Memphis	TN	USA	Backsiphonage of blue water from toilet into apartment's water supply.	Apartment Complex
*254	February-91	San Bernardino	CA	USA	Mercury backflowed into the potable water system.	Commercial
*396	February-91	San Antonio	TX	USA	Back-siphonage of blue water from commode tank due to non-approved ballcock.	Residential
*276	March-91	Johnson County	KS	USA	Boiler chemicals backflowed into water lines.	Institution
*396	April-91	Kelly AFB	TX	USA	Chilled water backpressures through system due to two leaking check valves.	
*141	April-91	Edmonton		Canada	Leaking check valve on fire sprinkler system allows backflow into washrooms.	Rail Transit Station
*269	April-91	San Antonio	TX	USA	Water from chiller backflowed into potable water through two malfunctioning single check valves. Contamination persisted for two weeks.	Institution
*396	May-91	San Antonio	TX	USA	Swimming pool water backpressured from a direct make-up connection to recirculation system.	Motel
*276	May-91	Johnson County	KS	USA	Backflow of CO2 from post-mix machine.	Commercial
*277	May-91	Gatlinburg	TN	USA	Recirculated water in the heating/cooling system backflowed into the potable water supply.	Commercial
*278	May-91	Selden	NY	USA	Antifreeze backflowed into the potable water system.	Institution
*396	June-91	San Antonio	TX	USA	Carbonated water backpressured into water supply through leaking check valve on soda dispenser.	Lounge

11 Backflow Incidents

Source	Month - Year	City	State	Country	Cause (What Happened?)	Type of Establishment
*279	June-91	Casa	AK	USA	Chicken house water backflowed into the potable water system through cross-connection where two single check valves were installed.	Commercial
*276	June-91	Johnson County	KS	USA	Backflow of CO2 from post-mix machine.	
*396	July-91	San Antonio	TX	USA	Stagnant water from looped fire sprinkler system backflowed into city main.	Motel
*276	July-91	Johnson County	KS	USA	Toilet tank contents that included a bowl cleaning chemical backflow to potable water.	Residential
*396	August-91	San Antonio	TX	USA	Backsiphonage of blue water from toilet.	Residence
*280	August-91	Newark	NJ	USA	Firefighting foam leaked into the city's water system due to a valve malfunction.	Institution
*281	August-91	Longmont	CO	USA	Water flowed into the gas line through a unique cross-connection.	Residential
*239	August-91	Northglenn	CO	USA	Coliform bacteria found in the water at a recreation center may have been caused by a cross-connection of a hose in a slope sink.	Institution
*162	September-91	Mission Viejo	CA	USA	Direct cross-connection between the potable water and a reclaimed water line irrigating slope.	Residence
*282	September-91	Mission viejo	CA	USA	Cross-connection between the domestic water line and a reclaimed water distribution system.	Residential
*283	September-91	Norridgewock	ME	USA	Backflow occurred when fire department refilled their trucks, causing the water pressure to drop.	Commercial
*409	September-91	Uintah	UT	USA	2.5 gallons of TriMec herbicide were backsiphoned into the Uintah Highlands water system.	Residential
*285	October-91	Southgate	MI	USA	Vacuum effect caused by a water main break forced some water containing parasites from a sprinkler system into the city water.	
*286	October-91	Louisville	KY	USA	Backflow occurred during a four-foot wide water main break.	Residential
*396	November-91	San Antonio	TX	USA	Backsiphonage of blue water from toilet spread several apartment's water supply.	Apartment Complex

Backflow Incidents

Source	Month - Year	City	State	Country	Cause (What Happened?)	Type of Establishment
*407	December-91	Salt Lake City	UT	USA	Propylene glycol backflows into potable water.	Building
*426	January-92	Santa Margarita	CA	USA	Cross-connection between recycled water irrigation system and house potable water irrigation system.	Residence
*287	February-92	Ansley	GA	USA	Cleaning solutions were back pressured into the potable water system through a cross-connection.	Commercial
*276	February-92	Johnson County	KS	USA	Backflow of toilet tank contents that included a bowl cleaning chemical entered the water system.	Residential
*239	March-92	Northglenn	CO	USA	Coliform bacteria found in the water at an elementary school may have been caused by a cross-connection on the school's boiler system.	Institution
*276	July-92	Johnson County	KS	USA	Backflow of toilet tank contents included a bowl cleaning chemical.	Residential
*288	September-92		ME	USA	Backsiphonage occurred during the winterizing of the water system, fecal coliform bacteria was distributed into the water system when system pressure was restored.	
*168	October-92	San Antonio	TX	USA	Blue water from a toilet backsiphoned into an apartment's water supply.	Apartment Complex
*415	October-92	Clifton	NJ	USA	Potable water was contaminated by a backflow from the boiler system.	Institution
*396	November-92	San Antonio	TX	USA	Soap backsiphoned into internal potable water system. City water not affected.	Soap Mfg. Plant
*426	January-93	Elsinore Valley	CA	USA	Cross-connection between residential potable water irrigation system and golf course recycled water.	Golf Course
*290	January-93	Tempe	AZ	USA	Orange brown water came from inside faucets due to backflow problems.	Residential

11 Backflow Incidents

Source	Month - Year	City	State	Country	Cause (What Happened?)	Type of Establishment
*396	April-93	San Antonio	TX	USA	Backsiphonage of blue water from commode tanks occurred due to non-approved ballcock.	Residential
*162	May-93	Fullerton	CA	USA	Industrial water backpressured to loop from the tissue manufacturing process.	Paper Mill
*409	July-93	Coos Bay	OR	USA	Water from drainage pond was pumped into potable water supply.	Residence
*139	August-93	Superior	AZ	USA	Backflow of antifreeze solution occurred through fouled single check valve on fire sprinkler system.	Arboretum
*437	September-93	Pinkham Notch	NH	USA	Cross-connection between potable water and river.	Lodge
*292	September-93	Brighton	MA	USA	Chemical drying agent was sucked into the potable water system by a pump in a malfunctioning soda system.	Commercial
*293	October-93	Peoria	AZ	USA	Dirty water backflowed into the water system when the water main was shut down.	Commercial
*294	November-93	Yorba Linda	CA	USA	Caustic soda backflowed into the drinking water line when the caustic soda line became crystallized.	Institution
*295	November-93	Wilson	NC	USA	Chemical mixer used for x-ray development backsiphoned into the water system during a repair.	Institution
*296	November-93	Melville	NY	USA	Antifreeze backflowed into the potable water system through a direct cross-connection.	Institution
*443	December-93	Tucson	AZ	USA	Several RP's are improperly by-passed, making it a susceptible for backflow conditions.	Building
*297	December-93	Seattle	WA	USA	Backflow from CO_2 dispenser resulted in bright blue water from drinking fountains in lobby and sinks in the concession areas at Seattle Center.	Institution

Backflow Incidents

Source	Month - Year	City	State	Country	Cause (What Happened?)	Type of Establishment
*426	January-94	Pomona	CA	USA	Cross-connection found between the recycled and potable water internal system at a dual plumbed facility.	School
*444	February-94	Marana	AZ	USA	An elementary school is surveyed, and found with several cross-connections.	School
*441	February-94	Tucson	AZ	USA	Cross-connection between the potable water and waste water.	Building
*298	February-94	Tucson	AZ	USA	Boiler treatment chemicals backflowed into the internal water system.	School
*299	February-94	Laredo	TX	USA	Soap backsiphoned into potable water supply through faulty check valve caused by a low pressure in the supply line.	Commercial
*442	March-94	Tucson	AZ	USA	Fecal matter was inserted into the bath tub faucet at an apartment complex.	Building
*300	March-94	Hilton Head IS	SC	USA	Wastewater flowed into the potable water system through a cross-connection at a golf course.	Commercial
*301	April-94	Fresno	CA	USA	Contaminated water backflowed into the community water system from the fire sprinkler system due to a failed single check valve.	Commercial
*276	April-94	Johnson County	KS	USA	Heavy usage on water district main causes the domestic water to backflow through meter.	
*274	April-94	Des Moines	IA	USA	Water from a chiller system backflowed into the domestic water system when the system was drained for maintenance.	Institution
*302	April-94	Dearborn	MI	USA	Water from cooling system backflowed into the potable water system.	Industrial
*375	April-94	Des Moines	IA	USA	Backflow occurred from a chilled water system into the potable water piping.	Building

11 Backflow Incidents

Source	Month - Year	City	State	Country	Cause (What Happened?)	Type of Establishment
*303	May-94	Franklin	NE	USA	Freon Backflowed into the city water system during repair of air-conditioning at a nursing home.	Institution
*386	May-94	Columbus	OH	USA	Ice machine connected to a sewer line sickened dozens of people.	Commercial
*304	June-94	Novato	CA	USA	100 gallons of diluted TRYCOL backflowed into the public water supply from a fire district pumper truck.	Residential
*305	June-94	Carmichael	CA	USA	Dyed water backflowed into water main from an amusement park.	Institution
*413	June-94	Ontario		Canada	Uranium contaminated the drinking water.	Factory
*306	July-94	Syracuse	NY	USA	Malfunctioning of an RPZ type backflow preventer on the main feed for incoming domestic water caused flooding in the sub-basement of Room CL55 in the Syracuse VA Medical Center. The Rubble gasket failure mode was most likely the cause.	Institution
*307	July-94	Canandaigua	NY	USA	Due to a broken valve, the hospital's basement was flooded with 8 feet water and the dirty water was pumped into the distribution system throughout the whole building.	Institution
*308	July-94	Tempe	AZ	USA	Irrigation water from a lake backflowed into the potable water system through cross-connections.	Residential
*410	July-94	Seattle	WA	USA	A hose faucet was directly connected to a non-potable cooling system.	Woodland Park Zoo
*309	October-94	Monterey	CA	USA	Water in the boiler and cooler backflowed into the water system due to a defective backflow device.	Institution
*276	October-94	Johnson County	KS	USA	Backflow of compressed air from a dry fire protection system resulted in milky, foamy, oily water in toilets and drinking fountains.	Commercial

Backflow Incidents

Source	Month - Year	City	State	Country	Cause (What Happened?)	Type of Establishment
*311	October-94	Bradbury	CA	USA	About 30 gallons of snow making Macroject 1 Concentrate backflowed into the water system through a connected garden hose.	Residential
*312	October-94	Tucson	AZ	USA	A private plumbing system contained a thick, green, greasy oil like substance due to cross-connections.	Commercial
*276	December-94	Johnson County	KS	USA	Water softener pellets backflowed into the water line during back flush cycle.	Residential
*396	December-94	San Antonio	TX	USA	Meter running backwards, belted services.	Residential
*426	January-95	Irvine Ranch	CA	USA	Cross-connection between recycled water and potable water systems.	
*313	January-95	Georgetown	SC	USA	Treated mold water back flowed into the city water system through a unprotected cross-connection.	Industrial
*314	March-95	La Verne	CA	USA	Sludge entered the drinking water line through a cross-connection.	Institution
*381	March-95	New Roads	LA	USA	Herbicide backsiphoned into the public water supply.	Residence
*276	May-95	Johnson County	KS	USA	Toilet tank contents that included a bowl cleaning chemical backflowed into water supply.	Residential
*315	June-95	Yorba Linda	CA	USA	Caustic soda flowed into the irrigation line through cross-connections.	Institution
*374	July-95	Gainsville	FL	USA	Cryptosporidium found on outdoor garbage can washer.	Day Camp
*316	August-95	San Diego	CA	USA	Chemically treated water form the heating and cooling water system backflowed into the potable water system through a failed single swing check when the distribution system was depressurized to replace the meter backflow preventer.	Institution

11 Backflow Incidents

Source	Month - Year	City	State	Country	Cause (What Happened?)	Type of Establishment
*299	August-95	Laredo	TX	USA	Softener brine filter residue contaminated internal water system. City's supply line was unaffected due to RP backflow preventer.	Institution
*299	September-95	Laredo	TX	USA	Toilet tank backsiphoned into internal water supply due to non-approved ball cock.	Commercial
*318	September-95	Newcastle	WY	USA	The city's waterlines were contaminated when a main was repaired and a hydrant was installed as part of the pool construction by private contractors.	Residential
*274	October-95	Cedar Rapids	IA	USA	Backflow from a cleaning solution dispenser contaminated the domestic water system.	Institution
*319	November-95	Philadelphia	PA	USA	Grass seed mixture backflowed into the potable water system when a subcontractor improperly attached a hose to a hydrant without backflow preventer.	
*320	December-95	San Diego	CA	USA	Carbonate water backflowed into the potable water system due to a failed dual check valve leading to a post-mix beverage machine.	Institution
*321	December-95	Charleston	SC	USA	Blue water backflowed into the potable water system from bathrooms during a line break.	Residential
*322	December-95	Detroit	MI	USA	Hot water in a heater backflowed into the cold water system due to high pressure.	Residential
*440	1996	Reno	NV	USA	Cross-connections discovered in a hotel where a Norwalk virus agent affected 1,000 guests and employees.	Building
*426	January-96	Lakewood	CA	USA	Drinking fountains connected to recycled water system.	School
*323	January-96	Atlanta	GA	USA	Compressed air was pushed back into the water distribution system due to a malfunction sprinkler system.	

Backflow Incidents

Source	Month - Year	City	State	Country	Cause (What Happened?)	Type of Establishment
*324	January-96	Yorba Linda	CA	USA	Thick solution was pumped into the potable water supply through cross-connections.	Commercial
*289	March-96		NJ	USA	Potable water was contaminated by a backflow from the boiler system.	Institution
*325	July-96	Honolulu	HI	USA	Bacterial contamination flowed through crevices in flexible pipe joints in the water line.	Residential
*299	September-96	Laredo	TX	USA	Private well used for irrigation, was interconnected with City water. Backflow operator broke city's water line and the wells water line connection.	Residential
*276	September-96	Johnson County	KS	USA	Backflow from beverage carbonator.	Institution
*327	November-96	Brighton	CO	USA	Ethylene Glycol in the hot water heating system backflowed into the drinking water system.	Institution
*328	November-96	Tucson	AZ	USA	5 floors of suspended water within City Hall backflowed to the public water system during a water main break.	Institution
*274	December-96	Iowa City	IA	USA	Backflow from the service bays of a Jiffy Lube and from a wax machine in a car wash contaminated the domestic water.	Commercial
*439	1997	Clark County	MS	USA	Diesel fuel backflows into the water main due to faulty RP which was untested for 5 years.	Residence
*426	January-97	Lakewood	CA	USA	Potable water tank filled with recycled water from fire hydrant.	Residence
*426	January-97	Lakewood	CA	USA	Direct cross-connection between potable water service line to a golf course and recycled water irrigation system.	Golf Course
*426	January-97	Las Virgenes	CA	USA	Illegal connection between irrigation system and a private residence's irrigation system.	Residence

11 Backflow Incidents

Source	Month - Year	City	State	Country	Cause (What Happened?)	Type of Establishment
*329	April-97	Boulder	CO	USA	Water from a chilled waterline flowed into the domestic water system due to high pressure.	Institution
*330	June-97	North Platte	NE	USA	Deposits from the water main flowed out of the taps when a major power outage dropped water pressure.	
*330	June-97	North Platte	NE	USA	Drinking water was contaminated due to cross-connections via a residential boiler or unprotected hose bibs.	Residential
*330	June-97	North Platte	NE	USA	The water had been contaminated during a boiler wash-down at an elementary school.	Institution
*330	June-97	North Platte	NE	USA	Direct cross-connection between a soft water machine and a sewer drain.	Residential
*331	June-97	La Verne	CA	USA	A cross-connection caused treated water to discharged into the filtration plant instead of the sewer.	Institution
*276	June-97	Johnson County	KS	USA	Backflow from beverage carbonator.	
*332	June-97	Westminster	CA	USA	Potable water was contaminated by a backflow from a fertilizer aspirating system.	Institution
*370	July-97	Tucson	AZ	USA	Chiller water with Freon backflowed into a building water system.	Office Bldg.
*333	July-97	Charleston	SC	USA	Blue water backflowed into the potable water system from toilets when the water service was shut down.	Commercial
*370	July-97	Tucson	AZ	USA	Chiller water with Freon backflowed into a building's water system.	Office Bldg.
*334	August-97	Guelph	Ontario	Canada	Chemical renosol 1515 backflowed into the city water system due to a failed backflow prevention device.	Industrial
*396	August-97	San Antonio	TX	USA	Air backpressured into potable water supply from water cooled air compressor.	Commercial
*335	September-97	Charlotte	NC	USA	Firefighting foam was mistakenly pumped into the public water main through a hydrant.	Institution

Backflow Incidents

Source	Month - Year	City	State	Country	Cause (What Happened?)	Type of Establishment
*336	September-97	Charlotte	NC	USA	Fire-smothering foam was backpressured into the public water distribution system through a fire hydrant by a fire truck.	Residential
*337	September-97	Reno	NV	USA	Carbonated water backflowed into the copper water lines due to a carbonation pump on a soft drink dispenser.	Institution
*338	October-97	Calabasas	CA	USA	Plumber mistakenly interconnected a recycled water line affecting 1600 homes and two schools.	Residential
*299	October-97	Laredo	TX	USA	Enclosed heating system backsiphoned into internal potable water supply, contaminating water fountain and room. City's supply was not effected due to proper protection.	Institution
*276	October-97	Johnson County	KS	USA	Backflow of car wash water softener entered residential water supply.	Commercial
*276	October-97	Johnson County	KS	USA	Backpressure from irrigation system being winterized with an air compressor.	Residential
*340	October-97	Charleston	SC	USA	Tidy Bowl Blue backsiphoned from an apartment toilet tank into the internal plumbing lines when the water was turned off for repairs.	Residential
*341	October-97	Fresno	CA	USA	Chemically treated boiler water backflowed into the potable water supply line.	Institution
*342	October-97	Northglenn	CO	USA	Rusty, muddy, dirty water and occasional pockets of air were exiting the pipes in toilets and sinks due to a cross-connection in the irrigation system.	Residential
*299	November-97	Laredo	TX	USA	Water heater contaminated internal potable water supply. Water connections were made with galvanized pipe. City's supply is protected by RP backflow preventer and was not affected.	Institution

11 Backflow Incidents

Source	Month - Year	City	State	Country	Cause (What Happened?)	Type of Establishment
*276	November-97	Johnson County	KS	USA	Backflow occurs from unused water softener connected to plumbing supply.	
*343	November-97	Sacramento	CA	USA	Water treatment chemical backflowed into the potable water system from the condenser water system.	Commercial
*344	December-97	San Bernardino	CA	USA	A cross-connection between the raw fire water system and the domestic water system induced coliform into the domestic water supply.	Residential
*345	December-97	Charleston	SC	USA	Tidy Bowl Blue backflowed into the potable water system.	
*426	January-98	San Diego	CA	USA	Four interconnections between a recycled water irrigation system and a potable water fire system.	
*346	January-98	Charleston	SC	USA	Blue water was backflowed into a private plumbing system.	Residential
*420	February-98	North Platte	NE	USA	CO_2 backflowed into a restaurant's water system.	Commercial
*347	March-98	McCormick	SC	USA	Chemicals backflowed into the water distribution system through cross-connections.	Industrial
*276	March-98	Johnson County	KS	USA	Toilet tank contents that included a bowl cleaning chemical backflowed into the water system.	
*276	March-98	Johnson County	KS	USA	Backpressure from water cooled air compressor.	Institution
*421	March-98	Washoe County	NV	USA	Blue water backsiphoned into the potable water system from toilet tank during a water outage.	Residential
*424	April-98	New Mexico	NM	USA	Ethylene glycol backsiphoned into the domestic water system through a garden hose.	Institution
*274	May-98	Perry	IA	USA	Ethylene glycol from an air cooling system backflowed into the plant domestic water.	Industrial

Backflow Incidents

Source	Month - Year	City	State	Country	Cause (What Happened?)	Type of Establishment
*376	May-98	Sac City	IA	USA	Broken water line causes backsiphonage of fertilizer.	Residential
*276	June-98	Johnson County	KS	USA	Debris accumulated in the irrigation system which backsiphoned into the potable water supply during repair to the lawn irrigation system.	Institution
*423	June-98	Arkansas	AR	USA	Fertilizer was accidentally pumped into the public water supply.	Institution
*348	July-98	Minden	NV	USA	Fecal material from the wastewater treatment plant backflowed into the drinking water system.	Institution
*349	July-98	Charleston	SC	USA	Blue water backflowed into the potable water system.	Institution
*276	August-98	Johnson County	KS	USA	Toilet tank contents that included a bowl cleaning chemical backflowed into the water system.	Residential
*276	September-98	Johnson County	KS	USA	Toilet tank contents that included a bowl cleaning chemical backflowed into the water system.	Residential
*274	September-98	Des Moines	IA	USA	Well water was pushed into the domestic water by booster pumps.	Industrial
*350	September-98	Tempe	AZ	USA	About 1.39 million gallons of landscape irrigation water backflowed into city mains through a cross-connection.	Institution
*404	September-98	San Antonio	TX	USA	Potable water system within trailer park contaminated with sewage caused by backsiphonage and below grade leaking hose bibs.	Trailer Park
*430	October-98	Lee's Summit	MO	USA	Fire suppression system discharged causing potable water to run yellow.	Building

11 Backflow Incidents

Source	Month - Year	City	State	Country	Cause (What Happened?)	Type of Establishment
*276	October-98	Johnson County	KS	USA	Air backpressured into water main via air compressor in a winterized irrigation system resulting in air & brown water.	
*351	December-98	San Antonio	TX	USA	Contaminated water and air leak into the potable water system from a pressure surge tank.	Residential
*405	December-98	San Antonio	TX	USA	Customer's private well backpressures well water into city water system. Bypassed backflow preventer found on-site.	Laundromat/Cleaners
*406	December-98	San Antonio	TX	USA	Fire line backpressured stagnant/oil laden water into city's water supply. Failed check on fire line.	Office Bldg.
*426	January-99	Irvine	CA	USA	Recycled water plumbed to 3 on-site trailers for toilet and wash sinks.	Trailer Park
*426	January-99	Santee	CA	USA	Drinking fountain supplied with recycled water.	Park
*352	January-99	Santa Fe Springs	CA	USA	Domestic water was contaminated by a cross-connection to the pressure side of two wash water pumps.	Industrial
*388	March-99	Toledo	OH	USA	CO_2 backflowed into a restaurant's water system.	Commercial
*408	March-99	Colonial Heights	VA	USA	Malfunctioning valve causes 200 gallons of latex to spill into city water system.	Paint Factory
*429	May-99	Orlando	FL	USA	Orange County Courthouse water supply contaminated.	Building
*276	May-99	Johnson County	KS	USA	Toilet tank contents that included a bowl cleaning chemical backflowed into water supply.	Residential
*353	June-99	Smyrna Beach	FL	USA	Two potable water services (one at a residence and one at a business) were mistakenly cross-connected to the reuse water lines for 18 months.	Residential & Commercial

Backflow Incidents

Source	Month - Year	City	State	Country	Cause (What Happened?)	Type of Establishment
*373	July-99	Stratford	CT	USA	Anti-corrosive chemical enters public water lines.	Industrial
*426	January-00	Newport Beach	CA	USA	Improper installation of recycled water and potable water lines impacts nearly 80 homes.	
*426	January-00	San Clemente	CA	USA	Recycled water irrigation system connected to potable water improperly.	
*354	January-00	Los Angeles	CA	USA	Water and debris from the Dominguez Storm Channel back-flowed into the pipeline through a cross-connection.	Institution
*371	January-00	Newport Beach	CA	USA	Reclaimed water gets into water supply.	Residential
*164	March-00	Martinez	CA	USA	Gas enters firefighting water, and ignites due to sparks.	Refinery
*378	March-00	Pineville	LA	USA	Sewer line mistakenly connected to underground water line.	Residential
*384	March-00	St. James	NC	USA	Fertilizer enters into water supply through water hydrant vacuum effect.	Residence
*372	May-00	Stanton	CA	USA	Water containing foaming agent backflows into distribution system through fire hydrant.	Industrial
*389	July-00	Oak Ridge	TN	USA	Cross-connections between creek water, fire system and potable water.	Factory
*412	July-00	Gulf of Taranto		Italy	Sewage contamination to drinking water.	Tourist Resort
*379	November-00	Rosemead	CA	USA	Water contaminated by chemical Phos-Chek through fire hydrant.	Residential
*380	November-00	R. Dominguez	CA	USA	Firefighting foam was backpressured into water supply line.	Residential
*426	January-01	El Dorado	CA	USA	Dual plumbed irrigation system cross-connected for 7 months.	Residence
*426	January-01	Marin	CA	USA	Dual plumbed irrigation system cross-connected for 2 months.	Residence
*426	January-01	Carlsbad	CA	USA	Illegal connection between irrigation system and recycled water piping system.	Residence

11 Backflow Incidents

Source	Month - Year	City	State	Country	Cause (What Happened?)	Type of Establishment
*355	February-01		NV	USA	Cooling towers at a hotel contaminate the drinking water.	Building
*355	June-01		AK	USA	Water well becomes contaminated with surface water.	Factory
*355	August-01		FL	USA	The potable water supply at a middle school became contaminated by ethylene glycol from the air conditioning system.	Institution
*355	August-01		IN	USA	Filtration system bypassed, causing several people to drink untreated water.	Residence
*356	December-01			Netherlands	AustraliaGrey water backflowed into the potable water system through cross-connections, resulting in an outbreak of gastroenteritis.	Residential
*357	March-02	W. Sacramento	CA	USA	Coolant backflowed into water lines after a backflow prevention device failed during maintenance on the building's cooling system.	Industrial
*358	March-02	San Antonio	TX	USA	A cross-connection occurred between recycled water and drinking water system.	Residential
*359	April-02	Reno	NV	USA	Antifreeze solution backflowed into the potable water and drinking fountains.	School
*355	April-02		NY	USA	A backsiphonage condition, allowed contaminated water to enter the system through a cross connection to sewage.	Residential
*360	April-02	Carson	NV	USA	A valve in cooling system failed to completely shut, allowing water treated with chemicals to flow into the fresh water supply.	Institution
*414	April-02	River Road	TX	USA	Cross-connection between drinking water line with a recycled water line at a Gold Course Park.	Golf Course
*355	June-02		MA	USA	A water main break might have caused water contamination.	Building
*355	July-02		AZ	USA	Drinking water was contaminated due to improper maintenance of ice and water dispensers.	Golf Course

Backflow Incidents

Source	Month - Year	City	State	Country	Cause (What Happened?)	Type of Establishment
*361	September-02	Santa Fe Springs	CA	USA	Liquid detergent from a metal treatment plant entered the water supply system through a cross-connection.	Industrial
*355	October-02		AZ	USA	Coliform contamination in residence kills two children.	Residence
*362	November-02	Altadena	CA	USA	Snow foam was pumped into water system due to lack of backflow protection.	Institution/Residential
*416	May-03	E. China Town	MI	USA	Steam containing hydrazine may have come in contact with food eaten by employees.	Power Plant
*363	July-03	Chicago	IL	USA	Chemical contamination was in the water after a pipe broke and a garden hose was used to continue the flow of cold water.	Commercial
*364	October-03	Aurora	IL	USA	Potable water was directly connected to the boiler and to a chemically fed tank.	Industrial
*365	July-04		GU	USA	Backflow occurred from a private well to the city water supply due to lack of backflow prevention.	Residential
*366	September-04	St. Ignatius	MT	USA	Illegally installed plumbing connection allowed backflow of water supplying a poultry pen.	Commercial
*367	March-05	Stratford	ON	Canada	A pressurized washer system at a car wash failed and released chemically polluted water into the city's drinking water.	Residential
*368	December-05	Fairhope	KA	USA	Plumber mistakenly interconnected a discharge line from a travel trailer into a subdivision's water main.	Residential
*369	May-06	Philadelphia	PA	USA	Backflow in a soda machine occurred downstream through a copper piping.	Institution
*428	August-07	Chula Vista	CA	USA	For two years several businesses have been drinking and washing their hands in treated sewage water.	Commercial

11 Backflow Incidents

Source	Month - Year	City	State	Country	Cause (What Happened?)	Type of Establishment
*425	August-09	Lenexa	KS	USA	Carbonate water backflowed into copper plumbing lines at a restaurant.	Restaurant
*445	October-09	Layton	UT	USA	Cross-connection between secondary water and the city's drinking water.	Residence

Reference Sources

Reference	Source
1	Wallace & Tiernan, Technical Bulletin #99.
2	Journal AWWA, Vol 31, No 2, Feb 1939, 365.
3	Ibid, 346.
5	Ibid, 345.
6	Ibid, 362.
7	East Bay Municipal Utility District. 1944. Cross-connections.
8	Ibid, 330.
9	Journal AWWA, Vol 31, No 2, Feb 1939, 328.
10	Ibid, 361.
11	Ibid, 347.
12	Journal AWWA, Vol 31, No 2, Feb 1939, 339.
13	Ibid, 353.
14	Ibid, 358.
15	Journal AWWA, Vol 31, No 2, Feb 1939, 348.
16	Ibid, 355.
17	Ibid, 349.
18	Ibid, 331.
19	Ibid, 341.
20	DWP File on Cross-Connections.
21	Journal AWWA, Vol 31, No 2, Feb 1939, 327.
22	Ibid, 329.
23	Ibid, 332.
24	Ibid, 350.
25	Government Printing Office. No. 62370. Hearing Before Senate Committee on Stream Pollution Control.
26	American Journal of Public Health 7., July 1939.
27	Los Angeles Times, March 25, 1939.
28	R. E. Dodson to R. L. Derby. January 17, 1951. Photostat in LADW&P Files.
29	Short, Don. 1939. Engineering News Record, Sept. 28., 389.
30	University of Michigan Bulletin, 1934., 84.
31	Journal AWWA., Vol 33, No 3, March 1941, 406.
32	DWP File on Cross-connections. Information on number effected not available due to war time.
33	W.E. Shaw. 1944. Regional Water Works Engineer. DWP File on Pollution Case History.
35	DWP File on Cross-Connections. Information on number effected not available due to war time.
36	DWP File on National Research Council. Chart prepared by Goudey.
37	February Water Pollution Instances. 1948. DWP Public Health News. , 38.
38	Public Works Magazine. 1946, 33.
39	El Segundo Water Company.
41	AWWA Water Transmission and Distribution, 2nd Ed., 326.
42	Fire Engineering, March 1968.
43	Water Turns Pastel as City Lines Cross. Atlanta Journal: 13-A, January 15, 1970.
44	Journal of Environmental Health. January-February 1965.
45	Journal of Environmental Health, July-August 1970.
46	Chicago Daily News, August 20, 1969.
48	Atlanta Journal. January 15, 1970.
49	Ohio Newspaper. Cincinnati. Dec 22, 1969.
50	Chicago Daily News.
51	Los Angeles Hearld Examinor. January 1970.
52	Faulkner, Peter. 1977. The Silent Bomb. New York: Vintage Books.
53	Willing Water, February 1973.
54	The Times. San Mateo. California. Dec 27, 1972.
57	Oregon State Health Department.
58	Staff Writer. 1972. Dangerous Pesticide Found in Bakersfield Area Water Supply. L.A. Times, May 3.
59	Los Angeles Times. September 23, 1972.
60	National Inquirer.
61	San Antonio Water Board Memo for Record. January 5, 1977.
63	San Mateo County Health Department.

Reference	Source
64	Los Angeles Times. October 14, 1975.
66	Analytical Control. Vol 2. No 1. 1977.
67	North Dakota Water & Pollution Conference. 1978. Official Bulletin. June.
68	City Activity Report. September 1977.
69	Chicago Tribune. December 11, 1977.
70	Hinsdale. 1978. Doings. January 16.
73	Cross-Connections Control Down on the Farm. 1979. AWWA Opflow. Vol.5 No. November 11.
74	Los Alamos Monitor. Aug 30, 1979.
75	Morbidity & Mortality Weekly Report. July 8, 1977.
76	Memo. Michigan State Health Dept. April 22, 1963, and AWWA OpFlow. February 1977.
77	Aquarius and Watts News. October 1974.
79	A Classic Car Wash Cross Connection. January 1985.
80	OpFlow AWWA. June 1980.
81	Manchester Water Department.
82	Pittsburgh Press. December 16, 1980.
83	AWWA OpFlow. Sept 1981.
84	Chicago Tribune. November 22, 1981.
85	Bangor Daily News. November 21, 1981.
86	Denver Post. May 5, 1982.
87	Water tainted by insecticide. 1982. Rocky Mountain News, May 6.
88	Buffalo Currier Express. June 6, 1982.
89	Water in plant now drinkable Western Electric assures employees. 1982. Sunday Eagle Tribune. Lawrence, MA., July.
90	Pasadena.1982. Star News, Sept 8.
91	AWWA Mainstream. Aug 1982.
92	Los Angeles Times. Feb 24, 1983.
94	Byland, Kathleen. 1983. Fighting bad water is round-the-clock battle. Akron Beacon Journal. Aug 21.
95	Florida. Sarasota Herald Tribune.
97	The Willets News. April 14, 1984.
98	Ventura County Environmental Health Department. Internal report. Nov 16, 1984.
99	Massachusett Department of Environmental Quality Engineering. 1687. Cross-Connection Control Manual.
100	University of Florida. Aligator Vol. 81 N. 28. October 1987.
102	Ethylene Glycol Intoxication due to Contamination of Water System CDC Morbidity and Mortality Weekly Report. September 18, 1987.
103	The Gridley Gleam Vol. 5, Issue 6. July 2, 1987.
104	Feil, Kenneth. 1986. Marin Municipal Water District. Backflow Incident Report. April 30.
105	Investigation Chronology of the Sand Point Marina Water Borne Desease Outbreak. December 5, 1979.
106	State of Idaho Division of Environmental. 1980. Meridian Case Report. December 2.
107	The Record-Chronicle Newspaper, Vol LXVII, No. 42.
108	Seattle Water Department & Pacific Northwest Section of AWWA (PNWS-AWWA)
109	Water & Sewage Work. & PNWS-AWWA.
110	WD&BP. October 1998.
111	Wilcox, Pat. 1976. Poison Found in Water of Homes In East Lake. 1976 Chattanooga Times, March 26.
112	Department of Public health. Seattle and King Company.
113	State of Washington. Department of Social and Health Service.
114	Vancouver Water Department & PNWS-AWWA.
115	Vancouver Water Department & PNWS-AWWA.
116	AWWA OpFlow. May 1979.
117	Portland Water Bureau & PNWS-AWWA.
118	Department of Health & Welfare, State of Idaho.
119	Seattle Water Department & PNWS-AWWA.
120	Tacoma Water Department.
121	Seattle Post Intelligence. October 18, 1980.
122	News Leaks. AWWA Indiana Section.
123	Springfield Utility Board & PNWS-AWWA.

Reference	Source
124	Lane County Health Division & PNWS-AWWA.
125	Tacoma Water Department & PNWS-AWWA.
126	Vancouver Water Department & PNWS-AWWA.
127	Oregon Health Division & PNWS-AWWA.
128	Seattle Water Department & PNWS-AWWA.
129	Bellevue Public Works/Utilities Department & PNWS-AWWA.
130	State of Washington. Department of Social and Health Services. 1995.
131	Rural Water Magazine. 1986. Volume 6. Number 2.
132	Backflow Prevention. 1986, Vol 3, Number 4.
133	Palmer Gillin, Susan. 1986. Town's water system poisoned by 2, 4-D. Wenatchee World, July 9.
134	County Engineering Department. San Luis Obispo County.
139	Backflow Incident-Arizona. DW&BP. October 1993.
140	Baird, Fred. 1979. Back-Siphonage, La Chateaux Apartments. San Antonio Water Board Memo for Record. August 9.
141	Western Canada Water and Wastewater Association. Bulletin Vol. 43, No. 4. December 1991.
142	Blue Water Complaints Continue. Crossfire Newsletter Vol.1 No.1. 1991.
143	Fluidmaster 200A Strikes Again. Crossfire Newsletter Vol.1 No.1. 1990.
144	Canadian Incidents. ABPA News Vol5. September/October 1993.
144	Canadian Incidents. ABPA News Vol5. September/October 1993.
145	King County Water District Number 107. Renton, Washington. Incident Report. Jan 3, 1991.
146	County of Orange, Santa Ana, CA. Incident Report. Feb 13, 1991.
147	Orr, Steve. 1990. Solvent mixed in Xerox water. Democrat & Chronicle, May 10.
148	Denver, Colorado. Rocky Mountain News. Jan 30, 1990.
149	Oxley, Donald R., Meyers M.D., Hildy, and Miller, Jack. 1989. County of Orange Health Care Agency. Cross Connection Incident - Restaurant Newport Beach. January 5.
150	Tonyan, Rick. 1988. Edgewater Lifts Ban on Water. ABPA News. June-August.
151	Elizabeth Gold-Rasmussen. 1988. Fire Retardent Chemical Pollutes Water. Backflow Prevention. August.
152	Clemings, Russel. 1988. FUSD Plugs Toxic Leak Valves Replaced After Chemical Found in Water. The Fresno Bee, September 29.
153	Oxley, Donald R., Meyers M.D., Hildy, and Miller, Jack. 1999. County of Orange Health Care Agency. Cross-Connection Incident - Orange County Health Care Agency Public Health Laboratory and Waste Management Modular. April 26.
154	City of Philadelphia. Incident Reports.
155	Han, Andrea. 1988. Citizens slam Gilbert over possible water leak. Mesa Tribune, SepTember 6.
156	Tucson, Arizona. Tucson Citizen, 30 March 1989.
157	Huachuca City, Arizona. 1972. Incident Report. March 23.
158	Incident Report, Peoria, Arizona. Sept 25, 1969.
159	Salt Lake City Public Utilities. 1966-1991. Incident Reports.
160	Tucson, Arizona. The Arizona Daily Star, May 1, 1990.
161	Flynn, Terry. 1989. Chemical in water of county building causes illnesses. Cincinnati Enquirer, July 29.
162	Oxley, Donald R., Meyers M.D., Hildy, and Miller, Jack. 1999. County of Orange Health Care Agency. Incident Reports. April 26.
163	Oxley, Donald R., Meyers M.D., Hildy, and Miller, Jack. 1999. County of Orange Health Care Agency. Backflow Incident: Newspaper Printing Plant. April 29.
164	Denis Cuff. "Tosco on hot seat for blaze." Contra Costa Times, (March 28, 2000), www.contracosta.com/aolo/aolfn/toscofolo_20000328.html.
165	Plumbing & Heating Business. August, 1940.
166	Water & Sewage Wks. July, 1949, 259.
167	Fire Engineering, March 1968.
168	San Antonio Water Board Case History Summary. February 1993.
169	Nassau County Department of Health. New York. 1969.
170	Philadelphia, Pennsylvania. Bulletin, September 13, 1969.
171	Boston Globe, October 5, 1989, and AWWA Water Transmission and Distribution 2nd Ed., 327.
172	AWWA. (IL), February 1971.
173	Dan Randall. City of Riverside Memo. October 11, 1996.
174	Globe Newspaper Co. Vol. 205, No. 172, June 21, 1974

Reference	Source
175	City of Riverside. 1996. Backflow Incidents Memo. October 11.
176	Dept. of Health & Environmental Control. South Carolina. October 13, 1975.
178	Dept. of health & Environmental Control South Carolina. March 1977.
179	"Makeshift" Cross Connection Resulting in Sickness Aboard U.S. Naval Ship. 1978. Watts News: Stop Backflow, October.
181	San Antonio Express News. February 26, 1978.
182	Residential Backflow-Its Not Going Away. 1981. Watts News: Stop Backflow, January.
184	Dept. of health & Environmental Control South Carolina. May 1979.
185	Des Moines Register. May 25, 1979.
186	Virginia Dept. of Health. 1980. Bulletin Volume 1. Spring Ed.
187	Dept. of health & Environmental Control South Carolina. July 18, 1980.
188	Fontana, Richard F. 1980. Contamiated water supply leaves Robinson taps dry. Post-Gazette, November 8.
189	Pennsylvania. Associated Press. December 10, 1980.
190	Watts News. January 1981.
191	Virginia PHC Image. January 29, 1981.
192	DW&BP. March 1996, 22.
193	Forida. City of Tampa. April 20, 1982.
194	Cross Connection and Backflow Incidents. 1999. Memphis Light, Gas and Water. April 29.
195	300 Forced From Class. 1982. Denver Post, May 5.
196	Michigan. 1982. Detroit Free Press, May 6.
197	Massachusetts. 1982. Eagle-Tribune, July.
198	Connecticut. 1982. The Bridgeport Post, August 5.
199	Backflow Prevention. November 1984.
200	Medical Center Incidents. DW&BP. January 1997, 16.
201	Texas. 1982. Alice Echo-News, September.
202	Maine. 1982. Brunswick Times-Record, September 30.
203	Florida. 1982. Air Conditioning, Heating & Refrigeration News, November 5.
204	Georgia. Gwinnett County Water System. October 4, 1982
205	North Carolina. 1982. Associated Press. October 8.
206	James, Kristi. 1994. DW&BP. Historical Perspective. April.
206	James, Kristi. 1994. DW&BP. Historical Perspective. April.
206	James, Kristi. 1994. DW&BP. Historical Perspective. April.
206	James, Kristi. 1994. DW&BP. Historical Perspective. April.
207	Virginia Dept. of Health. 1983. Winter.
208	Florida. 1983. St. Petersburg Times, January.
209	Baird, Fred L. 1983. Coleman Plant - Cross Connection/Backflow Situation. San Antonio Water Board Memo for Record. January 27.
210	Georgia. 1983. The Marietta Daily Journal, February 3.
211	Florida. 1983. St. Petersburg Times, April 2.
212	Tallman, Douglas. 1983. Woodsboro water contaminated. The Frederick Post, June 8.
213	Indiana. 1983. Indianapolis News, September 8.
215	Backflow Prevention. April 1986.
216	DW&BP. April 1997, 20.
217	Georgia Environmental Protection Division. Water Contamination Reported Near Perry. July 1984.
218	Backflow Prevention. July 1985.
219	Antifreeze in Solar Panel Contaminates Drinking Water. ABPA Colorado. March 26, 1985.
220	Wilkins, V. Harvey. 1985. Cross Connection Town of Harleyville Station Code Number 118002. Trident District EQC Incident. March 19.
221	Oklahoma. Tulsa Tribune. July 1985.
222	The Dallas Morning News. July 18, 1985.
223	Backflow Prevention. May 1986.
224	DW&BP. March 1996. 22.
225	New Jersey. Asbury Park Press. August 21, 1985.
226	Heller, Anne. 1985. Water contamination sparks probe in West Sacramento. Daily Democrat, September 6.
228	Froonjian, John. 1985. Harrah's Water Supply Tainted by Antifreeze. Atlantic City Press, November.

Reference	Source
229	Backflow Prevention. April 1986.
230	Georgia. Associated Press. January 1986.
231	Duncan, Howard. 1986. Malathion in a Factoty's Plumbing at Bonner Springs. Kansas Dept. of Health & Environment. January 13.
234	The Fayetteville Times. April 24, 1986.
235	Ashlock, Ken. 1986. Cross-Connection - Dairy. Report from Cross-Connection Control Inspector. April 17.
236	New York. 1986. Hudson Register - Record, October.
237	Alabama. The Birmingham News, October 1986.
238	Conway, Laura. 1986. Nevada water OK after nitrate leak. Nevada Journal, November 28.
239	Cross-Connections / Backflow Incidents. Laboratory Services Division of the Dept. of Natural Resources. April 16, 1999.240 *240 DW&BP. April 1994. 20.
243	DW&BP. April 1994. 18.
244	Magor, Ken. 1987. Worker's poisoned by nickel-contaminated water. Kitchener - Waterloo Record, June.
245	New Jersey. Hawthorne Press. July 2, 1987
246	DW&BP. 1994. 22.
247	New York. Times Union, October 18, 1987.
249	Cross Connection Closes Elementary School. Utah Department of Health, Bureau of Drinking Water/Sanitation.
251	Backflow Prevention. May 1988.
252	Selburg, Roger D. 1988. Bushnell - Pesticide Contamination of Distribuion System. Illinois Environmental Protection Agency. May 5.
253	Officials puzzled over pesticide death of worker. Florida Associated Press. January 1, 1988.
254	ICWC. Fax: Backflow Example No. 1. April 22, 1999.
256	Cross Connection Protection of the Public Water Supply System at 1111 Fulton Mall. City of Fresno Public Works Department. September 30, 1988.
257	DW&BP. December 1990.
258	Beauchamp, Lane. 1988. School's water tainted Teachers, students are tested at hospi tals. The Kansas City Times, November 12.
259	Backflow Prevention. January 1990.
261	Oberi, Kan. 1989. Cross Connections Found: Green Water. ABPA News Vol.4. September/November.
262	Exploding Courthouse Toilets Cause Big Blowup in Seattle. Associated Press. February 1989.
263	Redmond High School Backflow Incidents. Summary of Backflow Incidents Cross Connection Control Committee Pacific Northwest Section AWWA, R-89-003. June 1989.
264	*264 Lindsey, Donald G. Cross Connection Incident Involving Caustic Soda in Water Line at Henry J. Mills Filtration Plant. MWD Report. November 8, 1989.
265	*265 Report shows 3 deaths caused by water contamination. Kansas Rural Water Magazine. November 1990
266	Cross Connection/Backflow Incidents San Antonio Area. ABPA Texas. December 26, 1989.
267	The Denver Post. February 1, 1990.
268	Colorado. Rocky Mountain News. 1990.
269	Report. San Antonio Water System Inspection Division Backflow Prevention Section. December 1998.
270	Salkowski, Joe. 1990. Police station's cooling system contaminates its drinking water. The Arizona Daily Star, May 1.
271	DW&BP. January 1991.
272	Arthur, Tony. 1990. Comprehensive Coverage. Watts Regulator News. December 14.
273	Incident Summary. Water District No.1 of Johnson County Distribution Water Quality Department.
274	Des Moines Water Works. 1999. Backflow Incidents in Iowa 1977-1998. April 22.
275	School's Water Contaminated with Antifreeze. Indianapolis Star. November 3, 1990
276	Backflow Incidents in Johnson County, KS. Johnson County Distribution Water Quality Department Water District No.1 Fax. May 5, 1999.
277	Backflow Incident Space Needle, Gatlinburg, Tennessee. Gatlinburg Utility Depart ment.
278	DW&BP. May 1994. 22.

Reference	Source
279	Stone, Jeff. 1991. Chicken Hosue Cross-Connection. Watts News, Spring, 1991.
280	DW&BP. October 1991.
281	DW&BP. September 1991.
282	Tarman, Gene. Cross Connection 27197 Aurora. Santa Margarita Water District Memo. September 27, 1991.
283	DW&BP. December 1991. 18.
285	Baldwin, Carol. 1991. Unwanted Guests Residents find parasites in tap water. Heritage Newspapers/The News Herald. October 9.
286	DW&BP. January 1992. 8.
287	Clary, Dick. 1992. Near Backflow Disaster Averted by Alert Maintenance Crew. The Georgia Operator, February 6.
288	Maine Incident - Fecal Coliform in Private Supply. DW&BP. March 1993. 9.
289	Methemoglobinemia Attributable to Nitrite Contamination of Potable Water Through Boiler Fluid Additives - New Jersey, 1992 and 1996. MMWR. Vol.46. No.9. March 7, 1997. 203.
290	Cross-Connection Investigation Report. January 28, 1993.
292	Green, Ken. 1994. ABPA News, September/October 1994. 16.
293	Backsiphonage Incident. DW&BP. December 1993. 6.
294	Ochoa, Sergio. 1993 Cross-Connection Incident. Metropolitan Water District of Southern California. November 9.
295	Brown, Greg. 1994. NC Incident Report. DW&BP. February.
296	Marchon, Bob. 1994. Ice Machine Reveals Cross Connection. ABPA News. March/April.
297	Summary of Backflow Incidents. PNWS-AWWA Fax. December 29, 1993.
298	Boiler contaminates water at high school field. Tucson Citizen. March 1, 1994.
299	Incidents Report. Provided by Cross-Connection Control in Laredo, Texas. December 1997.
300	Cornett, Penny. 1994. Cross Connection Incident. April 1.
301	Nessl, Don. 1994. Backflow Incidents. Tuolumne Utilities District. May 16.
302	Chapman, Ron. 1994. Backflow Incident. DW&BP. July. 5.
303	Freon Burns Mouths. DW&BP. September 1994. 19.
304	Nelson, John O. Backflow Accident at Eagle Drive: Chronology of Events and Remedies. North Marin Water District. October 20, 1994.
305	Bowman, Chris. 1994. Blue tap water shocks homeowners, businesses. The Fresno Bee, June 30.
306	Incident Report for Utilities Management Program. Department of Veterans Affairs.
307	Gonzales, Daniel, and Reese, Tim. 1994. Veterans hospital, closed by flooding, may reopen Tuesday. Syracuse Herald-Journal Vol.118 No.35,306, July4.
308	Cross-Connection Investigation Report. City of Tempe. July 14, 1994.
309	Howe, Kevin. Backflow blamed for bad courthouse water. Monterey County Herald's, October 7.
311	Shigo, Barbara. Film Company Creates Backflow Situation. California-American Water Los Angeles Division. 17 October 1994.
312	Backflow Prevention Supplemental/Incident Report. Tucson Water. October 28, 1994.
313	Daily, Don B. 1995. Report. Georgetown Steel Corporation. January 17.
314	Ochoa, Sergio. 1995. Drinking Fountain Cross-Connection Incident. Metropolitan Water District of Southern California. March 16.
315	Ochoa, Sergio. 1995. Caustic Soda Cross-Connection Incident. Metropolitan Water District of Southern California. June 19.
316	Carlson, Richard. 1995. Prevention is Key. DW&BP. November.
318	DW&BP. December 1995. 18.
319	James, Kristi M. 1995. Green Water & Coliform Contamination. DW&BP. December.
320	Editorial Staff. 1996. San Diego Incident Investigated. DW&BP. April.
321	Incident Report. Commissioners of Public Works. December 28, 1995.
322	Brock, James. 1996. A Documented Backflow Incident. Michigan Newswire Drinking Water & Backflow Prevention.
323	Calhoun, Richard W. 1996. Investigators Solve Mystery of Air in Lines. Unknown magazine. June.
324	Ochoa, Sergio. 1996. Eye Wash Cross-Connection Incident. Metropolitan Water District of Southern California. January 24.
325	Waterweek. October 7, 1996. 6.
327	Lipsher, Steve. 1990. School antifreeze-water mix caused by error, faulty system. The Denver Post. February 1.

Reference	Source
328	Kalinski, Frank. 1996. Backsiphonage Incident. Tucson Water Supplemental/Incident Report. November 8.
329	Banuelos, Gil. 1997. Readers Response Backflow Incident. DW&BP. July.
330	Incidents Report. North Platte Water Department Backflow Prevention Division. June 1997.
331	Morel, Mike. 1997. Mobile Pilot Plant Cross-Connection at the F.E. Weymouth Filtration Plant. Memorandum. June 4.
332	Oxley, Donald R., Meyers M.D., Hildy, and Miller, Jack. 1999. Cross-Connection Incident. County of Orange Health Care Agency, Orange County Health Care Agency Public Health Division of Environmental Health. April 26.
333	Strong, Dale. 1997. Backflow Incident. Memorandum. July 31.
334	Bruce, Andrew. 1997. Half of Guelph goes without drinking water. The Guelph Mercury, August 30.
335	Carolinas/Connection. Carolinas Chapter ABPA Newsletter Vol.1 Issue 4. December 1997.
336	Krouse, Mark. 2001. Backflow Incidents Sparks Improvements. Opflow Vol. 27 No. 2. February 2001.
337	Svetich, E. Terri. 1997. Backpressure as a Result of a Carbonation Pump on a Softdrink Dispenser. Backflow Incident Report. September 24.
338	Oliande, Sylvia, and Scott Glover. 1997. Plumber's Mistake Contaminates Water for 2 Schools, 1,600 Homes. Los Angeles Times. October 31.
340	Strong, Dale. 1997. Cross Connection Incident. Commissioners of Public Works Memorandum. October 17.
341	Hubbs, Doyle. 1997. Backflow Incident at Yosemite International Airport. Cross-Connection Control Technician Report. October 23.
342	Report of Cross-Connection/Backflow Incident at Beacon House Apartments. October 1997.
343	Report Summary. Harding Lawson Associates. March 30, 1998.
344	Beuhler, Mark D., and Jay W. Malinowski. 1997. Domestic Water Quality Problem in the Iron Mountain Community Water System. Memorandum. December 10.
345	Fanning, Dixie H. 1997. Another Blue Water Complaint. Commissioners of Public Works email. December 11.
346	Strong, Dale. 1998. Backflow Incident. Commissioners of Public Works. Januray 30.
347	Field, Jeff S. 1998. McCormick Mill Cross-connection Incident Report. South Carolina Department of Health and Environmental Control Bureau of Water. April 28.
348	Drinking Water & Backflow Prevention, January 1999, p4
349	Commissioners of Public Works incident report, July 13, 1998
350	Cross-Connection Investigation Report. City of Tempe. September 29, 1998.
351	Baker, Vince. 1999. Backflow Cross Connection at 105 S. St. Mary's. Grinnell Fire Protection Systems Company. January 19.
352	Del Gann Jr., R.D. 1999. Cross-Connection Incident, Kelly Pipe. Department of Health Services Drinking Water Field Operations. January 8.
353	Turano, Genny. 1999. Cross-Connection Incident Report. New Symrna Beach Utilities Commission. June 30.
354	Morel, Mike. 2000. Executive Summary for Event Review 2000-2. Memorandum. February 28.
355	Appendix A. Surveillance of Waterborne-Disease Outbreaks Associated with Drinking Water - United States 2001-2002. CDC, http://www.cdc.gov/mmwr/preview/mmwrhtml/s5308a5.htm.
356	Setback for Netherlands Dual Supplies. Cooperative Reseach Centre for Water Quality and Treatment, (June 30, 2003), http://www.waterquality.crc.org.au.
357	Martineau, Pamela. 2002. Don't drink the water, some in West Sacramento told. The Sacramento Bee, (May 2), http://sacbee.com/content/news/story/2463985p-2915210s.html.
358	News 9 San Antonio Story, (July 3, 2003), http://news9sanantonio.com.
359	Hagar, Ray. 2002. Students cope with broken water system at Carson High. Reno Gazette-Journal, (April 23), http://www.rgj.com/news/printstory/php?id=12828.
360	http://rgj.com/news
361	Briscoe, Daren, and Anna Gorman. 2002. Chemical Spill Pollutes Town's Water System. Los Angeles Times, (September 1), http://www.latimes.com/news/local/la-me-nowater1sep01.story.
362	Dreuer, Howard. 2002. TV ad miscue causes. Unknow Newspaper. November 28.

Reference	Source
363	Contaminated Water Closes Loop Businesses. NBC5.com, (July 22, 2003), http://www.nbc5.com/news/2348714/detail.html?z=dp&dpswid=2265994&dppid=65172.
364	Thompson, Robert. 2003. Potable water cross connection. City of Aurora. October 21.
365	Worth, Katie. 2004. Water troubles vex south. Pacific Daily News. July 9.
366	Associated Press. 2004. Water poisoning likely came from poul try pen. Billings Gazzete, (September 6), http://www.billingsgazette.com/index/php?display=rednews/2004/09/22/build/state/70-poultry/pen.inc. September 6, 2004
367	Alphonso, Caroline. 2005. Spill from car wash contaminates water in Stratford. Globe andmail.com, (March 8), http://www.theglobeandmail.com/servlet/articlenews/tpprint/lac/20050308/stratfordsb08/tpnational/.
368	Dezember, Ryan. 2005. Sewer, water lines mixed. Everything Alabama, (December 4), http://www.al.com/printer/printer.ssf?/base/news/113334571926980.xml&coll=3.
369	Soda machine blamed for sickening Paul VI students. Philadelphia Inquirer, (May 19, 2006), http://www.philly.com/mld/inquirer/news/local/states/new_jersey.
370	Adams, Craig H. 1997. Backflow Prevention Supplemental Incident Report. Tucson Water. July 17.
370	Adams, Craig H. 1997. Backflow Prevention Supplemental Incident Report. Tucson Water. July 17.
371	Corbin, Laura. 2000. Reclaimed water gets into Newport supply. The Orange Country Register. 30 September.
372	Chang, David, and William C. Gedney. 2000. A Utility's Experience in Responding to a Backflow Incident. Southern California Water Company. May.
373	Ambrosini, Dana. 1999. Sikorsky chemical leak fouls water to 400 homes in 2 towns. Connecticut Post, (July 29), http://www.connpost.com/lead.html.
374	James, Kristi M. 1995. Gainesville Incident Investigated. DW&BP. November.
375	Magnant, Michael. 1994. Discolored, Bad Tasting Water at Capitol. State of Iowa Department of Public Health. April 20.
376	Sac City water tainted by fertilizer. The Des Moines Register, May 8, 1998.
377	Varin, George. 1979. Meridian's Water's Onion Odor. DW&BP Backflow Incidents. June 16.
378	Sewage water flowed in their faucets. CNN, (May 29, 2000), www.cnn.com/2000/US/05/29/drinking.sewage/.
379	Rosemead, El Monte water contaminated three days. NBC4.tv. November 2000.
380	Spath, David P. 2001. Fire Suppression Foam Contamination of the Public Water Supply. State of California Department of Health Services. May 18.
381	Frink, Chris. 1995. Water contamination spurs class action suit. The Advocate Regional News, March.
382	Boston Globe, October 5, 1989.
383	Bacflow Case Fast Food Chain Restaurant. 1975. Watts News: Stop Backflow, April.
384	Kolnitz, Cece von. 2000. Fertilizer sucked into Brunswick water system. Wilmington Star, (March 29), http://www.wilmingtonstar.com/temp/water29.html.
385	The Polio Cross-Connection. DW&BP. August 1992.
386	Mader, Robert P. 1995. Cross connection on ice machine sickens conventioneers in Ohio. Contractor. December.
387	Wine, Not Water, Ran From Faucets. The Cincinnati Post, December 24,1969.
388	Taylor, Chris. 1999. A Tip O'the Hat to Green Brew. Optflow. March.
389	Thomas, Susan. 2000. Oak Ridge site's water was tainted for decades. The Tennes sean. 30 July.
390	Baird, Fred L. 1976. Possible Backflow Lake Water. San Antonio Water Board Memo for Record. April 14.
391	Back-Siphonage. 1979. Memo for Record. October 19.
392	San Antonio. Texas. 1978. City Water Board Memo for Record. July 26.
393	San Antonio Water Board, Memo for Record. October 27, 1977.
394	San Antonio Water Board. July 11, 1977.
395	San Antonio Water Board. June 30, 1976.
396	San Antonio Water System Inspection Division Backflow Prevention Section, Cases of Cross-Connection/Backflow Situations within San Antonio. December 1998.
397	San Antonio Water Quality Inspector. February 1974. Memo for Record.
398	Filk, Tommy L. 1983. Cross-Connection Inspector Interoffice Memo. December 27.
399	Baird, Fred L. 1984. Lanier High School-chromates in water. San Antonio Water Quality Division Memo for Record. April 4.

Backflow Incidents

Reference	Source
400	Baird, Fred L. 1986. Schoenfield Materials - Contamination. San Antonio Water Quality Division Memo for Record. June 10.
401	Water Quality Complaint Investigation. San Antonio Water Board. March, 28 1989.
402	Water Quality Complaint Investigation. San Antonio Water Board. June 28, 1989.
403	Baird, Fred L. 1990. Unapproved Commode Ballcocks Marina Apartments Lawsuit. San Antonio Water Board Memo for Record. May 10.
404	Trailer Park's Water Shut-Off Due to Cross Connection Related Contamination. The Direct Connection ABPA Vol. 9 Issue 10. October 1998.
405	By-Pass Around Backflow Preventer Contributes to Backflow Situation. The Direct Connection ABPA Vol. 9 Issue 12. January 1999.
406	Fireline Back-Pressures Air/Oil Ladden Water Into City Supply.The Direct Connection ABPA Vol.9 Issue 2. February 1999.
407	Backflow Incident Report. Taylorsville-Bennion Improvement District. December 28, 1991.
408	Pope, Jon. 1999. Alert lets Heights curb spill/Valve at plant puts latex in water pipes. Richmond Times-Dispatch, (March12), http://www.timesdispatch.com/Backflow_Incidents/spill0312.shtml.
409	Accident Reports. Seattle Water Department: Seattle Post Intelligencer. February 1990.
410	Summary of Backflow Incidents. PNWS-AWWA. July 21, 1994.
412	Waterborne Outbreak of Norwalk-Like Virus Gastroenteritis at a Tourist Resort, Italy. CDC, (July 2000),http://www.cdc.gov/ncidod/EID/vol8no6/01-0371.htm.
413	Zircatec missed part a "costly mistake". Fort Hope Evening Guide, July 20, 1994.
414	Needham, Jerry. 2002. Tainted water sent to homes. MySanAntonio, (April 4), http://news.mysanantonio.com/story.cfm?xla=saen&xlb=290&xlc=674435.
415	McGraw, Seamus. 1992. Pupil's sickness is traced to water Clifton school's boiler blamed. New Jersey, October 23.
416	Weniger, Deanna. 2003. Edison finds hydrazine problem at St. Clair plant. The Timesherald.com, (May 24), http://www.thetimesherald.com/news/stories/20030524/localnews/359419.html.
417	AWWA. February 1971.
418	New Hampshire. 1983. The Union Leader, September 1.
419	Detergent in water line causes shutoff at LVHC. The Morning Call, May 12, 1989.
420	Kramer, Leroy. 1998. List of Incidents in the last four years. North Platte Water Department. February 26.
421	Svetich, Terri. 1998. Investigation of Backsiphonage Incident. Washoe County Utility Services Division. April 10.
422	WSBCWD. 1977. Incident Report. October.
423	Drinking Water & Backflow Prevention, August 1998, p6
424	Backflow Incident. DW&BP. September 1998.
425	22 People Sickened at Mexican Restaurant. 2009. KCTV5.com, (August 31),
426	Plumbing Code/Cross Connection Control Workshop of the 2002 Recycled Water Task Force. Draft White Paper. November 15, 2002.
427	San Antonio Water Board, Memo for Record. February 14, 1979.
428	Krueger, Anne. 2007. Chula Vista center connected to pipes carrying treated sewage. The Union Tribune, (August 22), http://www.sandiego.com/news/metro/200708229999-1n22otay.html.
429	Spears, Kevin, and Salamone Debbie. 1999. Courthouse's drinking water is no longer stinky. The Orlando Sentinel. May 25.
430	Douglas, Larry. 1998. Piggly Wiggly Fire Suppression System. City of Lee's Summit Water Utilities Department Memo. November 3.
431	King, Pat. 1977. Trott: improper plumbing endangered city's water. The Daily Mail. August 3.
432	Complaints Prompt Water Line Flushing. The Pittsburgh Press. December 10, 1974
433	Overfilled coolant turned water green at Hale Morgue. Lawrence Eagle-Tribune. September 6, 1984
434	Van, Jon. 1985. Report 'finds' 2 salmonella outbreaks. Chicago Tribune. September 15.
435	Bouchillon, Kim. 1987. State health officials test water in Yazoo County. The Clarion-Ledger. July 3.
436	Hackett, Ray. 1989. Test results expected in city water problem. Norwich Bulletin. September.
437	Braile, Robert. 1993. AMC agrees to alter Notch water system. The Boston Sunday Globe. September 19.

Reference	Source
438	Users Of Well Water Urged To Get Tests. The Greenville News. April 1, 1973.
439	Engle, Randy. 1998. Backflow Incident, Clark County, Harmony, Mississippi. Engsoft Solutions Letter. October 20.
440	Nevada. CA/NV AWWA Source Vol.7 Number 3. Autumn 1996.
441	Kalinski, Frank. 1994. Backflow Prevention Supplemental/Incident Report. Tucson Water. Februrary 2.
442	Kalinski, Frank. 1994. Backflow Prevention Supplemental/Incident Report. Tucson Water. March 10.
443	Adams, C.H. 1993. Backflow Prevention Supplemental/Incident Report. Tucson Water. December 16.
444	Davidson, Inga. 1994. Backflow Prevention Supplemental/Incident Report. Tucson Report. February 8.
445	445 cmayorga. 2009. Illegal connection causes four Layton cases of giardiasis. StandardNet, (October 15), http://www.standard.net/print/topics/health/2009/10/15/illegal-connection

Backflow Incidents

Appendix A
Field Test Guidance Documents

Appendix A

Inside Appendix A

General Information 484
    Line Pressure..48
    Bleed-off Valve Arrangement..48
    Field Test Results...486
Reduced Pressure Principle Backflow Prevention Assembly...................486
    Attaching hoses to self-actuating test cocks...486
    Check Valve No. 2: Direction-of- flow 487
Pressure Vacuum Breakers (PVB & SVB) 488
    Backpressure Evaluation 488
    Limited Space..488
Detector Assemblies: Operation of bypass 489
    RPDA: Verify Detection of Flow through Bypass 489
    RPDA Diagnostics: Operation of Bypass...490
    DCDA: Verify Detection of Flow through Bypass 491
    DCDA Diagnostics: Operation of Bypass..491
    RPDA-II: Verify Detection of Flow through Bypass................................493
    DCDA-II: Verify Detection of Flow through Bypass...............................493
Abbreviated Field Test Procedures: Training Aid 493
    RP 493
    DC 498
    PVB 499
    SVB..500
Periodic Test of Differential Pressure Field Test Kit 501

A.1 General Information

A.1.1 Line Pressure

Some jurisdictions require that the line pressure be recorded during the test of a backflow prevention assembly. To accomplish this a suitable pressure gage should be attached to the No. 1 or No. 2 test cock, and the value recorded.

A.1.2 Bleed-off Valve Arrangement

The bleed-off valve arrangement used in 9.3, 9.4, 9.5, 9.7 and 9.9 in the field test procedures requires one each of:

> One 1/4" ⌀ needle valve
> One brass 1/4" ⌀ nipple
> One brass 1/4" ⌀ tee
> One brass 1/4" ⌀ IPS × 45° SAE flare connector

constructed as shown in figure A.1.

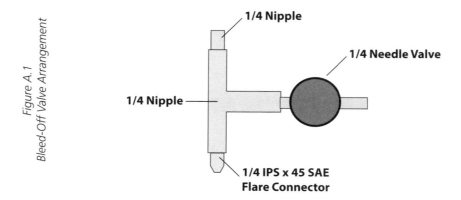

Figure A.1
Bleed-Off Valve Arrangement

The needle valve in the bleed valve arrangement is necessary for fine control of the flow during the diagnostic portions of the field test procedures. The arrangement as shown in Figure A.1 will allow for a greater volume of flow to compensate for shutoff valve leaks. The needle valve may be opened to the fully open position and accurate values recorded on the field test kit.

Swivel Connection

Common practice in the field has been to use a swivel connection attached to the 1/4"⌀ nipple of the bleed valve arrangement so that it may be attached to a 45° SAE flare connector inserted in the test cock of the assembly being field tested (Fig. A.2). However, the smaller inside diameter of the swivel connection will reduce the amount of flow through the bleed valve arrangement. To maintain accurate values on the field test kit with the swivel connection, it is recommended that the needle valve be opened no more than one-half turn.

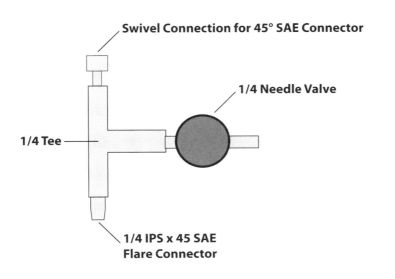

Figure A.2
Bleed-Off Valve Arrangement with Swivel Connection

Large Assemblies

Assemblies 2-1/2" ∅ and larger will use 1/2" ∅ or 3/4" ∅ test cocks, so it may be necessary to compensate for larger shutoff valve leaks. This may be accomplished with a bleed valve arrangement containing two 1/4" ∅ needle valves, and still maintain accurate values on the field test kit (Fig. A.3).

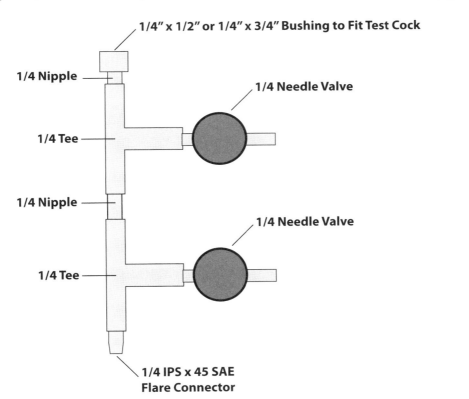

Figure A.3
Two needle valve bleed-off valve arrangement

With this arrangement directly threaded to a 1/4" x 1/2" or 1/4" x 3/4" bushing, both needle valves may be opened fully to compensate for shutoff valve leaks. This two needle valve arrangement should not be used with a swivel connection, since the smaller diameter of the swivel connection will restrict flow and cause inaccurate readings on the field test kit.

A.1.3 Field Test Results

When analyzing the field test results from any of the backflow prevention assemblies, it must be pointed out that there may be a distinct difference between a failing component versus a leaking or stuck closed component. The initial reaction may be that both situations are identifying the same unacceptable results but the level of backflow protection may vary significantly.

For example on a DC, if one of the check valve readings from the field test procedure is greater than 0.0 psid but less than 1.0 psid, this implies that the check valve is holding a differential reading below the acceptable reading of 1.0 psid. The key point is that the check valve is holding some differential pressure in the direction of flow, which implies that the check valve is preventing backflow. This is compared to a reading of 0.0 psid, which means that the check valve is not holding any differential pressure or in other words, the check is leaking. A leaking check valve would allow backflow to occur.

In the case of the RP, the minimum acceptable opening point of the differential pressure relief valve is 2.0 psid. If the field test results indicate that the relief valve opened at 1.5 psid, this is considered an unacceptable result because it is less than the required 2.0 psid. The relief did open; it just opened at an unacceptable value. On the other hand, if the differential reading on the field test kit goes to 0.0 psid and the relief valve does not open, this indicates that the relief valve is stuck shut and should be recorded as "did not open." In this instance, the relief valve is not capable of maintaining the pressure between the two check valves at a reduced pressure, therefore, reducing the backflow prevention capabilities of the RP.

For the PVB or SVB, if the air inlet valve opening point reading from the field test procedure is less than 1.0 psid but greater than 0.0 psid, this implies a failing air inlet valve. The air inlet failed to open at the minimum value of 1.0 psid but it did open to admit air into the body. This is contrasted to the air inlet valve not opening when the pressure in the body is lowered to 0.0 psid. In this case, the air inlet is not opening to allow air to enter the body. An air inlet valve, which does not open at all, significantly reduces the assembly's ability to break the vacuum or prevent backflow.

Administrative authorities tracking the field test results of backflow prevention assemblies in their jurisdiction may quote an overall annual failure rate. This data can be easily misconstrued to mean that any backflow preventer that failed its annual field test is failing to prevent backflow. However, as illustrated in the examples above, a failing reading from one of the field test procedures may cause the assembly to fail the criteria of the annual field test but the assembly is not, necessarily, allowing backflow to occur.

A.2 Reduced Pressure Principle Backflow Prevention Assembly

A.2.1 Attaching field test kit hoses to self-actuating test cocks

9.2.1.2: Test No. 1, step c, Five needle valve field test kit procedure
9.2.2.2: Test No. 1, step c, Two needle valve field test kit procedure
Some backflow prevention assemblies may contain a type of test cock that is opened when the hose from the field test kit is attached. There is no handle or screwdriver slot to open the test cock.

a. Attach hose from the low side of the field test kit to the No. 3 test cock, and then bleed all air from the low side of the field test kit by opening the low side bleed needle valve.
b. Maintain the low side bleed needle valve in the open position, then attach hose from the high side of the field test kit to the No. 2 test cock. Open the high side bleed needle valve to bleed all air from the high side of the field test kit.
c. Proceed to Test No. 1, step h.

9.2.3.2: Test No. 1, step c, Three needle valve field test kit procedure

Some backflow prevention assemblies may contain a type of test cock that is opened when the hose from the field test kit is attached. There is no handle or screwdriver slot to open the test cock.

a. Attach hose from the low side of the field test kit to the No. 3 test cock, and then bleed all air from the low side of the field test kit by opening the bypass needle valve approximately one turn, then open low side bleed needle valve.
b. Maintain the bypass needle valve and low side bleed needle valve in the open position, then attach hose from the high side of the field test kit to the No. 2 test cock. Open the high side bleed needle valve to bleed all air from the high side of the field test kit.
c. Proceed to Test No. 1, step h.

A.2.2 Check Valve No. 2: Direction of flow

LIMITATION: A direction of flow value may be obtained for check valve No. 2 as long as shutoff valve No. 2 is drip tight. Any leak in shutoff valve No. 2 will invalidate the direction of flow value. If shutoff valve No. 2 is drip tight and the direction of flow value for check valve No. 2 is below 1.0 psid, but greater than 0.0 psid, then the ability of the assembly to adequately prevent backflow is not affected.

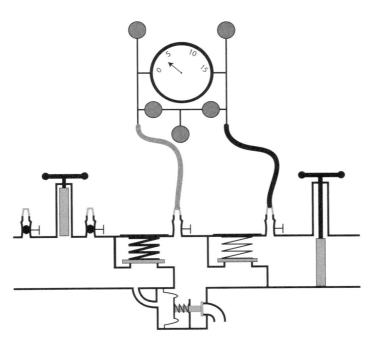

Figure A.4
Direction of Flow Test for Check Valve No. 2

Purpose: To determine the static pressure drop across check valve No. 2.

Requirement: The static pressure drop across check valve No. 2 shall be at least 1.0 psid. If shutoff valve No. 2 is found to be leaking this test cannot be performed accurately.

Steps:
a. With the field test kit attached as in 9.2.1.2 (9.2.2.2 or 9.2.3.2) Test No. 3, step a, close No. 2 test cock.
  - If the reading drops to zero, or rises above the actual pressure drop across check valve No. 1, this implies that the shutoff valve No. 2 is leaking. This test can not be completed accurately. Proceed to Test No. 3, step b.
  - If reading remains steady, then proceed to step b below.

b. Close test cocks No. 3 and No. 4. Remove test equipment.
c. Attach hose from the high side of the field test kit to the No. 3 test cock.
d. Attach hose from the low side of the field test kit to the No. 4 test cock.
e. Open test cocks No. 3 and No. 4, and bleed all air from the hoses and gage by opening the high side bleed needle valve and the low side bleed needle valve.
f. Close the high side bleed needle valve, then slowly close the low side bleed needle valve.
g. After the reading stabilizes, the differential pressure reading indicated is the static pressure drop across check valve No. 2 and is to be recorded as such.
h. Close all test cocks, slowly open shutoff valve No. 2, and remove all test equipment.

A.3 Pressure Vacuum Breaker (PVB & SVB)

A.3.1 Backpressure Evaluation

Purpose: To determine if the assembly (PVB or SVB) is subjected to continuous backpressure due to improper installation.

Requirement: Backpressure shall not be present. (Test cannot be performed if the assembly is part of a parallel installation.)

Steps:
After completing:
  PVB: 9.4, Test No. 2, step f
  or SVB: 9.5, Test No. 2, step e
proceed with the following:

a. Open shutoff valve No. 2. (No. 1 shutoff valve is still closed.)
b. For PVB: Open test cock No. 1, then open test cock No. 2.
   For SVB: Open vent valve, then open test cock.
c. Observe and record if a continuous discharge of water occurs from test cock No. 2 (or vent valve)

NOTE: Discharge of water may occur for many minutes since the downstream must be depressurize through the test cock No. 2 (or vent valve). If the downstream piping contains any air in the system, then the depressurizing may take a significant amount of time.

d. Close test cock No. 1 and No. 2 (test cock and vent valve for SVB). Close shutoff valve No. 2.
e. Open shutoff valve No. 1, then slowly open shutoff valve No. 2.
f. Replace the air inlet valve canopy.

A.4 Double Check Valve Backflow Prevention Assembly

A.4.2 Limited Space

Below grade installations of the double check valve assembly (DC) may provide limited clearance around the assembly, such as in a meter-box or pit, and the field test kit may be prevented from being held at the proper elevation during the field test procedure. Should this occur, the field test procedure can still be performed accurately providing that the downstream reference point (See 9.3.3.1) is raised to an elevation where the field test kit is maintained (Fig. A.5). This also applies to the field test of the DCDA (9.7) and DCDA-II (9.9).

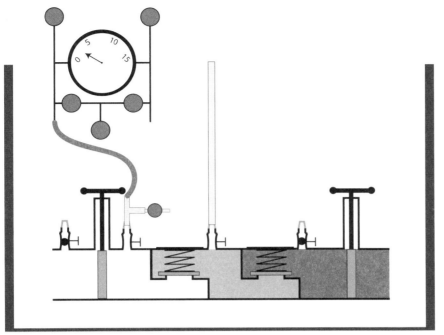

Figure A.5
Testing the Double Check with Limited Clearance

A.5   Detector Assemblies: Operation of bypass

A.5.1 Reduced pressure principle-detector backflow prevention assembly (RPDA)

   a. Conduct following steps after completing 9.6.2 Test No. 2, step a.
   b. Maintain No. 2 shutoff valve of the bypass reduced pressure principle backflow prevention assembly in open position.
   c. Maintain No. 2 shutoff valve of main-line assembly in closed position.

A.5.1.1 Verify Detection of Flow through Bypass

Purpose: To test the operation of the bypass by determining if a flow of water is detected by the water meter.

Requirement: The water meter in the bypass shall detect flow when the No. 4 test cock on the main-line reduced pressure principle backflow prevention assembly is opened.

   a. If the test cock No. 4 of the main-line assembly is located on the bypass piping, then it will be necessary to verify that the connection between the downstream bypass piping and main-line body will allow flow. See Diagnostics A.5.1.2. If test cock No. 4 is not located on the bypass piping, proceed to step b.
   b. Slowly open test cock No. 4 of the main-line assembly to create a small flow (approximately 1 to 2 gallons per minute).
   c. Verify that the water meter in the bypass indicates flow, and record as such.
   d. Open all shutoff valves of RPDA.

A.5.1.2 Diagnostics - Operation of Bypass

If test cock No. 4 attaches to main-line body at the same location as the bypass piping, it will be necessary to verify that the connection between the downstream bypass piping and main-line body is open to allow flow. Close shutoff valve No. 1 on the bypass reduced pressure principle assembly, then open test cock No. 4 on the bypass assembly. Water should pass through the check valves of the mainline assembly, then the connection of the main-line body to the bypass, and flow freely from the test cock No. 4.

- If water flows freely from test cock No. 4, then the connection between the downstream bypass piping and main-line body is open to flow. Open shutoff valve No. 1 on the bypass assembly, then proceed to A.5.1.1, step b.

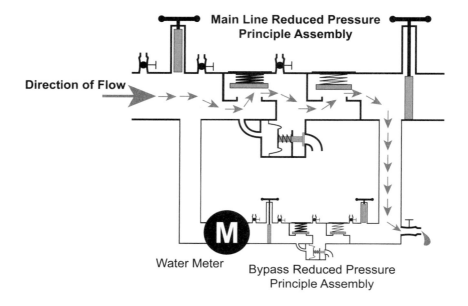

- If water does not continue to flow freely from test cock No. 4, then the connection between the bypass and the main-line body may be restricted or clogged. Repair or maintenance of the connection is necessary before an accurate test can be completed.

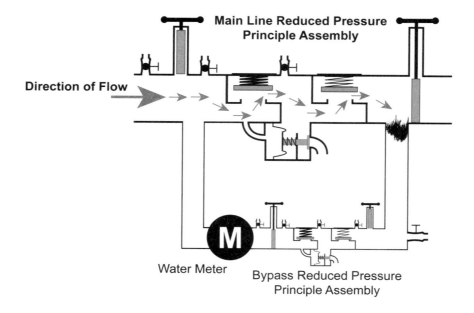

a. Proceed to 9.6.2, Test No. 2, step b.

A.5.2 Double Check –Detector Backflow Prevention Assembly (DCDA)

    a.   Conduct following steps after completing 9.7.2, Test No. 2, step a.
    b.   Maintain the No. 2 shutoff valve of the main-line assembly in the closed position.
    c.   Open No. 2 shutoff valve of the bypass assembly.

A.5.2.1 Verify Detection of Flow through Bypass

Purpose: To test the operation of the bypass by determining if a flow of water is detected by the water meter.

Requirement: The water meter in the bypass shall detect flow when the No. 4 test cock on the main-line double check valve assembly is opened.

    a.   If test cock No. 4 is located at the same location as the bypass piping, then it will be necessary to verify that the connection between the downstream bypass piping and main-line body will allow flow. See A.5.2.2. If test cock No. 4 is not located on the bypass piping, proceed to step b.
    b.   Slowly open test cock No. 4 of the main-line assembly to create a small flow (approximately 1 to 2 gallons per minute).
    c.   Verify that the water meter in the bypass indicates flow, and record as such.
    d.   Open all shutoff valves in the DCDA.

A.5.2.2 Diagnostics: Operation of Bypass

If test cock No. 4 attaches to main-line body at the same location as the bypass piping, it will be necessary to verify that the connection between the downstream bypass piping and main-line body is open to allow flow. Close shutoff valve No. 1 on the bypass double check valve assembly, then open test cock No. 4 on the bypass assembly. Water should pass through the check valves of the mainline assembly, then the connection of the main-line body to the bypass, and flow freely from the test cock No. 4.

- If water flows freely from test cock No. 4, then the connection between the downstream bypass piping and main-line body is open to flow. Return to A.5.2.1, step b.

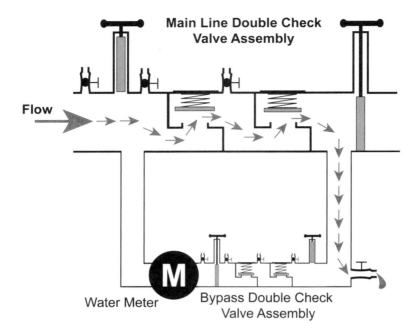

- If water does not continue to flow freely from test cock No. 4, then the connection between the bypass and the main-line body may be restricted or clogged. Repair or maintenance of the connection is necessary before an accurate test can be completed.

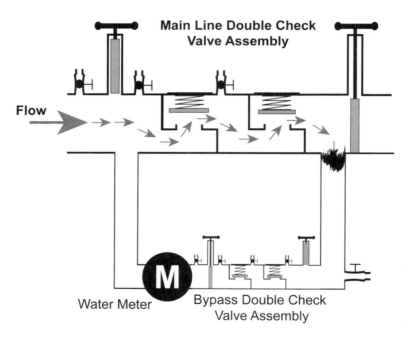

Proceed to 9.7.2, Test No. 2, step b.

A.5.3 Reduced Pressure Principle-Detector Backflow Prevention Assembly-Type II (RPDA-II)

a. Conduct following steps after completing Section 9.8.2, Test No. 2, step a.
b. Maintain the No. 2 shutoff valve of the main-line assembly in the closed position.
c. Open No. 2 shutoff valve of the bypass single check assembly.

A.5.3.1 Verify Detection of Flow through Bypass

Purpose: To test the operation of the bypass by determining if a flow of water is detected by the water meter.

Requirement: The water meter in the bypass shall detect flow when the No. 4 test cock on the main-line reduced pressure principle backflow prevention assembly is opened.

a. Slowly open test cock No. 4 of the main-line assembly to create a small flow (approximately 1 to 2 gallons per minute).
b. Verify that the water meter in the bypass indicates flow, and record as such.
c. Open all shut off valves of RPDA-II.

A.5.4 Double Check-Detector Backflow Prevention Assembly-Type II (DCDA-II)

a. Conduct following steps after completing 9.9.2, Test No. 2, step a.
b. Maintain the No. 2 shutoff valve of the main-line assembly in the closed position.
c. Open No. 2 shutoff valve of the bypass single check assembly.

A.5.4.1 Verify Detection of Flow through Bypass

Purpose: To test the operation of the bypass by determining if a flow of water is detected by the water meter.

Requirement: The water meter in the bypass shall detect flow when the No. 4 test cock on the main-line double check valve assembly is opened.

a. Slowly open test cock No. 4 of the main-line assembly to create a small flow (approximately 1 to 2 gallons per minute).
b. Verify that the water meter in the bypass indicates flow, and record as such.
c. Open all shutoff valves in the DCDA-II.

A.6 Abbreviated Field Test Procedures: Training Aid

A.6.1 Reduced Pressure Principle Backflow Prevention Assembly (RP)

A.6.1.1 RP - Five needle valve field test kit

Preliminary Steps
- Notify
- Identify
- Inspect
- Observe

Test No. 1 - Relief Valve Opening Point
1a. Bleed test cocks.
1b. Install fittings.
1c. Attach high hose to test cock No. 2.
1d. Attach low hose to test cock No. 3.
1e. Open test cock No. 3, open low bleed needle valve and leave open.
1f. Open test cock No. 2, slowly.
1g. Open high bleed needle valve.
1h. Close No. 2 shutoff valve.
1i. Close high bleed needle valve, then close low bleed needle valve. Observe reading.
- If reading drops to relief valve opening point, No. 1 check valve leaks. Go to 3b.
- If relief valve does not open, reading is apparent reading across No. 1 check.

1j. Open high control needle valve one turn and low control needle valve no more than ¼ turn.
- Record opening point of RV when first drip is detected.
- If low control needle valve must be opened more than ¼ turn, close low control needle valve and go to step T1.
- If reading drops to 0.0 and RV does not open, record as such.

1k. Close low control needle valve.

Test No. 2 - Tightness of No. 2 Check Valve
2a. Leave No. 2 shutoff valve closed.
2b. Open bypass needle valve to bleed air and close. (High control needle valve remains open.)
2c. Attach bypass hose to No. 4 test cock. Open No. 4 test cock.
2d. Open low bleed needle valve to raise reading above apparent, then close.
2e. Open bypass needle valve.
- If reading remains steady No. 2 check valve is OK. Go to 3a.
- If reading drops, but stabilizes above relief valve opening point, No. 2 check is OK. Go to 3a.
- If reading drops to relief valve opening point, bleed low to raise reading above apparent reading; then close.
  - If reading stabilized above relief valve opening point, No. 2 check is OK. Go to 3a.
  - If reading drops to relief valve opening point again, No. 2 check valve leaks. Go to 3b.
- If reading drops to 0.0 and relief valve does not open, bleed low, then close.
  - If reading stabilizes above 0.0 and relief valve does not open, No. 2 check is OK. Go to 3a.
  - If reading drops to 0.0 again and relief valve does not open, No. 2 check valve leaks. Go to 3b.
- If reading increases go to T2.

Test No. 3 - Tightness of No. 1 Check Valve
3a. Open low bleed needle valve (to raise reading above apparent reading), then close low bleed needle valve. When reading settles, record actual reading across No. 1 check valve.
3b. Close test cocks, open shutoff valve, remove equipment.

Diagnostic
T1. Attach temporary bypass hose from test cock No. 1 to test cock No. 4. Bleed air from hose and open test cocks No. 1 and No. 4.
- If reading stabilizes above relief valve opening point, re-open low control needle valve no more than 1/4 turn.
  - If reading drops, record relief valve opening point. Close low control needle valve, bleed low and close. Record No. 2 Check as OK, actual reading for No. 1 check and the No. 2 shutoff valve leaks.
  - If reading drops to 0.0 and the relief valve does not open, record as such.

Close low control needle valve, bleed low and close. Record No. 2 check as OK, actual reading for the No. 1 check and No. 2 shutoff valve leaks.
- If it is not possible to get the reading to drop, the No. 2 shutoff valve must be repaired.
- If reading drops, record relief valve opening point, bleed low and close.
  - If reading stabilizes above relief valve opening point, record No. 2 check as OK and actual reading for No. 1 check. Record No. 2 shutoff valve as leaks.
  - If reading drops to relief valve opening point again, repairs must be made. Go to 3b.

T2. Close Test Cock No. 2.
- If reading drops, there is no backpressure. Open No. 2 test cock. Go to 2e.
- If reading increases, backpressure is present. Close No. 4 test cock.
  - Record No. 2 check valve as OK.
  - Record apparent reading as actual reading for No. 1 check valve and No. 2 shutoff valve as leaks with backpressure. Go to 3b.

A.6.1.2 RP - Two needle valve field test kit

Preliminary Steps
- Notify
- Identify
- Inspect
- Observe

Test No. 1 - Relief Valve Opening Point
1a. Bleed test cocks.
1b. Install fittings.
1c. Attach high hose to test cock No. 2.
1d. Attach low hose to test cock No. 3.
1e. Attach bypass hose to low bleed needle valve. Open test cock No. 3, open low bleed needle valve and leave open.
1f. Open test cock No. 2, slowly.
1g. Open high bleed needle valve.
1h. Close No. 2 shutoff valve.
1i. Close high bleed needle valve, then close low bleed needle valve. Observe reading.
- If reading drops to relief valve opening point, No. 1 check valve leaks. Go to 3b.
- If relief valve does not open, reading is apparent reading across No. 1 check.
1j. Attach bypass hose to high bleed needle valve.
1k. Open high bleed needle valve one turn and low bleed needle valve no more than ¼ turn.
- Record opening point of RV when first drip is detected.
- If low bleed needle valve must be opened more than ¼ turn, go to step T1.
- If reading drops to 0.0 and RV does not open, record as such. Go to 3b.
1l. Close both needle valves. Detach hose from the low side bleed needle valve.

Test No. 2 - Tightness of No. 2 Check Valve
2a. Leave No. 2 shutoff valve closed.
2b. Attach bypass hose to No. 4 test cock.
2c. Open No. 4 test cock.
2d. Open low bleed needle valve to raise reading above apparent, then close.
2e. Open bypass needle valve.
- If reading remains steady No. 2 check valve is OK. Go to 3a.
- If reading drops, but stabilizes above relief valve opening point, No. 2 check is OK. Go to 3a.
- If reading drops to relief valve opening point, bleed low (to raise reading above apparent reading), then close.

- If reading stabilized above relief valve opening point, No. 2 check is OK. Go to 3a.
- If reading drops to relief valve opening point, No. 2 check valve leaks. Go to 3b.
- If reading drops to 0.0 and relief valve does not open, bleed low, then close.
  - If reading stabilizes above 0.0 and relief valve does not open, No. 2 check is OK. Go to 3a.
  - If reading drops to 0.0 again and relief valve does not open, No. 2 check valve leaks. Go to 3b.
- If reading increases go to T2.

Test No. 3 - Tightness of No. 1 Check Valve
3a. Open low bleed needle valve (to raise reading above apparent reading), then close low bleed needle valve. When reading settles, record actual reading across No. 1 check valve.
3b. Close test cocks, open shutoff valve, remove equipment.

Diagnostic
T1. Attach temporary bypass hose from test cock No. 1 to test cock No. 4. Bleed air from hose and open test cocks No. 1 and No. 4.
- If reading stabilizes above relief valve opening point, re-open low bleed no more than 1/4 turn.
  - If reading drops, record relief valve opening point. Close low bleed. Loosen low hose at No. 3 test cock, then tighten. Record No. 2 check as OK, actual reading for No. 1 check and No. 2 shutoff valve leaks.
  - If reading drops to 0.0 and relief valve does not open, record as such. Close low bleed, loosen low hose at No. 3 test cock, then tighten. Record No. 2 check as OK, actual reading for No. 1 check and No. 2 shutoff valve leaks.
  - If it is not possible to get the reading to drop, No. 2 shutoff valve must be repaired.
- If reading drops, record relief valve opening point. Loosen low hose at No. 3 test cock and then tighten.
  - If reading stabilizes above relief valve opening point, record No. 2 check valve as OK and actual reading for the No. 1 check valve. Record No. 2 shutoff valve as leaks.
  - If reading drops to relief valve opening point again, repairs must be made. Go to 3b.

T2. Close Test Cock No. 2.
- If reading drops, there is no backpressure. Open test cock No. 2. Go to 2e.
- If reading increases, backpressure is present. Close No. 4 test cock.
  - Record No. 2 check valve as OK.
  - Record apparent reading as actual reading for No. 1 check valve and No. 2 shutoff valve as leaks with backpressure. Go to 3b.

A.6.1.3 RP - Three needle valve field test kit

Preliminary Steps
- Notify
- Identify
- Inspect
- Observe

Test No. 1 - Relief Valve Opening Point
1a. Bleed test cocks.
1b. Install fittings.
1c. Attach high hose to test cock No. 2.
1d. Attach low hose to test cock No. 3.
1e. Open test cock No. 3, open bypass needle valve, then open low bleed needle valve and leave

open.
1f. Open test cock No. 2, slowly.
1g. Open high bleed needle valve.
1h. Close No. 2 shutoff valve.
1i. Close high bleed needle valve, then close low bleed needle valve. Close bypass needle valve. Observe reading.
- If reading drops to relief valve opening point, No. 1 check valve leaks. Go to 3b.
- If relief valve does not open, reading is apparent reading across No. 1 check.

1j. Open high bleed needle valve one turn and low bleed needle valve no more than ¼ turn.
- Record opening point of RV when first drip is detected.
- If low bleed needle valve must be opened more than ¼ turn, close low bleed needle valve. Go to step T1.
- If reading drops to 0.0 and RV does not open, record as such. Go to 3b.

1k. Close low bleed needle valve.

Test No. 2 - Tightness of No. 2 Check Valve
2a. Leave No. 2 shutoff valve closed.
2b. Attach bypass hose to bypass needle valve, open bypass needle valve to vent hose. Close bypass needle valve. (High bleed needle valve remains open.)
2c. Attach bypass hose to No. 4 test cock, then open No. 4 test cock.
2d. Loosen low side hose on No. 3 test cock to raise reading above apparent, then tighten hose.
2e. Open bypass needle valve.
- If reading remains steady No. 2 check valve is OK. Go to 3a.
- If reading drops, but stabilizes above relief valve opening point, No. 2 check is OK. Go to 3a.
- If reading drops to relief valve opening point, loosen low side hose on No. 3 test cock (to raise reading above apparent reading).
  - If reading stabilizes above relief valve opening point, No. 2 check is OK. Go to 3a.
  - If reading drops to relief valve opening point, No. 2 check valve leaks. Go to 3b.
- If reading drops to 0.0 and relief valve does not open, loosen low hose at No. 3 test cock, then tighten.
  - If reading stabilizes above 0.0 and relief valve does not open, No. 2 check is OK. Go to 3a.
  - If reading drops to 0.0 again and relief valve does not open, No. 2 check valve leaks. Go to 3b.
- If reading increases go to T2.

Test No. 3 - Tightness of No. 1 Check Valve
3a. Loosen low side hose on No. 3 test cock (to raise reading above apparent reading), then tighten low side hose. When reading settles, record actual reading across No. 1 check valve.
3b. Close test cocks, open shutoff valve, remove equipment.

Diagnostic
T1. Attach temporary bypass hose from test cock No. 1 to test cock No. 4. Bleed air from hose and open test cocks No. 1 and No. 4.
- If reading stabilizes above relief valve opening point, re-open low bleed no more than 1/4 turn.
  - If reading drops, record relief valve opening point. Close low bleed needle valve, loosen low hose at No. 3 test cock, then tighten. Record No. 2 check as OK, actual reading for the No. 1 check and the No. 2 shut off valve leaks.
  - If the reading drops to 0.0 and the relief valve does not open, record as such. Close low bleed needle valve, loosen low hose at No. 3 test cock, then tighten. Record no. 2 check as OK, actual reading for No. 1 check valve and No. 2 shutoff valve as leaks.
  - If it is not possible to get the reading to drop, the No. 2 shutoff valve must be

repaired.
- If reading drops, record relief valve opening point. Loosen low hose at No. 3 test cock, then tighten.
  - If reading stabilizes above relief valve opening point, record No. 2 check as OK and actual reading for the No. 1 check. Record No. 2 shutoff valve as leaks.
  - If reading drops to relief valve opening point again, repairs must be made. Go to 3b.

T2. Close Test Cock No. 2.
- If reading drops, there is no backpressure. Open test cock No. 2. Go to 2e.
- If reading increases, backpressure is present. Close No. 4 test cock.
  - Record No. 2 check valve as OK.
  - Record apparent reading as actual reading for No. 1 check valve and No. 2 shutoff valve leaks with backpressure. Go to 3b.

A.6.2 Double Check Valve Backflow Prevention Assembly (DC)

Preliminary Steps
- Notify
- Identify
- Inspect
- Observe

Test No. 1 - Tightness of No. 1 Check Valve
1a. Bleed test cocks.
1b. Install fittings and bleed valve arrangement.
1c. Attach sight tube to No. 3 test cock.
1d. Attach bleed valve arrangement and high side hose to No. 2 test cock.
1e. Bleed air from gage and fill tube.
1f. Close No. 2 shutoff valve, locate gage at proper elevation, then close No. 1 shutoff valve.
1g. Open test cock No. 3.
- If water level and reading are stable record reading and go to step 1h.
- If water continues to flow from tube go to step T1.
- If water recedes in the tube, go to step T4.

1h. Close test cocks, open No. 1 shutoff valve, and remove equipment.

Test No. 2 - Tightness of No. 2 Check Valve
2a. Attach bleed valve arrangement and high side hose to No. 3 test cock and attach sight tube to No. 4 test cock.
2b. Bleed air from gage and fill tube.
2c. Close shutoff valve No. 1.
2d. Open test cock No. 4.
- If water level and reading are stable, record reading and go to step 2e.
- If water continues to flow from tube go to step T6.
- If water recedes go to step T5.

2e. Close test cocks, remove equipment.
2f. Remove fittings, open shutoff valves.

Diagnostic
T1. Observe gage reading, open bleed valve.
- If water flows from the bleed valve but it can be adjusted so there is a drip at test cock No. 3, go to T2.
- If not possible to adjust bleed valve. Repair No. 1 shutoff. If that doesn't work repair

both check valves and No. 2 shutoff.
- If water stops flowing from bleed valve go to T3.

T2. Adjust bleed valve to drip at No. 3 test cock record the reading. Go to step 1h (less than 1.0 psi, the No. 1 check valve must be repaired before proceeding).

T3. Record observed reading from T1, No. 2 check valve leaking. No. 2 shutoff valve is leaking with backpressure. Go to 2e.

T4. Lower gage to centerline, record reading. No. 2 check valve is leaking and No. 2 shutoff valve is leaking. Go to 2e.

T5. Lower gage to centerline, record reading for No. 2 check valve. Record No. 2 shutoff valve is leaking. Go to 2e.

T6. Observe reading. Open bleed valve.
- If water stops flowing from bleed valve go to T7.
- If bleed valve can be adjusted go to T8.
- If bleed valve cannot be adjusted (water flows from test cocks 3 and 4), go to T9.

T7. Record observed reading from T6. No. 2 shutoff valve leaks with backpressure go to 2e.

T8. Record reading for No. 2 check valve, and No. 1 shutoff valve leaking. Go to 2e.

T9. If No. 1 check valve had a reading of less than 1.0 psi, repair No. 1 check and go to 1a, otherwise go to T10.

T10. Close bleed valve, open No. 2 test cock, open No. 4 test cock. Record reading for No. 2 check valve. Go to 2e.

A.6.3 Pressure Vacuum Breaker Backsiphonage Prevention Assembly (PVB)

Preliminary Steps
- Notify
- Identify
- Inspect
- Observe

Test No. 1 - Air Inlet Valve Opening Point
1a. Remove canopy.
1b. Flush test cocks.
1c. Install fittings and bleed valve arrangement.
1d. Attach high hose to No. 2 test cock. Open No. 2 test cock.
1e. Open high bleed needle valve to bleed air. Close high bleed needle valve.
1f. Close No. 2 shutoff valve, locate gage at proper elevation, and then close No. 1 shutoff valve.
1g. Open high bleed needle valve no more than ¼ turn.
- If air inlet valve opens, record opening point. Go to 1h.
- If air inlet reading drops to 0.0 and does not open, record as such. Close No. 2 test cock. Go to 1j.
- If reading will not drop to air inlet valve opening point, go to T1.
- If reading falls before opening bleed needle valve, record opening point. Go to 1h.

1h. Close high side bleed needle valve, remove hose from test cock No. 2. Record that air inlet valve fully opens.
1i. Close No. 2 test cock.
1j. Open No. 1 shutoff valve.

Test No. 2 - Check Valve Closing Point
2a. Attach high side hose to bleed valve arrangement on test cock No. 1. Open test cock No. 1.
2b. Open and close high bleed needle valve to bleed air from gage.

2c. Locate gage at proper elevation. Close No. 1 shutoff valve.
2d. Open test cock No. 2.
- If water flow stops (or drip) and readng stabilizes, record check valve reading. Go to 2e.
- If water does not stop flowing, go to step T2.

2e. Close test cocks.
2f. Remove equipment and fittings.
2g. Open No. 1 shutoff valve, then open No. 2 shutoff valve.
2h. Replace canopy.

Diagnostics

T1. Close high bleed needle valve. Open the No. 1 test cock to divert water leaking through No. 1 shutoff valve. Once the leak is diverted continue with step 1g.
T2. Open bleed valve arrangement until there is a drip at the No. 2 test cock.
- If the leak can be compensated, record reading on gage for the check valve. Go to 2e.
- If leak cannot be compensated, repair No. 1 shutoff valve and retest.

A.6.4 Spill Resistant Pressure Vacuum Breaker Backsiphonage Prevention Assembly (SVB)

Preliminary Steps
- Notify
- Identify
- Inspect
- Observe

Test No. 1 - Check Valve Closing Point

1a. Remove canopy.
1b. Flush test cock and vent valve.
1c. Install fitting and bleed valve arrangement.
1d. Attach high hose to bleed valve arrangement.
- Open test cock.
- Bleed air through high bleed needle valve.
- Close high bleed needle valve.

1e. Close No. 2 shutoff valve, locate gage at proper elevation, close No. 1 shutoff valve.
- If reading does not decrease, go to 1f.
- If reading decreases, and air inlet does not open, go to 1f.
- If reading decreases, and air inlet opens, go to T1.

1f. Open vent valve.
- If air inlet does not open, go to 1g.
- If air inlet opens, go to T2.

1g.
- If water stops draining through vent valve and gage stabilizes, record reading for check valve. Go to 2a.
- If water continues to flow through vent valve, go to T3.

Test No. 2 - Air Inlet Valve Opening Point

2a. Maintain gage at same level as vent valve.
2b. Open high side bleed needle valve no more than ¼ turn.
- If air inlet opens, record reading on gage. Go to 2c.
- If reading drops to 0.0 and air inlet does not open. Record as such. Go to 2d

2c. Close high side needle valve, and remove high side hose. Record that air inlet valve fully opens.
2d. Close test cock and vent valve.

2e. Remove test equipment and fittings.
2f. Open No. 1 shutoff valve, then No. 2 shutoff valve.
2g. Replace air inlet valve canopy.

Diagnostics
T1. Record reading when air inlet valve opens. Open vent valve, when water stops draining through vent valve and gage stabilizes, record reading on gage for check valve. Remove high side hose and record that air inlet valve fully opens. Go to 2d.
T2. Record reading on gage for air inlet valve opening point.
- If reading settles and water stops draining through the vent valve, record reading for check valve.
- If reading settles and water continues to flow through the vent valve, open bleed valve arrangement until there is a drip from the vent valve. Record reading for check valve. Remove high side hose, record that air inlet opens fully. Go to 2d.

T3. If water continues to flow from vent valve, record No. 1 shutoff valve as leaking. Open bleed valve arrangement.
- If water from vent is drip and reading stabilizes, record reading for check valve. Go to 2b.
- If it is not possible to reduce vent to drip, record that No. 1 shutoff valve is leaking too much. Go to 2d.

A.7 Guideline for the Periodic Test of the Different Pressure Field Test Kit

A properly calibrated field test kit is essential to ensure accurate data acquisition. Field test kits should be checked for accuracy at least once a year, and re-calibrated when inaccuracy is greater than ±0.2 psid. Local administrative authorities should be consulted regarding possible local accuracy verification requirements.

The following guidelines are recommended for the periodic inspection/test of the field test kit. Tests should be performed at ambient temperature. All gage hoses should be attached in their respective normal operating position, and all needle valves should be in the closed position unless otherwise noted.

A7.1 Accuracy verification

Field test kits should maintain an accuracy of ± 0.2 psid for decreasing differential pressure readings at each of the test points using either water or air/gas as the test medium. The field test kit should be verified against a reference source that has a calibration traceable to the National Institute of Standards and Technology (NIST). The verification reference source should have a maximum permissible error of ± 0.05 psig.

With the field test kit maintained in the vertical orientation, attach the verification reference source and the high side hose of the field test kit to a common pressure source. The low side of the field test kit should be maintained at atmospheric pressure. Increase the common pressure source to approximately 15 psig. Decrease the common pressure source until the verification reference source reads 12.0 psid. Record the field test kit reading. Repeat at values of 8.0 psid, 5.0 psid, 2.0 psid and 1.0 psid. Exceeding the accuracy of ± 0.2 psid should be cause for rejection.

A.7.2 Leakage Test

With both the high and low side hoses connected to a common water supply, flush all air from the field test kit by opening all needle valves. Close all needle valves. Increase the common water supply pressure to the maximum working water pressure (MWWP) of the field test kit, or a minimum of 175 psi, for a period of ten minutes. Evidence of leakage from the field test kit should be cause for rejection.

Appendix A

Appendix B: Evaluation Documents and Procedures

Inside Appendix B

Evaluation Forms 505
    Laboratory Submittal Form 505
    Field Site Application Form 508
    Letter of Acknowledgement for Field Evaluation Sites 512

Field Evaluation Phase of Approval Program 513
    3 PSI Buffer 513
    Field Test Procedure 513

B.1 Evaluation Forms

B.1.1 Laboratory Submittal Form

The form below is included on the accompanying CD-ROM and is available on the Foundation's website. For the most up to date version please visit the Foundation's website section entitled 'Approval Process.'

Disclaimer: Please be aware that the Foundation for Cross-Connection Control and Hydraulic Research at the University of Southern California DOES NOT GUARANTEE THAT A SUBMITTED ASSEMBLY(S) WILL BE GRANTED APPROVAL, as such assembly(s) may not comply with the required standard.

USC

Request for Evaluation
(Revised April 2011 - Page 1 of 3)
(Please use one sheet per model, size, and type)

Date : _____

Company name: _____ Project Contact Person: _____

Address: _____

Phone: (____) _____ Ext: _____ Fax: (____) _____

Email: _____ Web page address: _____

Submittal: ☐ Initial ☐ Re-submittal ☐ Production Review

Assembly submitted: Make _____ Model _____ Size _____
Type: ☐ DC ☐ RP ☐ AVB ☐ RPDA ☐ RPDA-II
       ☐ SVB ☐ PVB ☐ DCDA ☐ DCDA-II ☐ Other _____

Shutoff valves: _____ (see shutoff valves on List of Approved Backflow Prevention Assemblies)
If other shutoff valve please note make/model _____

Evaluate per the follow agency(s)* and their respective standard(s)*:
☐ USC ☐ CSA ☐ ASSE ☐ IAPMO
☐ UL   ☐ FM   ☐ AWWA ☐ Other _____

(Attach additional testing instructions if necessary)
* See page 2 of 3 for Information regarding Submittals for Multiple Standards - select ☐ Option #1
☐ Option #2

In the following orientation(s): *(See detailing of orientations - page 3 of 3)*
☐ H ☐ VU ☐ VD ☐ VUVD ☐ VDVU ☐ VDVD ☐ VUVU ☐ VUH ☐ HVD ☐ Other _____

For 2"⌀ and smaller assemblies we are submitting (check one):
☐ Three assemblies of each model and size per assembly type.
☐ A common body design; three assemblies of the largest size and one of the smaller size(s) for each model and size per assembly type.

For 2 1/2"⌀ and larger assemblies we are submitting (check one):
☐ One assembly of each model and size per assembly type, and an additional set of replacement parts for RP (1st and 2nd check valve seats and discs).
☐ A common body design with the internal components for each type, and an additional set of replacement parts for RP (1st and 2nd check valve seats and discs).

We are submitting a complete set of:

Enclosed	Previously Submitted	
☐	☐	Dimensioned drawings for the assembly and each of the components
☐	☐	Material specifications for each of the components

For Re-submittals:

☐	☐	Dimensioned drawings for each of the modifications/revisions
☐	☐	Material specifications for each of the modifications/revisions

Must be submitted before completion of Laboratory Evaluation:

☐	☐	Material non-toxicity certificates and documents
☐	☐	Installation instructions, repair and testing/maintenance instructions
☐	☐	Engineering specification sheets

[*Please enclose a copy of this Request for Evaluation form with each submittal*]

Foundation for Cross-Connection Control and Hydraulic Research
323 662 3536 | fcchrlab@usc.edu | fcchr.usc.edu

Appendix B

Request for Evaluation
(Revised April 2011 - Page 2 of 3)

Submittal for Multiple Standards

A "Request for Evaluation" form is required with each individual submittal for evaluation. A submittal may include a request to evaluate a backflow prevention assembly to multiple standards. For example, an assembly submitted may be evaluated not only under the Foundation's standard in the **Manual of Cross-Connection Control** - *10th edition*, but also to the latest version of standards from other agencies (i.e., American Water Works Association - AWWA, American Society of Sanitary Engineering - ASSE, Canadian Standards Association - CSA, International Association of Plumbing and Mechanical Officials - IAPMO, Underwriters Laboratory - UL or Factory Mutual - FM.)

Previous editions of some standards did not specify a particular order of the laboratory tests to be evaluated on the assembly, and as such, the laboratory test results of one standard might simultaneously be applicable for another standard(s). However, since many of the standards are now currently specifying the order of the laboratory tests, the assembly must be separately evaluated to each individual standard. Because of this, if only a single assembly for the 2 1/2"ø and larger assemblies is submitted for evaluation to multiple standards, the assembly that is being evaluated may be subjected to an excessive amount of laboratory tests. Subjecting the assembly to an excessive amount of tests may result in unacceptable test results (i.e., failures). These unacceptable test results may be due to the excessive amount of tests, and not the design of the assembly.

Please select one of the following options on the
Request for Evaluation Form:

Option #1. One (1) assembly for each make, model, and size may be submitted and evaluated to the multiple standards, or

Option #2. Multiple assemblies for each make, model and size may be submitted so that an individual assembly is evaluated to a particular standard.

Contact the Foundation Staff for any additional information.

Foundation Laboratory
3022 Riverside Drive
Los Angeles, CA 90039
EMAIL: fccchrlab@usc.edu
PHONE: 323-662-3536
FAX: 323-665-2055

Foundation for Cross-Connection Control and Hydraulic Research
323 662 3536 | fccchrlab@usc.edu | fccchr.usc.edu

Request for Evaluation
(Revised April 2011 - Page 3 of 3)

Backflow Prevention Assembly Orientations

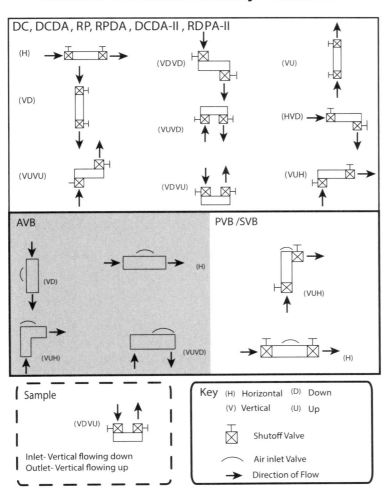

Foundation for Cross-Connection Control and Hydraulic Research
323 662 3536 | fccchrlab@usc.edu | fccchr.usc.edu

B.1.2 Field Site Application Form

The form below is included on the accompanying CD-ROM and is available on the Foundation's website. For the most up to date version please visit the Foundation's website section entitled 'Approval Process.'

Disclaimer: Please be aware that the Foundation for Cross-Connection Control and Hydraulic Research at the University of Southern California DOES NOT GUARANTEE THAT A SUBMITTED ASSEMBLY(S) WILL BE GRANTED APPROVAL, as such assembly(s) may not comply with the required standard.

USC   **Field Site Application Form**
(Revised April 2013- Page 1 of 4)

Instructions: Fill in form completely and attach letters of acknowledgment. Send application to:
    Field Evaluation Coordinator
    FCCC&HR Laboratory
    3022 Riverside Drive
    Los Angeles, CA 90039

    PHONE (323) 662-3536 or (213) 740-2032   FAX (323) 665-2055   EMAIL: fccchrlab@usc.edu

I. Field Test Assembly

Manufacturer _____

Model _____ (circle) RP RPDA DC DCDA PVB SVB DCDA-II RPDA-II

Orientation: (See detailing of orientations - page 4 of 4)

☐ H ☐ VU ☐ VD ☐ VUVD ☐ VDVU ☐ VDVD ☐ VUVU ☐ VUH ☐ HVD ☐ Other_____

Size _____ Serial Number (if known) _____

II. Field Site Information

Field site location (i.e., name of company, park, etc.):_____

Field site Address: _____
_____

Nearest Cross Street _____ GPS Coordinates _____

Map: Thomas Bros. Guide (if known): _____ (If necessary, provide sketch on separate page)

On site contact person _____ Phone number _____

Owner Name (owner of property or manager):_____

Owner Address: _____
_____

Phone number: _____

Foundation for Cross-Connection Control and Hydraulic Research
323 662 3536 | fccchrlab@usc.edu | fccchr.usc.edu

Field Site Application Form
(Revised April 2013- Page 2 of 4)

III. Type of Installation

Description of usage downstream of proposed field site (i.e., park irrigation; feeds into pump that irrigates school playground; supplies domestic water for 300 apartments; etc.):

_____

_____

_____

_____

if necessary, attach additional drawing/sketch

Field Site Flowing Condition: ☐ Static
　　　　　　　　　　　　　　　　☐ Flowing
　　　　　　　　　　　　　　　　　　　☐ * Designated high flowing [50-100%] of rated flow field site.
　　　　　　　　　　　　　　　　　　　　　(flow documentation required)
　　　　　　　　　　　　　　　　　　　☐ Flow Documentation Included

* Per Section 10.1.2.1.3 of the 10th Edition Manual; a minimum of one of the field evaluation site shall have flow rates reaching the range of 50-100% of rated flow.

Administrative Authority/Site Info:

Type of protection: 　☐ Service/Meter　Meter#_____　Service /Accnt. # _____
　　　　　　　　　　☐ Internal　　　　　　　　　　　　　Device ID # _____

Are there any hazardous materials or chemicals in use downstream from proposed field site? (i.e., chemical injection pump, aspirator, etc.)

　☐ Yes ☐ No

　if yes, please describe

　_____

　_____

　_____

Foundation for Cross-Connection Control and Hydraulic Research
323 662 3536 | fcchrlab@usc.edu | fcchr.usc.edu

Appendix B

Field Site Application Form
(Revised April 2013- Page 3 of 4)

IV. Administrative Authority (Water and/or Health Agency) having Cognizance

Name of agency: _____

Address: _____

_____

Phone number: _____ EMAIL _____

Contact person: _____ Title _____

V. Letters of Acknowledgment (See Sample Letter B.1.3 - revised December 2009)

A. The owner or manager of the property: (attach letter)

B. The water and/or health agency having cognizance: (attach letter)

VI. Applicant Information
To the best of our knowledge, the above proposed field test site location is in compliance with Section 10.1.2.1.3 "Selection of Field Locations," in the Manual of Cross-Connection Control, 10th Edition.

> **NOTE:** This completed application must be submitted, and the field site deemed acceptable, prior to the installation of the subject backflow prevention assembly.

Manufacturer representative (print name): _____

Title _____ Email _____

Phone number _____ FAX _____

Signature: _____ Date _____

— —

FOR OFFICE USE ONLY

ACCEPTED ☐
REJECTED ☐
COMMENTS: _____
_____
_____
_____
_____
_____
_____

EVALUATED BY: _____ DATE: _____

Foundation for Cross-Connection Control and Hydraulic Research
323 662 3536 | fcchrlab@usc.edu | fcchr.usc.edu

Appendix B

Field Site Application Form
(Revised April 2013- Page 4 of 4)

Backflow Prevention Assembly Orientations

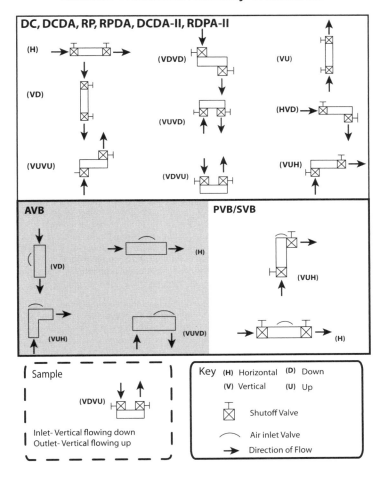

Appendix B

Foundation for Cross-Connection Control and Hydraulic Research
323 662 3536 | fccchrlab@usc.edu | fccchr.usc.edu

B.1.3 Letter of Acknowledgement for Field Evaluation Sites

The form below is included on the accompanying CD-ROM and is available on the Foundation's website. For the most up to date version please visit the Foundation's website section entitled 'Approval Process.'

Letter of Acknowledgment for Field Evaluation Sites
(Revised 13 April 2011)

To be submitted on company letterhead

_____ (date)

Field Evaluation Coordinator
Foundation for Cross-Connection Control
      and Hydraulic Research Laboratory
3022 Riverside Drive
Los Angeles, CA 90039

re: Field Test Site

_____ (manufacturer)

_____ (model and size)

It is the understanding of _____ (property owner, water purveyor, or health agency) that the subject test assembly(s) to be located at _____ (location address) will be in the field for at least twelve (12) months[1] and during that time the assembly(s) will be tested on a nominal thirty (30) day schedule by the Foundation for Cross-Connection Control and Hydraulic Research at the University of Southern California. A normal test will usually require a shutdown of water from five to ten minutes, however during these tests the water service may be turned off for an hour or more for inspection or maintenance. During the period of the Field Evaluation, only the Foundation Staff is permitted to test, inspect, or repair the assembly(s).

The subject test assembly(s) location shall be reviewed and deemed acceptable by the Foundation prior to installation. The Foundation Staff will only perform the initial field test after the assembly has successfully completed the Laboratory Evaluation. The manufacturer of the assembly(s) will be responsible for any required water/health agency field testing and maintenance of the assembly(s) installed prior to the initial field test.

At the successful conclusion of the Field Evaluation phase of the Foundation Approval program, the assembly(s) will gain Approval and become the property of _____ (the owner). Should the subject assembly(s) not complete the Field Evaluation and not gain Foundation Approval, then the manufacturer of the subject assembly(s) will be responsible for replacing the field test assembly(s) with an Approved assembly(s).

Disclaimer: Please be aware that the Foundation for Cross-Connection Control and Hydraulic Research at the University of Southern California DOES NOT GUARANTEE THAT A SUBMITTED ASSEMBLY(S) WILL BE GRANTED APPROVAL, as such assembly(s) may not comply with the required standard.

Sincerely,

_____ (signature)

_____ (printed name)

_____ (title)

cc.   (manufacturer)
       (property owner)
       (water purveyor)
       (health agency)

---

[1] Note: Field Evaluation does not start until the third assembly is installed and initial tested by the Foundation staff. If warranted, the Field Evaluation may be extended beyond the twelve (12) months.

B.2 Field Test Procedures: Field Evaluation Phase of Approval Program

B.2.1 Reduced Pressure Principle Assembly: Check Valve No. 1

In addition to the Field Evaluation protocol that is detailed in 10.1.2.1.4, the following field test data shall be required. During the initial, monthly and concluding field tests of the reduced pressure principle backflow prevention assembly, the No. 1 check valve differential pressure reading (9.2.1.2, Test No. 3, step a) shall be at least 3.0 psid greater than the differential pressure relief valve opening point.

B.2.2 Reduced Pressure Principle Assembly: Prevent backpressure

In addition to the Field Evaluation protocol that is detailed in 10.1.2.1.4, the following field test steps shall be conducted. These additional steps will prevent the No. 2 check valve from being backpressured during the closure of the No. 2 shutoff valve. After completing Test No. 1, step g of 9.2.1.2, the following steps shall be performed:

a. Maintain high side and low side bleed need valves in the open position
b. Open test cock No. 4
c. Close No. 2 shutoff valve
d. Close high side bleed needle valve
e. Close low side bleed needle valve
f. Close test cock No. 4
g. Proceed to Test No. 1, step j

Appendix B

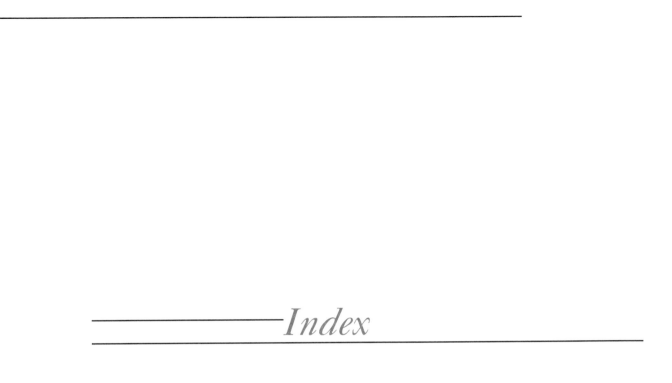

Index

Index

A

ABC's of a Cross-Connection Control Program 88
Absolute Pressure 2, 39
Accessibility of Internal Parts 334
Accessible 2
Administration 96
Administrative Authority 2, 88
Agricultural Chemicals 124, 126
Aircraft Manufacturing 121
Air Gap 2, 48
Air Gap, Periodic Test and Maintenance Report 161, 180
Air Inlet 75
Air Release 337
Air Scrubber 131-132
Alback, Walter H. 28
Alterton, E. D. 26
American Society of Sanitary Engineering 90
American Water Works Association 90
American Water Works Association Research Foundation 92, 94
Amoebic Dysentery 24
Amusement Parks 127
Approval 346-348, 410
Approvals/Listings 90
Approved Backflow Prevention Assemblies 3, 162, 191, 199
Aquaria 127
Aspirator 3
Aspirator Effect 3
Aspirators 131, 133
ASSE See American Society of Sanitary Engineering
Assembled Assemblies 333
Atmospheric Pressure 3, 40
Atmospheric Vacuum Breaker 4, 21, 54, 376
Authority 89
Autoclaves 134
Automatic Hood Wash Systems 123
Automotive Plants 121
Autopsy Table 135
Auxiliary Water Supply 4
Auxiliary Water Systems 96
AVB See Atmospheric Vacuum Breaker
AWWA See American Water Works Association
AwwaRF See American Water Works Association Research Foundation

B

Backflow 4, 38
Backflow Incident Report 197
Backflow Incidents 417, 418
Backflow Preventer See Backflow Prevention Assembly
Backflow Prevention Assembly 5, 89
Backflow Prevention Assembly Tester 5
Backflow Prevention Assembly Tester Certificate Applicaton 185
Backflow Protection 21
Backpressure 5, 46
Backsiphonage 5, 43
Backsiphonage Condition 72
Bakery 123
Battery Fill Station 123
Beauty Salon Sink 127
Bed Pan Washing Stations 124
Beverage Manufacturing 123
Biegler Hall 25
Bimini Baths 22
Bimini Water Company 22
Blanching 124-125, 128
Bluing 121
Bodies and Covers 340
Body and Bonnet 336
Boiler 47, 114, 120-124. 123-127, 136, 145, 158
Bottling 123, 137
Brewery 123
Bushings 336, 341
Bypass Arrangement 6, 51, 52, 53, 54

C

Cadmium 122, 342
Can and Bottle Washing Machines 124-125, 128, 137
Cannery 123
Carbonated Beverage Dispensers See Carbonators
Carbonators 139
Carlton, Frank 23
Car Washes 120, 138
Car Washing Machine 138
Caustic Solutions 122
Certification 92, 93, 97, 102, 103
Certified Cross-Connection Control Program Specialists 93
Certified Testers and Specialists 92
Change of Design, Materials or Operation 347, 411
Change of Occupancy or Use 97
Chatter 337
Check Valve 21, 49, 341
Chemical Dispenser 140

Chemical Plants 121
Chemicals 45, 115
Chlorine 123
Chromates 120-124, 126,-127
Chrome 122
Circulating Systems 120-121, 123-124, 126-127
Civil Works 126
Clappers 21
Classified 126
Classified Backflow Preventers 30
Clearance 336
Cloran, Everett 26, 30
Column of Water 5
Combined Service 98
Common-Cup 24
Community Water System 32
Compliance 115
Consumer 5
Containment 6, 93, 95
Contaminant 6, 24, 47, 53, 115
Control Piping 337, 342
Convalescent Hospitals 124
Cookers 141
Coolants 122-123, 125, 127
Cooling Tower 120-124, 126-128, 142
Copper 32, 120-127, 139, 340-343, 410
Corrosion 340
Counterbalances 341
Critical Level 6
Critical Services 6, 98
Cross-Connection 6
Cross-Connection Control Programs 88, 110
Cyanides 122

D

Dairy 123
DC See Double Check Valve Backflow Precention Assembly
DCDA See Double Check Detector Backflow Prevention Assembly
DCDA-II See Double Check Detector Backflow Prevention Assembly-Type II
Degree of Hazard 5-6, 47, 120
Dehydration Tanks 122, 124
Deionized Water Systems 125
Dental Offices 124, 143
Dental Vacuum Pump 143
Design of Waterway 335
Design Requirements for Backflow Preventers 334, 408
Design Requirements for Field Test Kits 408-409
Diaphragms 337, 341
Differential Pressure 52, 53, 65, 69, 70

Differential Pressure Relief Valve 335, 350
Direct Cross-Connection 6, 46, 47, 48, 57, 115
Dishwashing Equipment 127
Dissimilar Metals 340
Distribution 20
Documentation 100, 115
Double Check Detector Backflow Prevention Assembly 5, 7, 31, 50-52, 204, 320
Double Check Detector Backflow Prevention Assembly-Type II 5, 8, 50-52, 204, 324
Double Check Valve Backflow Prevention Assembly 5-6, 23, 50, 51, 59, 62, 64, 204, 277, 361
Downstream 46
Drawings and Requirements 343
Dual Services 98
Dyeing 121-122
Dye vats 121

E

E.C. #3 23
EC Valve Company 23
Education and Training 94
Elastomer 49
Elastomer Discs 341
Entriken, O.K. 23
Environmental Protection Agency 32
EPA See Environmental Protection Agency
Equipment 98
Evaluation of Design and Performance 343, 409
Existing Assemblies 98
Existing Systems vs. New Construction 111

F

FACA See Federal Advisory Committee Act
Factory Mutual 91
Federal Advisory Committee Act 33
Field Evaluation 31, 344
Field Locations 344
Field Survey Form 196
Film Laboratories 120
Film Processing 121, 125, 127-128
Firefighters 20
Fire Fighting Systems 122-123, 125, 127
Fireplug 20
Fire Sprinkler Systems 96, 98-99, 144
Fixtures 21
Flumes 124-125, 128
Food Processing 120, 123
Foot Baths 127
Foundation for Cross-Connection Control and Hydraulic Research 28, 90-91
Fountains 127

Frozen Foods Processing Plant 123

G

Gage Pressure 8, 39
Garbage Disposals 123, 127
Gate Valve 20, 114, 338
Glycerine 122-123, 125, 127
Grey Water 8
Guide Stems 341

H

Hayes, Harry 26
Hazard 25
Health Agency 8
Health Hazard 8, 47, 50, 57-58, 110
Heat Exchangers 145
Heavy Metals 122
History of Cross-Connection Control 20
Hospitals 98, 124
Hotels 127
Hydrant 44, 99
Hydraulically-operated equipment 121-122, 125, 127
Hydraulics 38
Hydraulics of the Backflow Prevention Assemblies 59-78
Hydrostatic Test 333, 339, 412

I

IAPMO See International Association of Plumbing and Mechanical Officials
Incidents 24, 26, 33, 99, 417
Indirect cross-connection 6, 46, 115
Industrial Fluids 8
Industrial Fluid Systems 122-123, 125, 127
Industrial Piping System 9
Inline Repairability 31
Inspection 21, 111-115
Installation and Maintenance Requirements 164
Installation Guidelines 92, 164-174
Interdependence of Components 334, 351, 369, 391
Internal Cross-Connections 110
Internal Plumbing 24-25
Internal Protection 9, 90, 93, 95, 110, 199
International Association of Plumbing and Mechanical Officials 86, 90, 495
Irrigation Systems 100-102, 146
Isolation See Internal Protection

J

Janitorial Sinks 124, 127
J. J. Lee. See Lee, J. J.

K

Kidney Dialysis Machine 147

L

Lab Faucets 124
Laboratory 24-29, 31, 90-91, 110 121-128
Laboratory Equipment 131, 148
Laboratory Evaluation of Backflow Prevention Assemblies 343-409
Laboratory Evaluation of Field Test Kits 410-413
Laboratory Sinks 121-128
LADWP See Los Angeles Department of Water and Power
Laundry Facilities 120, 149
Laundry Machines 121, 128, 149
Lavatories 121-128
Lee, J. J. 30
Lethal Hazard 47, 48
Linkage 21, 334
List of Approved Backflow Prevention Assemblies 89-91
List of Certified Backflow Prevention Assembly Testers 97, 187
Los Angeles Department of Water and Power 23, 25-26, 29
Low Water Pressure 102
Lubricants 229, 252, 276, 289, 317, 343
Lubrication Lines 123
Lubrication of Components 338

M

Manifold Assembly 9, 174, 339-340
Manufacturing 121
Markings 332
Material Requirements 340
Material Safety Data Sheets 115
Material Selection 338
Maximum Allowable Pressure Loss 330-331, 367
Maximum Contaminant Levels 24
Maximum Working Water Pressure 349, 351-353
MCL See Maximum Contaminant Level
Medical Facilities 124
Medical Laboratories 124
Medical Offices 124
Metallic Glucosides 120-127
Metal Works Facilities 121
Meter 13, 32, 51-54, 110, 112-113
Methods of Backflow Prevention 112
Minimum Requirements for Backflow Prevention Assembly Tester Certification Program 193
Minimum Requirements for Cross-Connection Control Specialist Certification Programs 195

Model Ordinance 189
Mortuary Equipment 124, 125, 128, 135
Mortuary Tables 135
Motels 127
Motion Picture Studios 127
MSDS See Material Safety Data Sheets
Multiple Services 98, 103

N
NAICS See North American Industry Classification System
Nameplate 332, 408
Negative Pressure 9, 40, 43
New York 91
Nickel 122
Non-Community Water System 32
Non-Compliance 103
Non-Compliance of Notice to Install Backflow Prevention Assembly 177
Non-Compliance of Periodic Test and Maintenance 178
Non-Health Hazard 6, 9, 47-50, 57-58
Non-Pressure Type Vacuum Breaker See Atmospheric Vacuum Breaker Vacksiphonage Prevention Assembly (AVB)
Non-Transient Non-Community Water System 32
North American Industry Classification System 114
Notice of Appointment of Water Supervisor 182
Notice of Discontinuance of Water Supply 179
Notice of Shutdown 188
Notification 111, 205

O
Oil/gas refineries 121
Ordinance 89, 96, 189

P
Packing House 123
Paper No. 5 27
Paraffin 122-127
Parallel Installation 9
Parallel Installation of Backflow Prevention Assemblies Guidelines 173
Pasturization Units 123
Patent 23
Pentachlorophenol 120-127
Periodic Test and Maintenance Report with Report Form 175
Period of Field Evaluation 345
Photographic Film Processing Machines 150
Pipette Washers 124

Piping Symbols 113-114
Plan Check 113
Plating Plants 121
Plating Solution Filtering Equipment 122
Plumbing Code 86, 101, 103
Plumbing Hazard 9
Point of Delivery 9
Policies and Procedures 95-105
Policy Regarding Design 334
Pollutant 6, 9, 47-50, 57-58
Pollution See Pollutant
Pools 124, 126, 127
Portable Cleaning Equipment 131, 151
Potable Water 9
Pounds Per Square Inch 9, 38-39
Power Plants 121
Preparing for a Cross-Connection Control Survey 111
Pressure 38-81
Pressure Cookers 124, 125, 128
Pressure Fluctuation 9
Pressure Gradient 10
Pressure Vacuum Breaker Backsiphonage Prevention Assembly 10, 37, 55, 74, 78, 81
Pressure Washers 123
Primacy 32
Prior Approval 347
Protective Coatings 341
PSI See pounds per square inch
Public Health Service Act 24
Public Potable Water System 10
Pulp, Bleaching, Dyeing and Processing Facilities 122
Pumps 20, 113, 121-128, 152
PVB See Pressure Vacuum Breaker Backsiphonage Prevention Assembly

R
Rated Flow 330-331, 351, 367
Readily Accessible 10
Reclaimed Water 10
Records 93
Recreational Vehicle Sewage Dump Station 127
Recycled Water 11
Recycled Water Systems 103, 124-128, 153
Reduced Pressure Principle Backflow Prevention Assembly 11, 24, , 65-73, 210-276, 349
Reduced Pressure Principle Detector Assembly 12, 31, 53
Reduced Pressure Principle Detector Assem-

bly-Type II 12, 53
Regulations 24, 86, 89
Relief Valve 11, 23, 52-53, 65-73
Renewal of Approval 347, 410
Repair Tools 340
Replacement Parts 348
Restaurants 123
Reservoirs 120-127
Resilient Seated 333, 336
Resolution Relative to Backflow Prevention Assembly Testers 183
Responsibility 86-87
Responsibility of the Certified Backflow Prevention Assembly Tester 88
Responsibility of the Consumer 87
Responsibility of the Health Agency 86
Responsibility of the Plumbing Official 87
Responsibility of the Repair and Maintenance Technician 88
Responsibility of the Water Purveyor 87
Restricted or Classified Services 104
Retention and mixing tanks 121-122, 124
Retorts 124-125, 128
Reused Water 13
Reynolds, Kenneth C. 27, 30
Romans 20
RP See Reduced Pressure Principle Backflow Prevention Assembly
RPDA See Reduced Pressure Principle Detector Assembly
RPDA-II See Reduced Pressure Principle Detector Assembly-Type II
Rubber Manufacturing Facilities 121

S

Safe Drinking Water Act 32
Sand and Gravel Washing Equipment 123
Sand/Gravel Facilities 121
Sanitary Sewer 13
Saunas 127
Schools and Colleges 127
Scope Washers 124
SCWUA See Southern California Water Utilities Association
SDWA See Safe Drinking Water Act
Seat Rings 340
Service Across Political or Water Agency Boundaries 105
Service Connection 13
Service Protection 13
Sewage 121-128
Sewage Ejector 154
Sewer-Connected Plumbing Fixtures 121-122, 125, 127
Shrinking 121

Shutoff Valves 338, 342
Single and Multiple Family Dwellings 104
Single Check 20, 49-50
Site Survey 93, 114-115
Snyder, Leonard 23
Southern California Water Utilities Association 28
Specifications 27
Specimen Tanks 124, 125, 128
Spill Resistant Pressure Vacuum Breaker Backsiphonage Prevention Assembly 13, 56, 78-81
Spring 49, 341
Springer, E. Kent 28, 30
Standard Industry Code 114
Standard for Field Test Kits 410-415
Standards for Backflow Prevention Assemblies 330-409
Static Condition 61, 63, 70, 72, 76, 79
Static Pressure 13
Steambaths 127
Steam Generating Facilities 120
Sub-Atmospheric Pressure 40, 43-44, 63
Sump and Lift Stations 104 123
Superior Pressure Principle Assembly 30
SVB See Spill Resistant Pressure Vacuum Breaker Backsiphonage Prevention Assembly
Swamp Coolers 127
Swing Arms 341
Swing Checks 21
Swing Pins 341
System Hazard 13
System Protection 90, 93, 95, 98, 110

T

Tanks 121-128, 155
TCR 33
Test Cocks 209, 336, 337, 342
Testing Laboratory 3, 90-91
Thermal Expansion 14, 168
Thermal Test 358, 359, 364, 373, 383, 387, 395, 399, 404
Three Questions 57
Tibbetts, William 26
Toilets 121-128, 156
Total Coliform Rule 33
Training 93-95
Transient Non-Community Water System 32
Truck wash 123
Typical Facilities 117-128
Typical High Hazard Services 103, 105

U

Underwriters Laboratories 91
University of Southern California 19, 25-27
Upstream 49
Urinals 121-128, 157
USC *See* University of Southern California
Used Water 14

V

Vacuum 2, 40-45
Vacuum Breakers 21, 121-122, 124-128
See also Atmospheric Vacuum Breaker, Pressure Vacuum Breaker and Spill-resistant Vacuum Breaker
Valve Seats 337
Van Meter, Roy 26
Vats 121-128, 155
Venturi 14
Venturi Effect 14
Veterinary Offices 124
Vibration 337
Viterbi School of Engineering 28

W

Waterborne disease 24
Water Cooled Equipment 120
Waterfront Facilities 127
Water Softener 158
Water Storage Tanks 121-128
Water Supervisor 14
Water Supplier 14
Water Supply 3
Whirlpool Tubs 124

Z

Zoos 127

Index

Index